Methods in Molecular Biology

Series Editor
John M. Walker
School of Life and Medical Sciences
University of Hertfordshire, Hatfield
Hertfordshire, UK

For further volumes:
http://www.springer.com/series/7651

For over 35 years, biological scientists have come to rely on the research protocols and methodologies in the critically acclaimed *Methods in Molecular Biology* series. The series was the first to introduce the step-by-step protocols approach that has become the standard in all biomedical protocol publishing. Each protocol is provided in readily-reproducible step-by-step fashion, opening with an introductory overview, a list of the materials and reagents needed to complete the experiment, and followed by a detailed procedure that is supported with a helpful notes section offering tips and tricks of the trade as well as troubleshooting advice. These hallmark features were introduced by series editor Dr. John Walker and constitute the key ingredient in each and every volume of the *Methods in Molecular Biology* series. Tested and trusted, comprehensive and reliable, all protocols from the series are indexed in PubMed.

Statistical Analysis of Proteomic Data

Methods and Tools

Edited by

Thomas Burger

Exploring the Dynamics of Proteomes, Proteomics French Infrastructure, Université Grenoble Alpes, CNRS, CEA, INSERM, Grenoble, France

Humana Press

Editor
Thomas Burger
Exploring the Dynamics of Proteomes
Proteomics French Infrastructure
Université Grenoble Alpes
CNRS, CEA, INSERM
Grenoble, France

ISSN 1064-3745 ISSN 1940-6029 (electronic)
Methods in Molecular Biology
ISBN 978-1-0716-1966-7 ISBN 978-1-0716-1967-4 (eBook)
https://doi.org/10.1007/978-1-0716-1967-4

This Humana imprint is published by the registered company Springer Science+Business Media, LLC part of Springer Nature.
The registered company address is: 1 New York Plaza, New York, NY 10004, U.S.A.

Preface

The ever-increasing throughput of mass spectrometry-based instrumental pipelines is a fantastic opportunity for proteomic researchers: Larger data volumes make it possible to answer more complex biological questions while respecting more stringent quality control guidelines. However, this also poses a challenge, which calls for new statistical methods and new data processing computational routines: Proteomics experts need to rely on reproducible bioinformatics workflow involving machine learning and high-dimensional statistics to extract the best of their data. To help them with this interdisciplinary challenge, this book traverses the most important processing steps of proteomics data analysis and presents practical guidelines as well as software tools, which are both user friendly and at the state-of-the-art in chemo- and biostatistics.

Chapter 1 leverages the recent theory of knock-off filters to propose a tutorial that bridges target–decoy competition (the universal approach to control the false discovery rate [FDR], when identifying fragmentation spectra from a reference database) and Benjamini–-Hochberg procedure (classically used to control the FDR associated to a list of differentially expressed proteins), hereby providing an insightful unified view on FDR controls in proteomics. Pushing the conceptual link between target–decoy competition and knock-off filters further, the authors of **Chapter 2** have developed a powerful approach to control the FDR at peptide level using multiple decoy databases. Their protocol describes how to practically benefit from these theoretical advances using their toolbox.

The three following chapters present software suites encompassing or wrapping the necessary tools to seamlessly construct the quantitation data table on which quantitative and differential analysis is usually conducted. While many such software suites exist in the literature, the focus on these three ones is justified by specific features. The one presented in **Chapter 3** proposes specific add-ons for post-translational modifications, while that of **Chapter 4** proposes routines to validate spectral identifications at various levels (peptide-spectrum match, peptide or protein) using Benjamini–Hochberg procedure in place of target and decoy searches. Finally, **Chapter 5** describes a complete software tool, which integrates all the processing steps (i.e., identification, quantification, and differential analysis). Such embedding provides a nice advantage: it helps preventing errors to propagate through the data processing workflow while simplifying the global error rate control.

Before focusing on the crucial step of differential analysis, two chapters provide specific and practical answers to missing value-related issues: **Chapter 6** proposes an original way of imputing quantitative values that are missing due to their too low abundances. While other types of missing values have focused many efforts, very few methods can be used for this one. To cope for this, a tool originally developed for mass spectrometry-based metabolomics is described. As for **Chapter 7**, it also focuses on an issue that is seldom addressed when heavily relying on imputation algorithms: how to preserve the data variance to avoid false positives in subsequent differential analysis.

The next three chapters present differential analysis software relying on various paradigms: protein-level (**Chapter 8**) and peptide-level (**Chapter 9**) processing of extracted ion chromatogram data as well as spectral count data (**Chapter 10**). Then, one focuses on three specific experimental designs, which are instrumental in many proteomics experiments: Immunoprecipitations (**Chapter 11**), post-translational modification analysis (**Chapter 12**), and constrained fold change (**Chapter 13**) using tools previously described in other chapters.

While univariate statistical test has been the default method so far to select differentially expressed proteins, **Chapter 14** describes an original alternative based on sparse regression models: The rationale underlying this increasingly popular approach in biostatistics is that a good set of candidate biomarkers should be of restricted size as well as sufficiently discriminant between the compared biological conditions. Following a similar trend, **Chapter 15** details how dimensionality reduction and sparse models can be leveraged for multi-omics data integration. Finally, the last two chapters focus on post-differential analysis approaches: **Chapter 16** addresses the issue of combining multiple proteomic assays to increase statistical power as well as to conduct meta-analyses in a proteomic context, and **Chapter 17** describes a tool to contextualize the results of a differential analysis using Gene Ontology enrichment and visualization.

To summarize, this book contains a series of protocols that covers the needs of any proteomics researcher wanting to extract the best of their data with the state-of-the-art tools while deepening their understanding of data analysis.

Grenoble, France ***Thomas Burger***

Contents

Preface ... *v*
Contributors ... *ix*

1 Unveiling the Links Between Peptide Identification and Differential Analysis FDR Controls by Means of a Practical Introduction to Knockoff Filters ... 1
Lucas Etourneau, Nelle Varoquaux, and Thomas Burger

2 A Pipeline for Peptide Detection Using Multiple Decoys ... 25
Syamand Hasam, Kristen Emery, William Stafford Noble, and Uri Keich

3 Enhanced Proteomic Data Analysis with MetaMorpheus ... 35
Rachel M. Miller, Robert J. Millikin, Zach Rolfs, Michael R. Shortreed, and Lloyd M. Smith

4 Validation of MS/MS Identifications and Label-Free Quantification Using Proline ... 67
Véronique Dupierris, Anne-Marie Hesse, Jean-Philippe Menetrey, David Bouyssié, Thomas Burger, Yohann Couté, and Christophe Bruley

5 Integrating Identification and Quantification Uncertainty for Differential Protein Abundance Analysis with Triqler ... 91
Matthew The and Lukas Käll

6 Left-Censored Missing Value Imputation Approach for MS-Based Proteomics Data with GSimp ... 119
Runmin Wei and Jingye Wang

7 Towards a More Accurate Differential Analysis of Multiple Imputed Proteomics Data with mi4limma ... 131
Marie Chion, Christine Carapito, and Frédéric Bertrand

8 Uncertainty-Aware Protein-Level Quantification and Differential Expression Analysis of Proteomics Data with seaMass ... 141
Alexander M Phillips, Richard D Unwin, Simon J Hubbard, and Andrew W Dowsey

9 Statistical Analysis of Quantitative Peptidomics and Peptide-Level Proteomics Data with Prostar ... 163
Marianne Tardif, Enora Fremy, Anne-Marie Hesse, Thomas Burger, Yohann Couté, and Samuel Wieczorek

10 msmsEDA & msmsTests: Label-Free Differential Expression by Spectral Counts ... 197
Josep Gregori, Àlex Sánchez, and Josep Villanueva

11 Exploring Protein Interactome Data with IPinquiry: Statistical Analysis and Data Visualization by Spectral Counts ... 243
Lauriane Kuhn, Timothée Vincent, Philippe Hammann, and Hélène Zuber

12 Statistical Analysis of Post-Translational Modifications Quantified by Label-Free Proteomics Across Multiple Biological Conditions with R: Illustration from SARS-CoV-2 Infected Cells 267
Quentin Giai Gianetto

13 Fast, Free, and Flexible Peptide and Protein Quantification with FlashLFQ 303
Robert J. Millikin, Michael R. Shortreed, Mark Scalf, and Lloyd M. Smith

14 Robust Prediction and Protein Selection with Adaptive PENSE 315
David Kepplinger and Gabriela V. Cohen Freue

15 Multivariate Analysis with the R Package mixOmics 333
Zoe Welham, Sébastien Déjean, and Kim-Anh Lê Cao

16 Integrating Multiple Quantitative Proteomic Analyses Using MetaMSD 361
So Young Ryu, Miriam P. Yun, and Sujung Kim

17 Application of WGCNA and PloGO2 in the Analysis of Complex Proteomic Data 375
Jemma X. Wu, Dana Pascovici, Yunqi Wu, Adam K. Walker, and Mehdi Mirzaei

Index *391*

Contributors

FRÉDÉRIC BERTRAND • *Laboratoire de Modélisation et Sûreté des Systèmes, Institut Charles Delaunay, Troyes, France*

DAVID BOUYSSIÉ • *Institut de Pharmacologie et de Biologie Structurale (IPBS), Université de Toulouse, CNRS, UPS, Toulouse, France*

CHRISTOPHE BRULEY • *University of Grenoble Alpes, CEA, Inserm, BioSanté U1292, CNRS ProFI FR2048, Grenoble, France*

THOMAS BURGER • *University of Grenoble Alpes, CNRS, CEA, INSERM, ProFI FR 2048, Grenoble, France*

CHRISTINE CARAPITO • *Laboratoire de Spectrométrie de Masse BioOrganique, Institut Pluridisciplinaire Hubert Curien, Strasbourg, France*

MARIE CHION • *Institut de Recherche Mathématique Avancée; UMR 7501, Laboratoire de Spectrométrie de Masse BioOrganique; Institut Pluridisciplinaire Hubert Curien, Strasbourg, France*

GABRIELA V. COHEN FREUE • *The University of British Columbia, Department of Statistics, Vancouver, BC, Canada*

YOHANN COUTÉ • *University of Grenoble Alpes, CEA, Inserm, BioSanté U1292, CNRS ProFI FR2048, Grenoble, France*

SÉBASTIEN DÉJEAN • *Institut de Mathématiques de Toulouse, Université Paul Sabatier, Toulouse, France*

ANDREW W DOWSEY • *Department of Population Health Sciences and Bristol Veterinary School, Faculty of Health Sciences; University of Bristol, Bristol, UK*

VÉRONIQUE DUPIERRIS • *University of Grenoble Alpes, CEA, Inserm, BioSanté U1292, CNRS ProFI FR2048, Grenoble, France*

KRISTEN EMERY • *School of Mathematics and Statistics, University of Sydney, Sydney, Australia*

LUCAS ETOURNEAU • *University of Grenoble Alpes, CEA, Inserm; BioSanté U1292, CNRS ProFI FR2048, Grenoble, France*

LUCAS ETOURNEAU • *University of Grenoble Alpes, CNRS; TIMC, Grenoble, France*

ENORA FREMY • *University of Grenoble Alpes; CEA, Inserm; BioSanté U1292; CNRS ProFI FR2048, Grenoble, France*

QUENTIN GIAI GIANETTO • *Institut Pasteur, Université Paris Cité, Bioinformatics and Biostatistics Hub, Paris, France; Institut Pasteur, Universiteé Paris Cité, Proteomic Platform, Mass Spectrometry for Biology Unit, CNRS, UAR 2024, Paris, France*

JOSEP GREGORI • *Vall Hebron Research Institute, Barcelona, Spain*

PHILIPPE HAMMANN • *Plateforme Protéomique Strasbourg Esplanade du CNRS, Université de Strasbourg, Strasbourg, France*

SYAMAND HASAM • *School of Mathematics and Statistics, University of Sydney, Sydney, Australia*

ANNE-MARIE HESSE • *University of Grenoble Alpes, CEA, Inserm, BioSanté U1292, CNRS ProFI FR2048, Grenoble, France*

SIMON J HUBBARD • *School of Biological Sciences, Faculty of Biology, Medicine and Health, University of Manchester, Manchester, UK*

LUKAS KÄLL • *Science for Life Laboratory, KTH Royal Institute of Technology, Solna, Sweden*
URI KEICH • *School of Mathematics and Statistics, University of Sydney, Sydney, Australia*
DAVID KEPPLINGER • *George Mason University, Volgenau School of Engineering, Department of Statistics, Fairfax, VA, USA*
SUJUNG KIM • *School of Community Health Sciences, University of Nevada Reno, Reno, NV, USA*
LAURIANE KUHN • *Plateforme protéomique Strasbourg Esplanade du CNRS, Université de Strasbourg, Strasbourg, France*
KIM-ANH LÊ CAO • *School of Mathematics and Statistics, Melbourne Integrative Genomics, The University of Melbourne, Melbourne, VIC, Australia*
JEAN-PHILIPPE MENETREY • *University of Grenoble Alpes; CNRS, CEA; INSERM; ProFI FR 2048, Grenoble, France*
RACHEL M. MILLER • *University of Wisconsin-Madison, Department of Chemistry, Madison, WI, USA*
ROBERT J. MILLIKIN • *University of Wisconsin-Madison, Department of Chemistry, Madison, WI, USA*
MEHDI MIRZAEI • *Australian Proteome Analysis Facility, Macquarie University, Sydney, NSW, Australia*
WILLIAM STAFFORD NOBLE • *Department of Genome Sciences, Department of Computer Science and Engineering, University of Washington, Seattle, WA, USA*
DANA PASCOVICI • *Insight Stats, Croydon Park, NSW, Australia*
ALEXANDER M PHILLIPS • *Department of Electrical Engineering and Electronics and Computational Biology Facility, Faculty of Health and Life Sciences, University of Liverpool, Liverpool, UK*
ZACH ROLFS • *University of Wisconsin-Madison, Department of Chemistry, Madison, WI, USA*
SO YOUNG RYU • *School of Community Health Sciences, University of Nevada Reno, Reno, NV, USA*
ÀLEX SÀNCHEZ • *Vall Hebron Research Institute, Department of Genetics Statistics and Microbiology, Barcelona, Spain*
MARK SCALF • *University of Wisconsin-Madison, Department of Chemistry, Madison, WI, USA*
MICHAEL R. SHORTREED • *University of Wisconsin-Madison, Department of Chemistry, Madison, WI, USA*
LLOYD M. SMITH • *University of Wisconsin-Madison, Department of Chemistry, Madison, WI, USA*
MARIANNE TARDIF • *University of Grenoble Alpes, CEA, Inserm; BioSanté U1292, CNRS ProFI FR2048, Grenoble, France*
MATTHEW THE • *Chair of Proteomics and Bioanalytics, Technische Universität München, Freising, Germany*
RICHARD D UNWIN • *Stoller Biomarker Discovery Centre and Division of Cancer Sciences, School of Medical Sciences, Faculty of Biology, Medicine and Health, The University of Manchester, Manchester, UK*
NELLE VAROQUAUX • *University of Grenoble Alpes, CNRS; TIMC, Grenoble, France*
JOSEP VILLANUEVA • *Tumor Biomarkers Lab, Vall Hebron Institute of Oncology, Barcelona, Spain*
TIMOTHÉE VINCENT • *Institut de biologie moléculaire des plantes; CNRS, Université de Strasbourg, Strasbourg, France*

ADAM K. WALKER • *Neurodegeneration Pathobiology Laboratory, Queensland Brain Institute; The University of Queensland, St. Lucia, QLD, Australia*
JINGYE WANG • *Verisk Analytics, Houston, TX, USA*
RUNMIN WEI • *The University of Texas MD Anderson Cancer Center, Department of Genetics, Houston, TX, USA*
ZOE WELHAM • *Kolling Institute, Sydney, NSW, Australia*
SAMUEL WIECZOREK • *University of Grenoble Alpes; CEA, Inserm; BioSanté U1292; CNRS ProFI FR2048, Grenoble, France*
JEMMA X. WU • *Australian Proteome Analysis Facility, Macquarie University, Sydney, NSW, Australia*
YUNQI WU • *Australian Proteome Analysis Facility, Macquarie University, Sydney, NSW, Australia*
MIRIAM P. YUN • *Department of Psychology, Institute for Neuroscience, University of Nevada Reno, Reno, NV, USA*
HÉLÈNE ZUBER • *Institut de biologie moléculaire des plantes; CNRS, Université de Strasbourg, Strasbourg, France*

Chapter 1

Unveiling the Links Between Peptide Identification and Differential Analysis FDR Controls by Means of a Practical Introduction to Knockoff Filters

Lucas Etourneau, Nelle Varoquaux, and Thomas Burger

Abstract

In proteomic differential analysis, FDR control is often performed through a multiple test correction (i.e., the adjustment of the original *p*-values). In this protocol, we apply a recent and alternative method, based on so-called knockoff filters. It shares interesting conceptual similarities with the target–decoy competition procedure, classically used in proteomics for FDR control at peptide identification. To provide practitioners with a unified understanding of FDR control in proteomics, we apply the knockoff procedure on real and simulated quantitative datasets. Leveraging these comparisons, we propose to adapt the knockoff procedure to better fit the specificities of quantitative proteomic data (mainly very few samples). Performances of knockoff procedure are compared with those of the classical Benjamini–Hochberg procedure, hereby shedding a new light on the strengths and weaknesses of target–decoy competition.

Key words FDR control, Knockoff filters, Benjamini–Hochberg procedure, Proteomics data analysis

1 Introduction

Controlling the false discovery rate (FDR) is a well-established practice in most -omic approaches, as it answers a pervasive question: Considering quantitative measurements for many covariates (be they genes, transcripts, metabolites, or proteins) in a set of samples split in at least two different biological conditions, how can we shortlist some differentially expressed ones, while controlling the risk of false positives (i.e., wrongly selected covariates due to their looking differentially expressed while they are not)? To cope with this, the most commonly used procedure is without a doubt the Benjamini-Hochberg one (BH) [1]. However, due to its large field of application, FDR control has focused a lot of additional efforts in biostatistics, and many authors have proposed to improve upon BH FDR control [2, 3], or have proposed alternative frameworks to do so [4–6].

Thomas Burger (ed.), *Statistical Analysis of Proteomic Data: Methods and Tools*,
Methods in Molecular Biology, vol. 2426, https://doi.org/10.1007/978-1-0716-1967-4_1,

In the specific case of proteomics, FDR control is not only used in the aforementioned biomarker selection problem. It is also an essential quality control metric when matching experimental fragmentation spectra onto *in silico* spectra (i.e., derived from reference database of protein sequences). However, for historical reasons, the associated FDR control is not performed using classical tools from biostatistics. On the contrary, a rather empirical approach termed target–decoy [7] is almost exclusively used. It consists in searching two databases: the first one, referred to as target, containing the genuine protein sequences, and another one, referred to as decoy, containing artifactual sequences. Under the assumption that target mismatches and decoy matches are equally likely, the number of decoy matches can be used to estimate the number of target mismatches, thus opening the door to FDR control.

For a long time, FDR control for peptide identification and for protein differential analysis has been considered as largely independent. However, theoretical connections exist: Notably, it has long been established [8] that if target and decoy databases are searched independently, then the procedure is broadly equivalent to relying on empirical null theory to estimate the FDR in a BH-related way [3]. More recently, it has been shown ([9] and Chapter 4) that BH procedure could be a user-friendly and computationally attractive alternative to target–decoy competition (TDC) (*see* **Note 1**). However, recent developments in theoretical biostatistics have made the links between both approaches to FDR control even tighter. Notably, the authors of [4] have proposed to tackle the biomarker research FDR control using an algorithmic procedure akin to that of TDC. Broadly, this novel approach, denoted as "knockoff filter," works as follows. First, knockoff variables are simulated to be as independent as possible from conditions of samples, but yet preserve the covariance structure of the original variables (*see* **Note 2**). Second, a competition is organized between each original variable and its associated knockoff. Third, the proportion of retained knockoffs is used to estimate the proportion of wrongly selected original covariates (*see* Table 1 for a more detailed comparison with TDC). Conversely, authors have recently leveraged the theory underlying knockoff filters to propose improved TDC strategies (*see* [10] as well as Chapter 2).

Overall, the framework of knockoff filters is particularly insightful to provide a global understanding of FDR control in proteomics, and the purpose of this protocol is to root such unified view on empirical comparisons using both real and simulated data. Interestingly, the results of these comparisons are compliant with empirical knowledge about the various strengths and weaknesses classically associated with each FDR control method.

Table 1
Comparison of the target–decoy and knockoff filter procedures for FDR control (PSM stands for Peptide-Spectrum Match)

Target–decoy competition	Knockoff filter (2nd order approximation)
1. Construct peptide decoys such that decoy PSMs have same score distribution than erroneous target PSMs.	1. For each protein, generate knockoff abundances with same mean and correlation matrices as original abundances.
2. For each real spectrum obtained, find the best match among all targets and decoys, and retain its score.	2. For each protein, compute a score describing whether the original abundances vector or its knockoff best predict the condition.
3. The number of selected PSMs from decoys at a given cutoff enables to estimate the FDR on selected target PSMs.	3. The number of selected knockoffs at a given cutoff enables to estimate the FDR on selected original proteins deemed differentially abundant.

2 Notations

We first start by reviewing commonly used yet conflicting notations in biostatistics and proteomics.

2.1 Classical Notations in Biostatistics

In biostatistics, the false discovery rate (FDR) and the false discovery proportion (FDP) are distinct notions. The FDP corresponds to what was classically and informally referred to as the "true FDR" in proteomics, i.e., the exact proportion of false positives among the proteins that passed the user-defined selection threshold, and therefore deemed as differentially abundant. Of course, except for benchmark artificial or simulated datasets, this quantity is unknown in practice.

The FDR reads as $\text{FDR} = \mathbb{E}[\text{FDP}]$, where $\mathbb{E}$ stands for the expectation, which broadly amounts to the long run average of the FDP on an infinite number of related experiments subject to stochastic fluctuations. This quantity is also unknown but it is much easier to estimate, and such estimate is classically noted $\widehat{\text{FDR}}$. Estimating the FDR is insightful, but unfortunately, not always sufficient [11]. An unbiased FDR estimate is expected to provide a value close to $\mathbb{E}[\text{FDP}]$. However, on a given dataset, this value may be larger or smaller than the FDP. While a slightly too large estimate implies a conservative behavior (there will be less false positives than expected among the shortlisted biomarkers), a too small FDR implies a too liberal quality control and subsequent risks in post-proteomics experiments.

To cope with weaknesses of FDR estimation, FDR control procedures have been developed: They rely on more conservative assumptions that yield slightly lesser selected discoveries at a given cutoff parameter. If we note as $\widehat{\text{FDR}}_\alpha$ the FDR estimate resulting from controlling the FDR at level α (α being classically tuned to 1%), it is likely that

$$\widehat{\text{FDR}}_\alpha \leq \alpha.$$

In other words, if one cuts-off a list of putative biomarkers according to an FDR controlled at 1%, the FDR estimate on this very list is likely to be slightly lower than 1%. However, as the FDP remains unknown, it is the only way to safely assume that the FDP is equal to or lower than 1%.

2.2 Classical Notations in Proteomics

In proteomics, most of the notions described above (*see* Subheading 2.1) are conflated. Since the mid-2010s, discriminating between the FDP and the FDR has progressively become standard. However, distinction between FDR (as equal to $\mathbb{E}[\text{FDP}]$), $\widehat{\text{FDR}}$, $\widehat{\text{FDR}}_\alpha$, and α is scarce. The reason is obvious: Except for specific methodological publications, most of them are not useful to the community. Indeed, in practice, a proteomic researcher only needs to manipulate α, the cutoff parameter, and to understand that after applying the FDR control accordingly, the FDP is not necessarily strictly equal to α, but possibly slightly smaller. However, the everyday language is error-prone: when one says or writes "We selected the putative biomarkers at an FDR of 1%," what is referred to as FDR is not $\mathbb{E}[\text{FDP}]$, $\widehat{\text{FDR}}$, or $\widehat{\text{FDR}}_\alpha$, but α.

To cope with this, it is possible to rely on other notations. They are not as formal as those of mainstream biostatistics (*see* Subheading 2.1) although they are sometimes reported in mathematics works [12]. However, they are sufficient for a rigorous everyday work in a proteomic lab. Essentially, it amounts to conflate the FDR estimate with α, and to define the FDR control as a procedure which provides the following guarantee with a sufficiently high probability:

$$\text{FDR} \geq \mathbb{E}[\text{FDP}] \ . \qquad (1)$$

This formulation can be misleading in the sense it gives the impression that the FDR control procedure indeed controls the FDP (*see* **Note 3**). However, it has two advantages: First, it makes the everyday language compliant with the minimum amount of statistical notions possible; second, it simplifies the understanding of other statistical notions such as "q-value" or "adjusted p-value," as using this formalism, they are simply equivalent to the FDR, as detailed in [13]. In the rest of the protocol, the naming conventions resulting from Eq. 1 are used, so that FDR refers to α, the FDR level tuned by the practitioner to perform FDR control.

2.3 Other Notations Used in This Protocol

Hereafter, the following mathematical notations are used:

1. n: The number of biological samples.
2. p: The number of proteins to include in differential analysis.
3. $X \in^{n \times p}$: The matrix of protein abundances, where each row corresponds to a sample and each column corresponds to a protein.
4. X_j: The vector of abundance of the j-th protein, i.e., the j-th column of X.
5. $x_{i,j}$: The abundance value of j-th protein for the i-th replicate.
6. y: The vector representing the condition label (numerical value) of biological samples, of length n. For example, the i-th coefficient of y is 1 if the i-th sample comes from the healthy condition, and -1 if it comes from the disease condition.
7. $X^{\text{Ko}} \in^{n \times p}$: The knockoff dataset, generated from original dataset matrix X.
8. X_j^{Ko}: The knockoff vector of abundance of the j-th protein.
9. W: The vector of scores of all proteins (only the original ones, not the knockoff), of length p.

10. W_j: the score associated with the j-th protein. A large positive value W_j is evidence that the protein j is differentially expressed. It is typically constructed by comparing the predictive power of X_j and X_j^{Ko} of the sample conditions. Swapping X_j and X_j^{Ko} should swap the sign of W_j. A null W_j means that both X_j^{Ko} and X_j^{Ko} bring the same amount (or lack thereof) of information on the condition.

3 Material

3.1 R version

R version 4.0.3 (or above) is required to use the following packages. We recommend using an integrated development environment like Rstudio to execute the commands of this protocol. It can be downloaded from https://www.rstudio.com/.

3.2 Packages

The following packages are necessary:

1. The packages `knockoff`, `lars` [14], and `glmnet` [15] must be installed from the CRAN:

```
install.packages("knockoff")
install.packages("lars")
install.packages("glmnet")
```

2. `cp4p` [16] provides two datasets with controlled ground truth: They result from analysis of samples containing different abundance of 48 human proteins spiked in a yeast background [17]. The p-values from a Welch t-test associated with each protein are also provided, along with functions to apply Benjamini–Hochberg procedure for differential analysis. To install `cp4p` package, it is first necessary to install the BioConductor [18] packages it depends on:

```
if (!requireNamespace("BiocManager", quietly = TRUE))
    install.packages("BiocManager")
BiocManager::install("multtest")
BiocManager::install("limma")
BiocManager::install("qvalue")
```

3. Then `cp4p` can be installed from the CRAN:

```
install.packages("cp4p")
```

4. Finally, load the packages in the environment:

```
library(cp4p)
library(knockoff)
library(lars)
```

3.3 Data Format

This protocol relies on a data format which is quite uncommon in proteomics (*see* **Note 4**). The input data X on which FDR control is applied should have at least 3 rows, i.e., at least biological 3 samples in total are needed. The number of proteins to include in differential analysis can be arbitrary. Values of abundance in X should be $\log_2$-scaled.

For conveniency, we use two datasets in this protocol: A dataset resulting from real mass-spectrometry output, called `LFQRatio25` (*see* Subheading 3.4), and a simulated dataset with adjustable parameters (*see* Subheading 3.5).

3.4 Data Loading from cp4p

The following commands enable to load and prepare `LFQRatio25` dataset [16]:

1. Load the dataset with the following command:

```
data("LFQRatio25")
```

2. Then, abundances values for all 6 samples are extracted to form the rows of the `X_yups` variable:

```
X_yups = t(LFQRatio25[,1:6])
```

3. Similarly, vector `y_yups` contains the condition labels of these samples:

```
y_yups = c(1,1,1,-1,-1,-1)
```

4. For this particular dataset, differentially abundant proteins (or in statistical language, variables under the alternative hypothesis H_1) are known. It is possible to display their name and their index in the list of proteins. These are the 46 first proteins, as the output of this code chunk suggests (*see* **Note 5**):

```
mask_human = LFQRatio25$Organism == "human"
names_diff_yups = LFQRatio25$Majority.protein.IDs[mask_human]
idx_diff_yups = which(mask_human)
idx_diff_yups
```

```
[1]   1  2  3  4  5  6  7  8  9 10 11 12 13 14 15 16 17
[18] 18 19 20 21 22 23 24 25 26 27 28 29 30 31 32 33 34
[35] 35 36 37 38 39 40 41 42 43 44 45 46
```

5. Check the dataset to make sure the same dataset is obtained:

```
head(X_yups[,1:5])
```

```
            [,1]     [,2]     [,3]     [,4]     [,5]
A.R1 31.27392 29.48101 29.80982 29.10410 26.85626
A.R2 31.27147 29.46032 29.84163 29.22384 27.11535
A.R3 31.26327 29.45797 29.83771 29.00945 26.94358
B.R1 29.83022 28.04973 28.41002 27.45505 25.71735
B.R2 29.81413 28.02686 28.38101 27.58463 25.74196
B.R3 29.84867 28.00774 28.42514 27.52028 24.62264
```

3.5 Data Simulation

The following commands enable to prepare a simulated dataset:

1. The code below randomly generates a dataset broadly akin to LFQRatio25. Due to randomness, it will be different from one run to another. To ensure the results are reproducible and to obtain same results as in the remaining of the protocol, use the following optional command to set the random seed (*see* **Note 6**):

```
set.seed(1234)
```

2. Tune the parameters of the dataset:

```
n_h1 = 50              # Number of proteins differentially abundan
n_rep = 3              # Number of replicates of each condition
p=1500                 # Number of proteins
mu = runif(p, 24, 32)
sigma1 = diag(runif(p,0,0.02))
sigma2 = diag(runif(p,0,0.02))
mu_diff = c(runif(n_h1, 0.5, 2)*sign(runif(n_h1, -1, 1)),
    rep(0, p-n_h1))
```

3. Create and concatenate arrays of both conditions:

```
p = length(mu)
X1 = matrix(rnorm(n_rep*p),n_rep)
X2 = matrix(rnorm(n_rep*p),n_rep)
X1 = t(t(X1)+mu+mu_diff/2)
X2 = t(t(X2)+mu-mu_diff/2)
X_sim = rbind(X1,X2)
y_sim = c(rep(1, n_rep), rep(-1, n_rep))
idx_diff_sim = 1:n_h1
```

4. Check the dataset to make sure there are no mistakes:

```
head(X_sim[,1:5])
```

```
          [,1]     [,2]     [,3]     [,4]     [,5]
[1,] 23.96519 28.55181 27.95301 29.64470 31.05108
[2,] 24.04396 28.21652 27.74679 29.56570 31.51248
[3,] 24.05717 28.39634 27.90406 29.74762 31.56869
[4,] 25.65308 29.55612 29.92890 28.21821 30.47228
[5,] 25.74846 29.63377 29.77653 28.30777 30.32441
[6,] 25.89306 29.58248 29.80624 28.33325 30.52107
```

4 Methods

This section falls into the following subsections:

1. We explain how to apply the original knockoff filter procedure to control the FDR for differential expression analysis. Precisely, we show how to (1) generate knockoff variables; (2) compute a score for each protein/knockoff pair; (3) select differentially abundant proteins for a predefined target FDR.
2. We detail how to replace the default scoring strategy with other ones, and compare these alternative knockoff procedures to the classical Benjamini-Hochberg (BH) procedure.
3. We propose some code to illustrate the sensitivity of the knockoff filter procedure to the random generation of knockoffs.

4.1 Original Knockoff Procedure

1. Choose the dataset on which applying the knockoff procedure:
   (a) To apply it on the LFQRatio25 dataset, use:

```
X_data = X_yups
y_data = y_yups
idx_diff = idx_diff_yups
```

   (b) Alternatively, to apply it on the simulated dataset, use:

```
X_data = X_sim
y_data = y_sim
idx_diff = idx_diff_sim
```

   For the rest of this section, we will use the LFQRatio25 dataset.

2. Rescale the data to have null mean and unitary variance for each protein abundance vector (i.e., for each X_j) (*see* **Note 7**):

```
X_data = scale(X_data)
```

3. Execute these commands to generate the knockoff dataset from original data with a fixed seed (*see* **Note 8**):

```
set.seed(1234)
X_data_k = create.second_order(X_data)
```

4. For each protein, compute a score based on the Lasso path of covariates (*see* **Note 9**). An inevitable warning concerning the lack of replicates appears: "one multinomial or binomial class has fewer than 8 observations; dangerous ground."

```
set.seed(1234)
W_lasso = stat.lasso_lambdasmax_bin(X_data, X_data_k, y_data)
```

5. Set the value of targeted FDR, compute the resulting threshold, and select proteins for which their score is above this threshold. The `target_fdr` parameter must be a number between 0 and 1. The `offset` parameter determines which FDR estimator to use, it can be set to either 0 or 1 (*see* **Note 10**). When `offset` is 0, a biased FDR estimate is used, and when `offset` is 1, a non-biased, yet more conservative estimate is used.

```
target_fdr = 0.05
thres = knockoff.threshold(W_lasso, fdr=target_fdr, offset=0)
selected_lasso = which(W_lasso >= thres)
```

6. **This step and the following ones are optional, as they can only be applied for a dataset endowed with a ground truth, such as `LFQRatio25` or a simulated dataset.** Display the names of proteins selected as differentially abundant at the FDR tuned with the `target_fdr` parameter (here 0.05).

```
names_diff_yups[selected_lasso]
```

```
[1] P02768upsedyp ALBU_HUMAN_upsedyp - CON__P02768-1
[2] O00762upsedyp UBE2C_HUMAN_upsedyp
[3] P00709upsedyp LALBA_HUMAN_upsedyp
[4] P02788upsedyp TRFL_HUMAN_upsedyp
[5] P06396upsedyp GELS_HUMAN_upsedyp
[6] P12081upsedyp SYHC_HUMAN_upsedyp
```

7. This code instantiates useful functions to compute the FDP and power from ground truth data. For a certain selection level α, the power is defined as

$$\text{Power}_\alpha = \frac{\#\text{ofselectedoriginalvariablesunder } H_1}{\#\text{oforiginalvariablesunder } H_1}.$$

Where H_1 denotes the alternative hypothesis, i.e., "the protein is differentially abundant." The power gives a measure of how well our selection covers all the proteins differentially expressed:

```
compute_fdp = function(selected, nonzero) {
    if (length(selected) != 0) {
        return(1-sum(nonzero
    }
    return(0)
}

compute_power=function(selected, nonzero) {
    if (length(selected) != 0) {
        return(sum(nonzero
    }
    return(0)
}
```

8. The following code computes the FDP and power of the procedure for a user-defined range of target FDRs (for both offset values):

```
FDR = seq(0,0.5,0.04)
template = rep(0,length(FDR))
FDP = list(template, template)
POWER = list(template, template)

for (t in 1:length(FDR)) {
  for (offs in 1:2) {
    thres = knockoff.threshold(W_lasso, fdr=FDR[t],
        offset=offs-1)
    selected = which(W_lasso >= thres)
    FDP[[offs]][t] = compute_fdp(selected, idx_diff)
    POWER[[offs]][t] = compute_power(selected, idx_diff)
    }
}
```

9. Using the results computed at the previous step, the following code displays the FDP and power as a function of the FDR (*see* Fig. 1 for LFQRatio25 and Fig. 2 for simulated dataset):

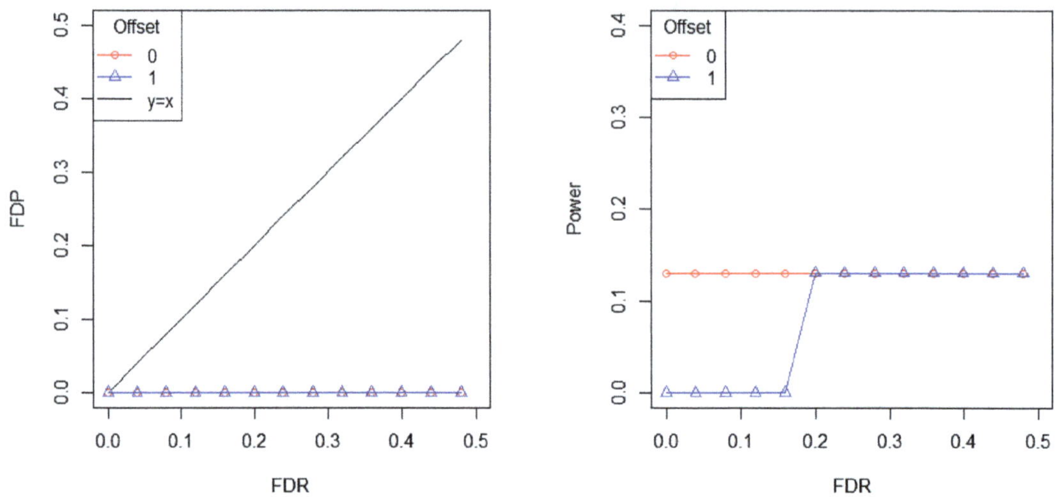

Fig. 1 FDP and power *vs.* FDR for LFQRatio25 dataset, with and without offset, for the knockoff filter procedure with Lasso-based scores

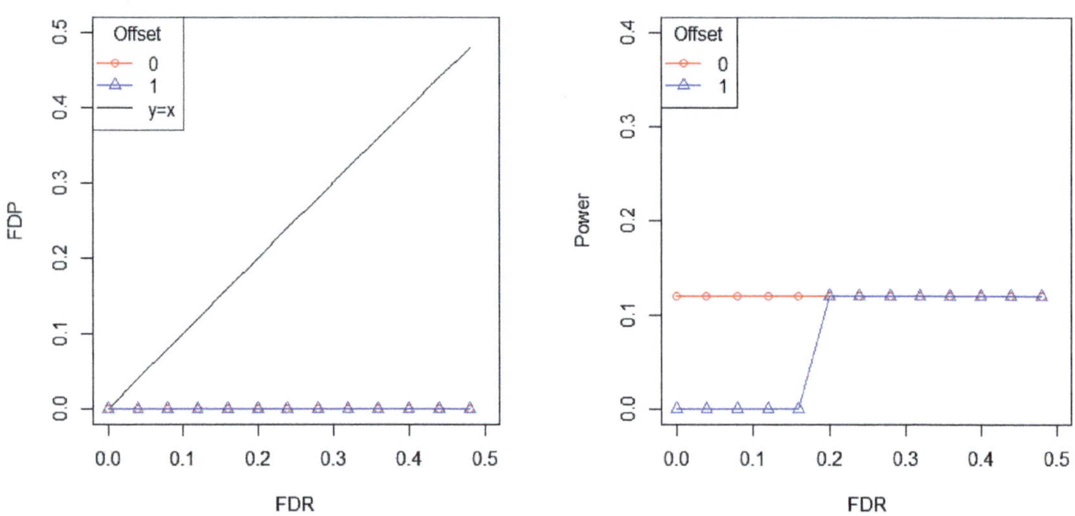

Fig. 2 FDP and Power *vs.* target FDR for the simulated dataset, with and without offset, for knockoff procedure with Lasso-based scores

```
par(pty='s')
cols = c("red","blue", "black")
plot(FDR, FDR, type='l', ylab = "FDP", xlab = "FDR",
    ylim=c(0,0.5), xlim=c(0,0.5))
lines(FDR, FDP[[1]], col="red")
lines(FDR, FDP[[2]], col="blue")
points(FDR, FDP[[1]], col="red", pch=1)
points(FDR, FDP[[2]], col="blue", pch=2)
legend("topleft", legend=c(0,1, "y=x"), col=cols,
    pch=c(1,2,-1), lty = 1, title="Offset")

plot(1, type="n", ylab = "Power", xlab = "FDR",
    ylim=c(0,0.4), xlim=c(0,0.5))
lines(FDR, POWER[[1]], col="red")
lines(FDR, POWER[[2]], col="blue")
points(FDR, POWER[[1]], col="red", pch=1)
points(FDR, POWER[[2]], col="blue", pch=2)
legend("topleft", legend=c(0,1), col=cols, pch=c(1,2),
    lty = 1, title="Offset")
```

We notice that FDP and power curves on Figs. 1 and 2 are almost always horizontal. This means that variables selected remain the same whatever the FDR target chosen. When the offset equates 1 (unbiased estimator), no proteins are deemed differentially expressed below a certain value of FDR. Thus, even though there are no false positive, there are no true positive either, making the FDR control through knockoff filters practically useless.

We mainly explain this over-conservativeness by the usage of variable selection with the Lasso algorithm, at the step of W scores computation. In fact, in the setting $n << p$, the Lasso algorithm will only select n variables. This is problematic for differential expression analysis where the total number of samples rarely exceeds the number of *a priori* differentially expressed proteins. On top of that, as very few covariates are selected, and some original variables are much more differentially abundant than all the others, knockoff variables are almost never selected. Thus, estimating the number of false discoveries from the number of selected knockoffs is not appropriate in our cases. This efficiency of variable selection with Lasso is thoroughly discussed in [19].

4.2 Scoring Methods Based on Forward Stagewise Regression and t-Test

Preliminary experimental comparisons highlighted the knockoff procedure accuracy highly depends on the chosen feature selection algorithm. We hereafter describe two procedures that we found to address the issue described above (*see* Subheading 4.1). The first scoring method consists in using forward stagewise selection (FS) algorithm (*see* **Note 11**). The second one is derived from the variable selection procedure classically used in proteomics: It amounts to computing a t-test p-value for both original and knockoff variables; then, the final score (i.e., W_j) is defined by the log difference of p-values (LDP) obtained between each original variable and its knockoff.

1. To instantiate the functions that compute the W_i's for the FS and LDP methods, use the following chunks of code (it is advised to run them both, so as to allow subsequent comparisons):

   (a) For the FS method:

```
stat_forward_sel=function(X, X_k, y) {
    Xconcat = cbind(X, X_k)
    res = lars(Xconcat, c(1,1,1,-1,-1,-1), type="for",
        use.Gram = FALSE)
    lambdas = rep(0, 2*ncol(X))
    lambdas[res$entry != 0] = res$lambda[res$entry]
    W_fs = lambdas[1:ncol(X)]-lambdas[-(1:ncol(X))]
    W_fs
}
W_fs = stat_forward_sel(X_data, X_data_k, y_data)
```

   (b) For the LDP method:

```
stat_log_diff_pval=function(X, X_k) {
    Xconcat = cbind(X, X_k)
    pvals = apply(Xconcat, 2, function(x){res =
        t.test(x[1:3], x[4:6]);return(res$p.value)})
    pvals_or = pvals[1:(length(pvals)/2)]
    pvals_k = pvals[(length(pvals)/2+1):length(pvals)]
    W_pvals = (-log(pvals_or)+log(pvals_k))
    W_pvals
}

W_ldp = stat_log_diff_pval(X_data, X_data_k)
```

2. Plot the histogram of W_i's to better visualize the selection process (*see* Fig. 3 for `LFQRatio25` dataset):

```
hist(W_ldp[W_ldp!=0], col=c(rep("red", 2), rep("grey", 4),


    rep("blue", 11)), main="Histogram of W", xlab="W")
axis(1, at=c(-5, -2, 0, 2, 5, 10))
```

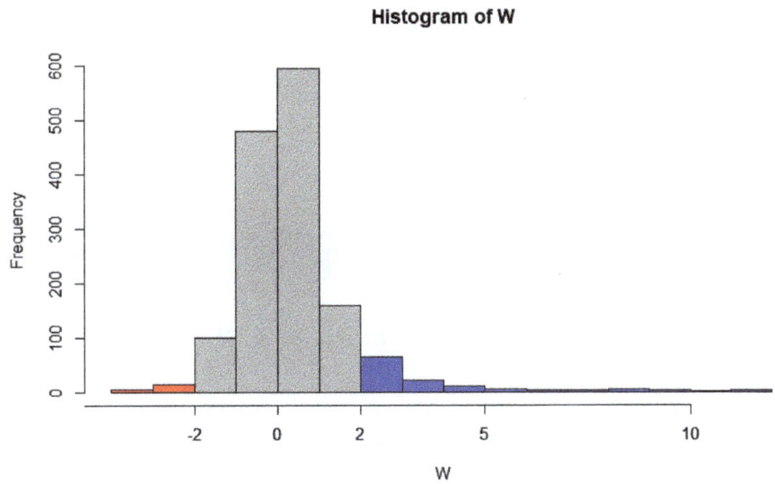

Fig. 3 Histogram of scores W_i's obtained with log diff of p-values scoring method, on `LFQRatio25` dataset. The blue area corresponds to original variables that are selected, and the red area represents knockoff variables selected, both at a threshold of 2 (hence, a conservative FDR estimate at a selection threshold of 2 reads $\widehat{FDR} = \frac{\text{redarea}+1}{\text{bluearea}}$)

3. To illustrate the interest of using FS and LDP within the knockoff filter procedure, we compare those two approaches with the classically used Benjamini-Hochberg (BH) procedure. Depending on the dataset being `LFQRatio25` or the simulated one, the code differs:

   (a) With `LFQRatio25`, the p-values resulting from Welch t-test are provided in the dataset:

```
pvals = LFQRatio25[,7]
res = adjust.p(pvals, pi0.method = 1)
```

   (b) With the simulated dataset, p-values must be computed beforehand (a Welch t-test is also used here):

```
pvals = apply(X_data, 2, function(x){res = t.test(
    x[1:n_rep], x[(n_rep+1):(2*n_rep)]);
    return(res$p.value)})
res = adjust.p(pvals, pi0.method = 1)
```

4. Compute the FDP and power for BH and knockoff filter procedure with LDP and FS methods (with offset=1), at different FDR levels:

```
FDP = list(template, template, template)
POWER = list(template, template, template)
W_list = list(W_fs, W_ldp)

for (t in 1:length(FDR)) {
  for (W_idx in 1:2) {
    thres = knockoff.threshold(W_list[[W_idx]], fdr=FDR[t],
        offset=1)
    selected = which(W_list[[W_idx]] >= thres)
    FDP[[W_idx]][t] = compute_fdp(selected, idx_diff)
    POWER[[W_idx]][t] = compute_power(selected, idx_diff)
  }
  selected_bh = which(res$adjp$adjusted.p<=FDR[t])
  FDP[[3]][t] = compute_fdp(selected_bh, idx_diff)
  POWER[[3]][t] = compute_power(selected_bh, idx_diff)
}
```

5. Finally plot the FDP and power *vs.* FDR level, as illustrated on Figs. 4 and 5, respectively for the `LFQRatio25` and simulated datasets):

```
par(pty='s')
cols = c("red", "blue", "orange")
leg = c("Knockoff w F.S.", "Knockoff w log diff.", "B-H.")
plot(FDR, FDR, type='l', ylab = "FDP", xlab = "FDR",
    ylim=c(0,0.6), xlim=c(0,0.15))
for (i in 1:3) {
  lines(FDR, FDP[[i]], col=cols[i])
  points(FDR, FDP[[i]], col=cols[i], pch=i)
}
legend("topleft", legend=leg, col=cols, pch=1:3,
    title="Procedure")

plot(1, type="n", ylab = "Power", xlab = "FDR",
    ylim=c(0,1.2), xlim=c(0,0.15))
for (i in 1:3) {


  lines(FDR, POWER[[i]], col=cols[i])
  points(FDR, POWER[[i]], col=cols[i], pch=i)
}
legend("topleft", legend=leg, col=cols, pch=1:2,
    title="Procedure")
```

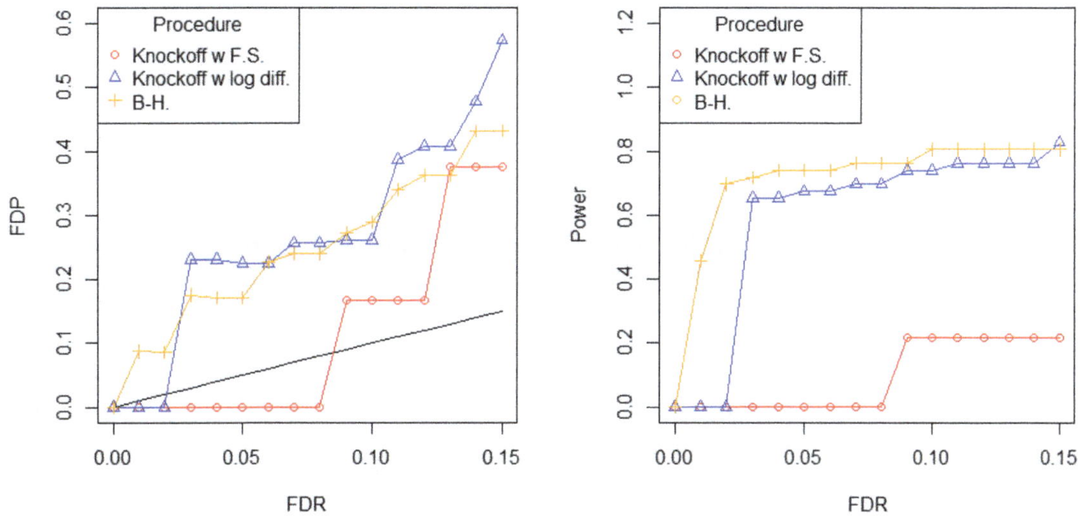

Fig. 4 FDP and power *vs.* target FDR for knockoff filter procedure with offset=1 applied with forward stagewise selection and log diff of *p*-values scoring, and Benjamini–Hochberg procedure, obtained with `LFQRatio25`

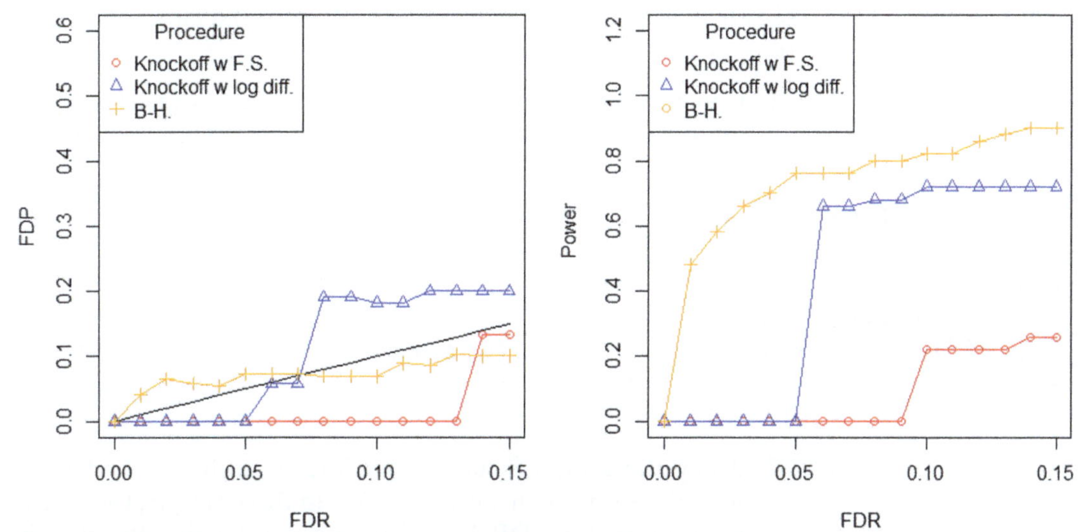

Fig. 5 FDP and power *vs.* target FDR for knockoff procedure with offset=1 applied with forward stagewise selection and log diff of *p*-values scoring, and Benjamini–Hochberg procedure, obtained with simulated data

We observe that the knockoff filter procedure with LDP broadly follows the same trend as the BH one on `LFQRatio25` (*see* Fig. 4). By construction, the LDP scores is never null, yielding a rather symmetric distribution (*see* Fig. 3). The largest positive scores (depicted in the right hand tail) result from differentially abundant proteins, while the left hand one amounts to selected knockoff proteins. The distribution being more symmetric than when using the Lasso, it is possible to select a larger subset of proteins at a given FDR. However, when using the FS based scores, knockoff filters roughly behaves as with the Lasso, yielding a greater but yet insufficient power.

Finally, the BH procedure also yields anti-conservative results on `LFQRatio25`, as the FDP is always higher than the FDR. However, this can be explained by other preprocessing steps (match between runs, normalization, imputation, etc.) which tends to shrink the within-condition variance prior to differential analysis as well as to increase the risk of false positives that are not accounted by FDR control. Indeed, Benjamini-Hochberg is conservative on simulated data (*see* Fig. 5).

4.3 Sensitivity of FDR Control to Knockoff Used

Knockoff generation with `create.second_order` function (*see* Subheading 4.1, **Step 3**) involves the random draw of a knockoff matrix (similarly to the random generation of decoy sequences). Hence, on a given dataset, running two consecutive FDR control procedures with knockoff filters should lead to slightly different results. We hereafter propose several experiments to illustrate the sensitivity of the knockoff filter procedure to the knockoff generation, as well as to evaluate its magnitude.

1. Generate 30 knockoff datasets and store them in a list (depending on the machine, this step may last between 30 min to an hour):

```
set.seed(3456)
n_k = 30
l_k = list()
for (i in 1:n_k) {
  l_k[[i]] = create.second_order(X_data)
}
```

2. Apply the knockoff filter procedure to each knockoff series, with FDR varying from 1% to 15%. In this example, the scoring method used is LDP. For all the knockoff series, the effective FDP *vs.* FDR curves are iteratively plotted, leading to a display akin to that of Fig. 6. The proteins selected at an FDR of 5% for each knockoff series are retained in a matrix referred to as `scores`:

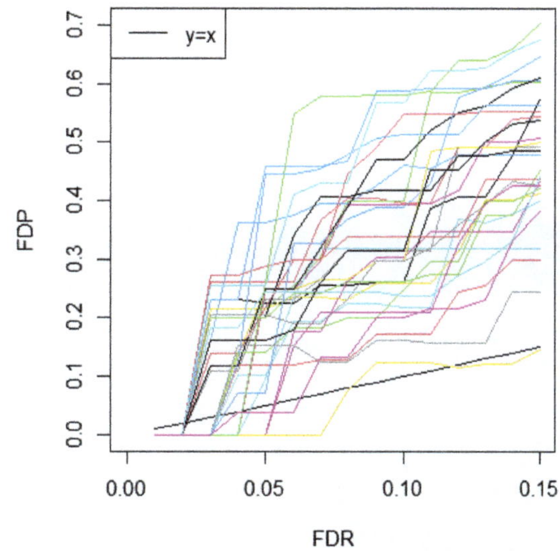

Fig. 6 Curves of FDP *vs.* FDR for 30 different Knockoff procedures, applied with log diff of values score on `LFQRatio25` dataset

```
par(pty='s')
FDR = seq(0.01,0.15,0.01)
FDP <- POWER <- matrix(rep(0,n_k*length(FDR)), nrow=n_k)
scores = matrix(rep(0, n_k*ncol(X_sim)), nrow=n_k)

plot(FDR, FDR, type='l', ylab = "FDP", xlab = "FDR",
    ylim=c(0,0.7), xlim=c(0,0.15))
for (i in 1:n_k) {
  W = stat_log_diff_pval(X_data, l_k[[i]])
  for (t in 1:length(FDR)) {
    thres = knockoff.threshold(W, fdr=FDR[t], offset=1)
    selected = which(W >= thres)
    FDP[i,t] = compute_fdp(selected, idx_diff)
    if (FDR[t] == 0.05) {
      scores[i,selected] = 1
    }
  }
  lines(FDR, FDP[i,], col=i)
}
legend("topleft", legend = "y=x", lty=1, col="black")
```

3. Finally, plot a heatmap featuring the `scores` matrix which highlights with different colors the selected proteins under H_0 and H_1 for each knockoff filter series, at an FDR target of 5%. (*see* Fig. 7):

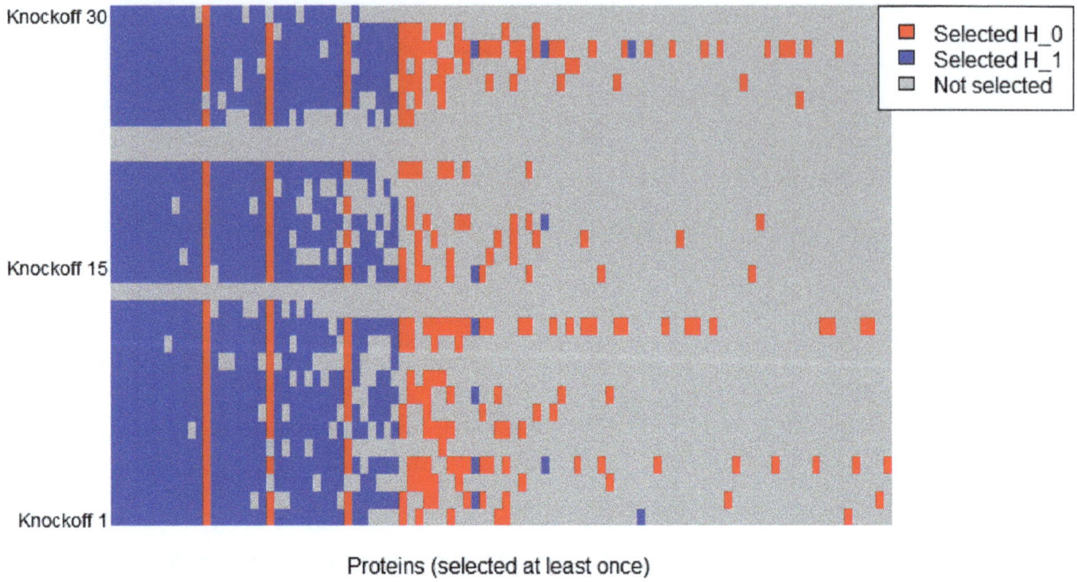

Fig. 7 Proteins selected according to 30 different knockoff procedure iterations (using LDP score) on the `LFQRatio25` dataset. Blue cells depict original differentially abundant proteins (human proteins) that were selected using a given knockoff. Similarly, red cells depict non-differentially abundant proteins (yeast proteins) mistakenly selected. Proteins are sorted from the most selected one (left hand side) to the least selected one (right hand side)

```
par(mar=c(5, 5, 2, 8), xpd=TRUE, mgp=c(1,1,0))
heights = sort(colSums(scores), decreasing = T,


    index.return = T)
heights_in_plot = heights$ix[heights$x>0]
submat = scores[,heights_in_plot]
submat[(submat == 1) & t(matrix(rep(heights_in_plot>46,
    nrow(scores)), ncol=nrow(scores)))] = 2

image(t(submat), col=c("grey", "blue", "red"),
    xlab="Proteins (selected at least once)", axes=F)
mtext(text=c(paste("Knockoff",c(1,15,30))), side=2, line=0.1,
    at=seq(0.0,1,1/2), las=1, cex=0.9)
legend("topright",inset=c(-0.23, 0),
    legend=c("Selected H_0","Selected H_1", "Not selected"),
    fill=c("red","cyan", "grey"))
```

Figures 5 and 7 emphasize the important variability resulting from the random nature of knockoff filters. To counter this variability, [20] proposes a method to aggregate multiple knockoffs. In fact, similar sensitivity has already been commented upon with target–decoy procedures [21], so it seems to be a problem ubiquitous to FDR control procedures which involve simulating artifactual data under the null hypothesis. Finally these observations provide an intuitive support for the tools described in Chapter 2, which relies on multiple decoy databases to construct a knockoff-like score.

5 Notes

1. *Target–Decoy Competition.* TDC is a specific target–decoy strategy where both databases are concatenated, so that each spectrum can only match to either a decoy or a target spectrum; in other words, both databases are competing for the matches.
2. *On the generation of knockoffs.* Knockoff variables, under second order approximation, are simulated (*see* Subheading 2.3 for mathematical notations) such that: (1) a knockoff variable X_i^{Ko} has the same mean as the original variable X_i; (2) the covariance between knockoff variables is equal to the covariance between the original variables: $cov(X_i^{\mathrm{Ko}}, X_j^{\mathrm{Ko}}) = cov(X_i, X_j)$; (3) the covariance between knockoff variables and original variables is equal to the covariance between the original variables: $cov(X_i^{\mathrm{Ko}}, X_j) = cov(X_i, X_j) \quad \forall i \neq j$; (4) but the variance between a knockoff variable and the original variable is null: $cov(X_i^{\mathrm{Ko}}, X_i) = 0$. Fulfilling all of those constraints in the data simulation process is impossible. Thus, an optimization procedure is used to fulfill them to the best extent possible. Note that knockoff variables are generated without looking at the condition of samples. This ensures that knockoff variables are independent from response *y* conditionally to original variables, as explained in [5].
3. *FDR or FDP control?* Directly controlling the FDP is more difficult than controlling its expectation, the FDR. However, some papers have tackled this challenge [22, 23]. The notion of "control" is tightly defined in statistics and various FDRs have been defined to induce different forms of FDR control (for instance exact, strong, or weak controls, such as discussed in [24]). However, Eq. 1 does not suggest that the procedure controls the FDR (as it now refers to α) but the FDP, although indirectly, through its expectation. We acknowledge this could be a source of confusion, and tentatively propose to coin the term "indirect FDP control."
4. *Data format.* In proteomics, data tables are generally structured with proteins as rows and replicates in columns. However, `knockoff` and `lars` packages were designed for more general use cases. Hence, they adopt another convention widely used in statistics, with features as columns and samples as rows.

5. *Missing proteins in cp4p*. Th UPS1 mixture contains 48 human proteins. However, according to [17], only 46 are confidently identified and quantified by mass spectrometry. Therefore, when processing the `LFQRatio25`, only 46 differentially abundant UPS1 proteins are sought.
6. *Seed in R*. Setting the seed of the random number generator should give you the same sequence of random numbers as presented here. However, different versions of R may yield different sequences of random number due to changes in the pseudo-random number generator.
7. *Data scaling*. This step is particularly important when variable variances span over a large range. To give an order of magnitude, in the `LFQRatio25` dataset, the lowest variance among all covariates equates 0.001 while the largest one equates 9. If the dataset is not scaled, the program used to generate knockoff converges after 10 min, while 40 s are sufficient with scaled data.
8. *Warnings in knockoff generation*. We have observed that the following warnings "Reached upper boundary" and "only 0 eigenvalue(s) converged, less than $k = 1$," may appear in some environments. We assume these warnings come from the too high dependence between the columns which corresponds to differentially abundant proteins, yet, the algorithm can still operate.
9. *Scoring methods*. The `knockoff` package already provides the functions to compute scores according to different methods. A classical scoring method used in the original knockoff procedure is based on the Lasso path of variables, thus we try to apply it first. However, other methods using Lasso are proposed in the package, such as `stat.lasso_lambdadiff_bin` and `stat.lasso_coefdiff_bin` (this last one is not applicable on our data by lack of samples). Also, a method based on random forests is proposed, but during our preliminary experiments, it gave poorer results on our datasets.
10. *Offset parameter*. The offset parameter corresponds to the difference between the natural FDR estimate and the conservative estimate yielding FDR control. The former reads

$$\frac{\#\text{ofknockoffsselectedat } \alpha \text{ level}}{\#\text{oforiginalvariablesselectedat } \alpha \text{ level}}$$

while the second reads:

$$\frac{\#\text{ofknockoffsselectedat } \alpha \text{ level} + 1}{\#\text{oforiginalvariablesselectedat } \alpha \text{ level}}.$$

This distinction also exists in the TDC procedure [11] where the natural estimate reads d/t and the conservative ones $(d+1)/t$, where t and d respectively denote the number of selected target and decoy PSMs. In the knockoff literature, the "+1" is termed "offset," and when equal to 0 (respectively 1), it leads to the biased (respectively, non-biased) estimate. The non-biased estimate is then more conservative than the biased estimate.

11. *Lasso vs forward stagewise.* From a theoretical point of view, the forward stagewise algorithm behaves very much like the Lasso [25]. However, it copes with the issue of the Lasso being unable to select more variables than the number of samples, an essential feature in proteomics. The FS score we use is based on the norm of predictors coefficients when a given variable enters the model, similarly to the Lasso method. This method is already proposed in the `knockoff` package and the associated paper [5] through the `stat.forward_selection` function. However, we use slightly modified approach relying on the `lars` package to obtain an equivalent regularization term, instead of variable selection ranking.

Acknowledgements

This work was supported by grants from the French National Research Agency: ProFI project (ANR-10-INBS-08), GRAL project (ANR-10-LABX-49-01), DATA@UGA and SYMER projects (ANR-15-IDEX-02) and MIAI @ Grenoble Alpes (ANR-19-P3IA-0003).

References

1. Benjamini Y, Hochberg Y (1995) Controlling the false discovery rate: a practical and powerful approach to multiple testing. J R Stat Soc Ser B (Methodological) 57(1):289–300. http://www.jstor.org/stable/2346101
2. Benjamini Y, Krieger AM, Yekutieli D (2006) Adaptive linear step-up procedures that control the false discovery rate. Biometrika 93(3): 491–507. https://doi.org/10.1093/biomet/93.3.491
3. Efron B (2012) Large-scale inference: empirical Bayes methods for estimation, testing, and prediction, vol 1. Cambridge University Press, Cambridge
4. Barber RF, Candès EJ, et al (2015) Controlling the false discovery rate via knockoffs. Ann Stat 43(5):2055–2085. https://doi.org/10.1214/15-AOS1337
5. Candès E, Fan Y, Janson L, Lv J (2018) Panning for gold: 'model-x' knockoffs for high dimensional controlled variable selection. J. R Stat Soc: Ser B (Stat Methodol) 80(3): 551–577. https://rss.onlinelibrary.wiley.com/doi/pdf/10.1111/rssb.12265
6. Stephens M (2017) False discovery rates: a new deal. Biostatistics 18(2):275–294. https://doi.org/10.1093/biostatistics/kxw041
7. Elias JE, Gygi SP (2007) Target-decoy search strategy for increased confidence in large-scale protein identifications by mass spectrometry. Nat Methods 4(3):207–214, https://doi.org/10.1038/nmeth1019
8. Käll L, Storey JD, MacCoss MJ, Noble WS (2008) Assigning significance to peptides identified by tandem mass spectrometry using decoy databases. J Proteome Res 7(01): 29–34. https://doi.org/10.1021/pr700600n
9. Couté Y, Bruley C, Burger T (2020) Beyond target-decoy competition: stable validation of peptide and protein identifications in mass spectrometry-based discovery proteomics.

Anal Chem 92(22):14898–14906. https://doi.org/10.1021/acs.analchem.0c00328

10. Emery K, Hasam S, Noble WS, Keich U (2019) Multiple competition-based FDR control for peptide detection. Preprint. https://arxiv.org/abs/1907.01458
11. He K, Fu Y, Zeng WF, Luo L, Chi H, Liu C, Qing LY, Sun RX, He SM (2015) A theoretical foundation of the target-decoy search strategy for false discovery rate control in proteomics. Preprint. https://arxiv.org/abs/1501.00537
12. Bouret P, Bastien F (2018) Erreurs et tests statistiques (40 min). https://hal.inria.fr/medihal-01774420/
13. Burger T (2018) Gentle introduction to the statistical foundations of false discovery rate in quantitative proteomics. J Proteome Res 17(1):12–22. https://doi.org/10.1021/acs.jproteome.7b00170
14. Hastie T, Efron B (2013) LARS: Least Angle Regression, Lasso and Forward Stagewise. R package version 1.2. https://CRAN.R-project.org/package=lars
15. Friedman J, Hastie J, Tibshirani R (2010) Regularization paths for generalized linear models via coordinate descent. J Stat Softw 33(1): 1–22. http://www.jstatsoft.org/v33/i01/
16. Giai-Gianetto Q, Combes F, Ramus C, Bruley C, Couté Y, Burger T (2019) cp4p: calibration plot for proteomics. R package version 0.3.6. https://CRAN.R-project.org/package=cp4p
17. Ramus C, Hovasse A, Marcellin M, Hesse AM, Mouton-Barbosa E, Bouyssié D, Vaca S, Carapito C, Chaoui K, Bruley C, Garin J, Cianférani S, Ferro M, Van Dorssaeler A, Burlet-Schiltz O, Schaeffer C, Couté Y, Gonzalez de Peredo A (2016) Benchmarking quantitative label-free LC–MS data processing workflows using a complex spiked proteomic standard dataset. J Proteom 132:51–62. https://www.sciencedirect.com/science/article/pii/S187439191530186X
18. Huber W, Carey VJ, Gentleman R, Anders S, Carlson M, Carvalho BS, Bravo HC, Davis S, Gatto L, Girke T, Gottardo R, Hahne F, Hansen KD, Irizarry RA, Lawrence M, Love MI, MacDonald J, Obenchain V, Ole's AK, Pag'es H, Reyes A, Shannon P, Smyth GK, Tenenbaum D, Waldron L, Morgan M (2015) Orchestrating high-throughput genomic analysis with Bioconductor. Nat Methods 12(2): 115–121. http://www.nature.com/nmeth/journal/v12/n2/full/nmeth.3252.html
19. Zou H, Hastie T (2005) Regularization and variable selection via the elastic net. J R Stat Soc: Ser B (Stat Methodol) 67(2): 301–320. https://rss.onlinelibrary.wiley.com/doi/pdf/10.1111/j.1467-9868.2005.00503.x
20. Nguyen TB, Chevalier JA, Thirion B, Arlot S (2020) Aggregation of multiple knockoffs. In: International conference on machine learning, PMLR, pp 7283–7293. http://proceedings.mlr.press/v119/nguyen20a.html
21. Keich U, Tamura K, Noble WS (2019) Averaging strategy to reduce variability in target-decoy estimates of false discovery rate. J Proteome Res 18(2):585–593. https://doi.org/10.1021/acs.jproteome.8b00802
22. Romano JP, Shaikh AM, et al. (2006) On step-down control of the false discovery proportion. In: Optimality, Institute of Mathematical Statistics, pp 33–50
23. Luo D, He Y, Emery K, Noble WS, Keich U (2020) Competition-based control of the false discovery proportion. Preprint. https://arxiv.org/abs/2011.11939
24. Ge Y, Dudoit S, Speed TP (2003) Resampling-based multiple testing for microarray data analysis. Test 12(1):1–77. https://doi.org/10.1007/BF02595811
25. Efron B, Hastie T, Johnstone I, Tibshirani R (2004) Least angle regression. Ann Stat 32(2): 407–499. https://doi.org/10.1214/009053604000000067

Chapter 2

A Pipeline for Peptide Detection Using Multiple Decoys

Syamand Hasam, Kristen Emery, William Stafford Noble, and Uri Keich

Abstract

Target–decoy competition has been commonly used for over a decade to control the false discovery rate when analyzing tandem mass spectrometry (MS/MS) data. We recently developed a framework that uses multiple decoys to increase the number of detected peptides in MS/MS data. Here, we present a pipeline of Apache licensed, open-source software that allows the user to readily take advantage of our framework.

Key words FDR control, Multiple decoys, Peptide detection, Tandem mass spectrometry

1 Introduction

Proteins are the primary functional molecules in living cells, and tandem mass spectrometry (MS/MS) currently provides the most efficient means of studying proteins in a high-throughput fashion. Knowledge of the protein complement in a cellular population provides insight into the functional state of the cells. Thus, MS/MS can be used to functionally characterize cell types, differentiation stages, disease states, or species-specific differences. For this reason, MS/MS is the driving technology for much of the rapidly growing field of proteomics.

A typical MS/MS experiment will generate thousands of spectra and canonically, each observed spectrum is generated by a single peptide. Thus, the first goals of the downstream analysis are to identify which peptide generated each of the observed spectra (the spectrum-ID problem) and to determine which peptides and which proteins were present in the sample (the peptide and protein detection problems).

Spectrum-ID is typically done by scanning each input spectrum against a *target* peptide database looking for the optimal peptide–spectrum match (PSM) for the given spectrum. Sometimes, this optimal PSM is correct—the peptide assigned to the spectrum was present in the mass spectrometer when the spectrum was

Thomas Burger (ed.), *Statistical Analysis of Proteomic Data: Methods and Tools*,
Methods in Molecular Biology, vol. 2426, https://doi.org/10.1007/978-1-0716-1967-4_2,

generated—and sometimes the PSM is incorrect. Ideally, we would report only the correct PSMs, but obviously we do not know which PSMs are correct and which are incorrect: all we have is the score of the PSM, indicating its quality.

Target–decoy competition (TDC) is commonly used to determine which PSMs to report while controlling the false discovery rate (FDR) [1]. The FDR is the expected value of the false discovery proportion (FDP): the proportion of incorrect/false PSMs among all reported PSMs. TDC works by comparing the results of two searches: a search against a database containing real ("target") peptides and a search against a database containing randomly shuffled or reversed ("decoy") versions of the target peptides. More precisely, each optimal target PSM is compared with its corresponding optimal decoy PSM, and only the higher of the two is kept together with a label indicating whether the PSM involves a target or a decoy peptide. Note that in practice this TDC is often done implicitly by searching the concatenated target–decoy database.

Subsequently, based on the assumption that an incorrect PSM is equally likely to be a target win or a decoy win—an incorrect PSM is a random match and for those the target database should offer no advantage over the decoy—the number of decoy wins among the top scoring PSMs is used to estimate the number of false discoveries (and therefore the false discovery rate) in the list of top target wins. To control the FDR at level α, we choose the largest list of top target PSMs for which the above FDR estimate (using the +1 correction of [2]) is still $\leq \alpha$.

The same TDC can be used for the peptide detection problem: instead of looking at PSMs, we are now considering peptides, and each target or decoy peptide is assigned a score that is the maximum of all the PSM scores that were optimally matched to this peptide [3]. We next compare each target peptide score with the score of its corresponding decoy, and we keep only the higher of the two together with a target/decoy label. The rest continues exactly as in the spectrum-ID problem.

We recently developed tools that allow us to take advantage of multiple decoy databases (obtained by multiple random shuffles of the target database) in order to reduce the variability and/or increase the number of reported PSMs or peptides while controlling the FDR. In doing so, we found that when it comes to using multiple decoy databases (multiple decoys for short), the spectrum-ID and the peptide detection problem are somewhat different. Specifically, our recently developed multi-decoy competition framework presented in [4] is applicable to peptide detection but not to spectrum-ID ([4, 5] provide further details). Here, we offer a hands-on protocol for how to use this framework that is implemented in the `multicomp` `R` package [4]. For the spectrum-ID problem, however, we recommend using our "average target–decoy competition" (aTDC) [6, 7], which is implemented in the Crux mass spectrometry analysis toolkit [8].

Our protocol is described in detail starting with the next section (*see* Subheading 2.1), and the rest of this introduction is dedicated to give some intuition on how our method works. In our multi-decoy competition framework, each target peptide competes against multiple randomly shuffled peptides rather than a single one as in TDC. Each of those target and decoy peptides is assigned a score as in the single decoy case: each spectrum is searched against each database, its optimal PSM is noted, and the maximum of all PSM scores involving any specific peptide is assigned to that peptide.

To understand how we use multiple decoys to control the FDR, it is worth looking at the so-called "mirror method," which is one of several different such procedures that our multiple decoy framework encompasses [4]. The mirror method declares a target peptide win if its score is higher than more than half of the scores of the decoy peptides derived from that target peptide (assume for simplicity that the number of decoys is odd); otherwise, the method declares a decoy win. Like in TDC, in case of a target win, the winning score is that target peptide score. However, in case of a decoy win, the winning score is the score of the decoy peptide whose rank mirrors that of the target win (Fig. 1). In this way, every peptide is once again associated with a label (target or decoy win) and a (winning) score just as in TDC (which is equivalent to the mirror method with a single decoy). The rest follows exactly like in TDC with the justification being that once again we can expect the incorrectly detected peptides to split evenly between the target and decoy wins (*see* **Note 1**).

More generally, our multiple decoy framework allows us to define different thresholds for a target win (how many decoy scores does the target score need to be higher than) and a decoy win (how many decoy scores does the target score need to be smaller than). Choosing these thresholds has to be done carefully, and our `multicomp` package implements an intricate bootstrap strategy called "labeled bootstrap monitored maximization" (LBM) that tries to select the optimal thresholds while still controlling the FDR [4].

This protocol introduces an `R` script that allows the user to readily apply LBM for peptide detection. The script relies on Crux to generate the multiple decoy databases and uses the Tide search engine to find the optimal PSM for each spectrum in each database, which it then converts to the peptide score of each peptide. The script then invokes `multicomp` that applies LBM to these generated target and decoy peptide scores, and it outputs the peptides detected at the desired level of FDR.

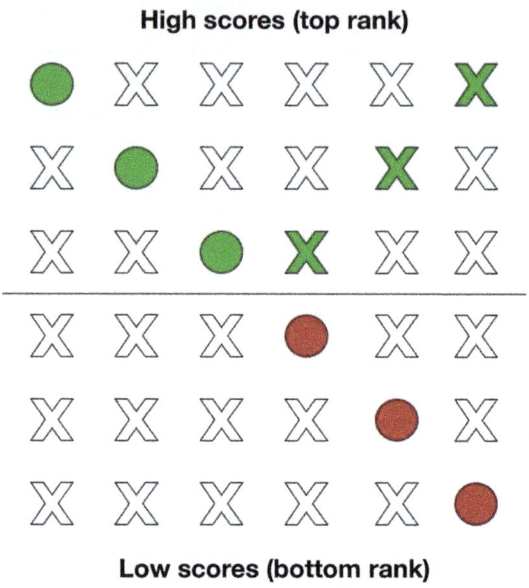

Fig. 1 Schematic description of the mirror method using 5 decoys. Each column of symbols represents one possible ordering of the target (circle) and its 5 decoy (crosses) peptide scores. The midline is the mirror method's target-win cutoff, and the green color represents the winning scores (red circles correspond to decoy wins)

2 Material

2.1 Software Requirements

1. The operating system can be Linux, Mac OS X, or Windows 10 (the latter using the built-in Unix subsystem).
2. A recent version of `R` should be installed in a directory in which the user has read and write permissions.
3. The `multicomp R` package [9] can be downloaded from https://bitbucket.org/noblelab/multi-competition-fdr/src/master/. Direct instructions for installation can be found below (*see* Subheading 2.2).
4. The Crux mass spectrometry analysis toolkit [8] should be downloaded into the working directory.
5. The interface script, called `multidecoy_peptide.R`, that takes the mass spectrometry files and runs `Crux` as well as the relevant algorithms from `multicomp` should also be in the working directory.

2.2 Software Installation

1. To install the `multicomp` package, run the following command on the `R` command line:

```
install.packages("https"://bitbucket.org/noblelab/
    multi-competition-fdr/raw
    /40693d05b381497df360f4f15d0d2da75d471b47/
    multicomp_0.2.0.tar.gz", repos = NULL)
```

2. To install the Crux mass spectrometry analysis toolkit [8], visit the `crux.ms` website and download the latest version of Crux for your operating system. Unzip the downloaded folder and place it in the same working directory where the R script below and the data will reside.
3. The `R` interface script, `multidecoy_peptides.R`, can be downloaded from https://bitbucket.org/noblelab/multi-competition-fdr.

2.3 Data Format

The data is first run through Crux, which requires a FASTA file containing the database being searched, and a file consisting of the mass spectra which may be in MS2, MGF, or mzML/mzXML. For a full description of the different files that the Crux system takes, visit http://crux.ms/fileformats.html. The interface script (*see* Subheading 2.1) makes reference to the `tide-search` function, and so only files accepted by that function will be accepted by this script.

3 Methods

The technique requires specification of parameters in a text file that acts as an interface with `crux` and `multicomp`. The parameters file consists of a number of lines that define variables to be used in the script. There should not be any whitespace except in the names of the variables. An example of such a parameters file is given below.

3.1 Parameter Specification

1. On a new line in the parameters file, specify the path to the FASTA file for the database (*see* **Note 2**):

```
fasta_file="[INSERT PATH TO FASTA FILE HERE]"
```

2. On a new line in the parameters file, specify the path to the spectrum file(s) for the dataset. Multiple files may be specified using standard regular expression (*see* **Note 3**) notation:

```
spectra_file="[INSERT PATH TO SPECTRUM FILE HERE]"
```

3. On a new line in the parameters file, specify the scoring method used to measure the quality of peptide–spectrum match. When using XCorr, then “xcorr” must be specified. When using p-values, then “pvalue” must be specified:

```
scoring="[EITHER xcorr OR pvalue]"
```

4. On a new line in the parameters file, specify the number of decoys used for the multiple decoy FDR competition algorithm:

```
n_decoys=[NB OF DECOYS USED FOR MULTIPLE FDR COMPETITION]
```

5. On a new line in the parameters file, specify the FDR threshold for which to report the discovered peptides:

```
fdr_threshold=[INSERT VALUE OF FDR THRESHOLD]
```

6. On a new line in the parameters file, specify the path to the output file with which to report the peptide discoveries:

```
output_file="[PATH TO OUTPUT FILE]"
```

7. (OPTIONAL)
   (a) On a new line in the parameters file, specify whether you want to keep the PSMs that are the output of Crux (T); otherwise, they are ordinarily deleted after the output is printed (default = F). If you are planning to run the script on the same dataset multiple times, keeping this option as (T) will avoid having to search the spectra for the PSMs again:

```
keep_psms=[EITHER T (TRUE) OR F (FALSE)]
```

   (b) On a new line in the parameters file, specify a name for your dataset. This name is only relevant should you choose to keep the PSM output from Crux. If you are running this script on different sets of data in the same working directory, you should use different names for each dataset:

```
name="[INSERT YOUR DATA NAME HERE]"
```

8. (OPTIONAL) On a new line in the parameters file, specify any extra parameters to be passed to the tide search function of the Crux package. The parameters must be specified in the same way they would be specified when running Crux directly from the command line. For a full list of the parameters that tide search can take, visit http://crux.ms/commands/tide-search.html. You can refer to the example at the end of this protocol to see how these extra parameters may be specified:

```
extra_tide_search_parameters="[INSERT EXTRA PARAMETERS]"
```

9. (OPTIONAL) On a new line in the parameters file, specify the seed to be used when running the script. This is mainly useful for running the script on the same dataset multiple times to verify results:

```
seed=[INSERT NUMERICAL SEED]
```

3.2 Obtaining the Discoveries

1. Place the parameters file that is appropriately specified (*see* Subheading 3.1) into the working directory.
2. Place the `multidecoy_peptides.R` script into the working directory.
3. Run the following command in the command line.

```
Rscript multidecoy_peptides.R data_parameters.txt
```

where `data_parameters.txt` is the parameters file specified in step 1.

4. The peptide discoveries are found as a newline separated file with the name specified in the parameters file in step 1.

3.3 Example

In this example, we use publicly available spectra and proteome data for *Saccharomyces cerevisiae* to test our script, to find the peptide discoveries at an FDR of 0.01 using the multi-decoy competition algorithm described in this protocol with 5 decoys. The spectrum data, derived from [10], can be downloaded from the PRIDE [11] database, project PXD002726, or via the following link: ftp://ftp.pride.ebi.ac.uk/pride/data/archive/2016/04/PXD002726/Yeast_In-gel_digest_2.mgf. A FASTA file associated with the same paper is found on the same PRIDE website. It can also be downloaded via the following link: ftp://ftp.pride.ebi.ac.uk/pride/data/archive/2016/04/PXD002726/KFSwYeastCrTry_150506.fasta. If either of the above links is not available, then this data can be accessed on the repository from which the script is derived (*see* Subheading 2.2).

Below, an example is shown of a parameters file which searches the above spectrum file against the peptide database derived from the above FASTA file (both of which are found within the working directory). There are 5 decoys used for the multi-decoy competition, and the peptide discoveries at FDR 0.01 are reported in `"yeast_discoveries.txt"`:

```
fasta_file="KFSwYeastCrTry_150506.fasta"
spectra_file="Yeast_In-gel_digest_2.mgf"
scoring="xcorr"
n_decoys=5
fdr_threshold=0.01
output_file="yeast_discoveries.txt"
keep_psms=T
name="yeast"

# Next parameter setting (extra_tide_search_parameters)
# should be single lined, as in:
# extra_tide_search_parameters="--precursor-window 10 --pr...
# (replace new lines/tabs by spaces)
extra_tide_search_parameters="--precursor-window 10
                              --precursor-window-type ppm
                              --mz-bin-width 0.02
                              --pm-min-peak-pairs 100
                              --pm-charges 2"

seed=123
```

Saving the above parameters to the file `yeast_parameters.txt` (*see* **Note 4**) and running the script

```
~/working_directory$ Rscript multidecoy_peptides.R
yeast_parameters.txt
~/working_directory$ head yeast_discoveries.txt
```

will yield the following:

```
LGEHNIDVLEGNEQFINAAK
GGSGGSHGGGSGFGGESGGSYGGGEEASGSGGGYGGGSGK
FSSCGGGGGSFGAGGGFGSR
GGGGSFGYSYGGGSGGGFSASSLGGGFGGGSR
GGGGGGYGSGGSSYGSGGGSYGSGGGGGGGR
...
```

Finally, the following command

```
~/working_directory$ wc -l yeast_discoveries.txt
```

indicates there are 202 peptide discoveries made at FDR 0.01 saved to `yeast_discoveries.txt`:

```
202 yeast_discoveries.txt
```

Note that in the preceding example, the seed was set to a value of 123 (*see* Subheading 3.2 for how to set the seed). The version of Crux used was the Linux 64-bit 33.2.ea362c4 version (*see* **Note 5**).

4 Notes

1. The theoretical foundation that ensures our multi-decoy setup controls the FDR was laid by Barber and Candés in [12].
2. In all the commands in this section (*see* Subheading 3.1), the square brackets and text inside them should be replaced, but the quotation marks should be kept.
3. Regular expressions are ways to search for and specify multiple different strings. For example, using "dataset_*.mgf" will specify all datasets of the form "dataset_1.mgf", "dataset_2.mgf", and so on as the shell will look for any character set to substitute in for the asterisk.
4. Make sure each line contains a single parameter, and in particular, note the line starting with "`extra_tide_search_parameters`" extends here on multiple lines due to formatting constraints but should be a *single* line in the file.
5. One may find small differences in the total number of discoveries based on the version of Crux and OS used due to differences in the random generation of the decoy databases.

References

1. Elias JE, Gygi SP (2007) Target-decoy search strategy for increased confidence in large-scale protein identifications by mass spectrometry. Nat. Methods 4(3):207–214. https://doi.org/10.1038/nmeth1019
2. He K, Fu Y, Zeng WF, Luo L, Chi H, Liu C, Qing LY, Sun RX, He SM (2015) A theoretical foundation of the target-decoy search strategy for false discovery rate control in proteomics. https://arxiv.org/abs/1501.00537v1
3. Granholm V, Navarro JF, Noble WS, Käll L (2013) Determining the calibration of confidence estimation procedures for unique peptides in shotgun proteomics. J Proteom. 80: 123–131. https://doi.org/10.1016/j.jprot.2012.12.007
4. Emery K, Hasam S, Noble WS, Keich U (2020) Multiple competition-based FDR control and its application to peptide detection. In: Lecture notes in computer science. Springer International Publishing, Cham, pp 54–71. https://doi.org/10.1007/978-3-030-45257-5_4
5. Keich U, Noble WS (2017) Controlling the FDR in imperfect database matches applied to tandem mass spectrum identification. J Amer Statist Assoc 113:973–982. https://doi.org/10.1080/01621459.2017.1375931
6. Keich U, Noble WS (2017) Progressive calibration and averaging for tandem mass spectrometry statistical confidence estimation: Why settle for a single decoy. In: Sahinalp S (ed) Proceedings of the international conference on research in computational biology (RECOMB). Springer, Lecture Notes in Computer Science, vol 10229. Springer, Berlin, pp 99–116. https://doi.org/10.1007/978-3-319-56970-3_7
7. Keich U, Tamura K, Noble WS (2018) Averaging strategy to reduce variability in target-decoy estimates of false discovery rate. J Proteome Res 18(2):585–593. https://doi.org/10.1021/acs.jproteome.8b00802
8. Park CY, Klammer AA, Käll L, MacCoss MJ, Noble WS (2008) Rapid and accurate peptide identification from tandem mass spectra. J Proteome Res 7(7):3022–3027. https://doi.org/10.1021/pr800127y
9. Emery K (2019) Multicomp: Multiple Competition FDR Control. R package version 0.2.0
10. Schittmayer M, Fritz K, Liesinger L, Griss J, Birner-Gruenberger R (2016) Cleaning out the litterbox of proteomic scientists' favorite pet: optimized data analysis avoiding trypsin artifacts. J Proteome Res 15(4): 1222–1229. https://doi.org/10.1021/acs.jproteome.5b01105

11. Perez-Riverol Y, Csordas A, Bai J, Bernal-Llinares M, Hewapathirana S, Kundu DJ, Inuganti A, Griss J, Mayer G, Eisenacher M, Pérez E, Uszkoreit J, Pfeuffer J, Sachsenberg T, Yılmaz Ş, Tiwary S, Cox J, Audain E, Walzer M, Jarnuczak AF, Ternent T, Brazma A, Vizcaíno JA (2018) The PRIDE database and related tools and resources in 2019: improving support for quantification data. Nucleic Acids Res 47(D1):D442–D450. https://doi.org/10.1093/nar/gky1106
12. Barber RF, Candès EJ (2015) Controlling the false discovery rate via knockoffs. Ann Statist 43(5):2055–2085. https://doi.org/10.1214/15-AOS1337

Chapter 3

Enhanced Proteomic Data Analysis with MetaMorpheus

Rachel M. Miller, Robert J. Millikin, Zach Rolfs, Michael R. Shortreed, and Lloyd M. Smith

Abstract

MetaMorpheus is a free and open-source software program dedicated to the comprehensive analysis of proteomic data. In bottom-up proteomics, protein samples are digested into peptides prior to chromatographic separation and tandem mass spectrometric analysis. The resulting fragmentation spectra are subsequently analyzed with search software programs to obtain peptide identifications and infer the presence of proteins in the samples. MetaMorpheus seeks to maximize the information gleaned from proteomic data through the use of (a) mass calibration, (b) post-translational modification discovery, (c) multiple search algorithms, which aid in the analysis of data from traditional, crosslinking, and glycoproteomic experiments, (d) isotope-based or label-free quantification, (e) multi-protease protein inference, and (f) spectral annotation and data visualization capabilities. This protocol provides detailed descriptions of how use MetaMorpheus and how to customize data analysis workflows using MetaMorpheus tasks to meet the specific needs of the user.

Key words Proteomics, Tandem mass spectrometry, Bottom-up, Database search, Open-source, Post-translational modification discovery, Crosslink, Glycopeptides

1 Introduction

Bottom-up proteomics is the foremost method for in-depth identification and characterization of proteins from biological systems. In bottom-up proteomics, proteins are digested generating peptides, which are separated, typically by reverse phase high-performance liquid chromatography (RP-HPLC), before mass spectrometric analysis [1]. As peptides elute from the HPLC, they are emitted into the mass spectrometer (MS) using electrospray ionization. Inside the MS, an MS1 spectrum is acquired to determine the mass-to-charge ratio of the eluting intact peptides. These peptides are then isolated and fragmented to produce MS2 spectra, which serve as "bar-codes" to identify each peptide's amino acid sequence. MS data acquisition is a repetitive process, wherein the acquisition of an MS1 spectrum is followed by acquisition of several

Thomas Burger (ed.), *Statistical Analysis of Proteomic Data: Methods and Tools*,
Methods in Molecular Biology, vol. 2426, https://doi.org/10.1007/978-1-0716-1967-4_3,

MS2 spectra. These experiments generate too much data for reliable and efficient manual analysis, thus database search software programs are commonly employed [2]. These programs compare the observed fragmentation data (MS2 spectra) with the predicted fragment ions of theoretical peptides derived from a reference database.

The principal goal of database search software programs is to correctly identify as many peptide sequences as possible from the acquired fragmentation spectra [3–6]. As the number of quality peptide identifications increases, the ability to accurately and comprehensively characterize the proteins present in the sample also improves. This is important for both discovery-based studies seeking to identify the proteins present, and quantitative studies seeking to determine protein abundance changes.

MetaMorpheus is a free and open-source database search software program designed to be user-friendly and maximize the information extracted from bottom-up proteomic experiments. MetaMorpheus contains several features to facilitate extensive data analysis, including (a) mass calibration, (b) post-translational modification (PTM) discovery [7], (c) specialized search algorithms to aid in the analysis of data from various proteomic experiments (traditional, crosslink [8], and glycoproteomic [9]), (d) isotope-based or label-free quantification [10], (e) multi-protease protein inference [11], and (f) spectral annotation and data visualization capabilities. This protocol is devoted to presenting how to construct MetaMorpheus data analysis workflows, enabling users to customize their experience based on their specific experiment. MetaMorpheus version 0.0.310 was used in the development of this protocol (*see* **Note 1**).

2 Material

2.1 Mass Spectra Requirements

MetaMorpheus accepts spectra in .mzML (centroided), .mgf, or Thermo .raw formats. Other formats can be converted to .mzML with MSConvert (*see* **Note 2**). Regardless of the format, all mass spectra must contain MS2 scans. MetaMorpheus was originally designed for the analysis of high-resolution MS2 data but has since been adapted for the analysis of low-resolution MS2 data (*see* **Note 3**).

2.2 Protein Database Requirements

Protein databases for analysis can be supplied in UniProt .XML (*see* **Note 4**) or .FASTA formats in either their compressed (.gz) or uncompressed states.

2.3 System Requirements

1. MetaMorpheus can be installed and operated on any desktop computer with a Windows, MacOS, or Linux 64-bit operating system via the command-line interface (CLI). However, the

graphical user interface (GUI) version is currently Windows-only.

2. There is no formal RAM requirement for MetaMorpheus, but a minimum of 8 GB of RAM is recommended (*see* **Note 5**).
3. MetaMorpheus requires the installation of .NET Core 3.1 (*see* **Note 6**).

2.4 Download and Installation

MetaMorpheus can be installed and operated as a graphical user interface (GUI) program in the Microsoft Windows environment or as a command-line interface (CLI) in Microsoft Windows, Apple MacOS, or Linux environments. The latest release of MetaMorpheus can be retrieved from GitHub (https://github.com/smith-chem-wisc/MetaMorpheus/releases). To download the GUI from GitHub, click on the `MetaMorpheusInstaller.msi` for the latest, or desired release. Follow the directions provided by the installer to complete the installation process. The command-line version of MetaMorpheus can be downloaded by selecting `MetaMorpheus_CommandLine.zip` for the latest or desired release. Once the zip file has downloaded, contents must be extracted to use the program. A Docker image of the command-line version of MetaMorpheus can be retrieved from Docker Hub by using the following Docker Pull Command:

```
docker pull smithchemwisc/metamorpheus
```

3 Methods

MetaMorpheus, similar to many other search software tools, takes in user-supplied spectra files and protein databases. However, a distinguishing feature of MetaMorpheus is the construction of custom data analysis workflows from individual analysis modules called tasks. Tasks, once added to a workflow, are run sequentially, and when appropriate, use information from previous tasks to improve the overall results. The use of individual tasks enables the user to mix and match different analyses to best meet their specific needs. The following protocol will explain the basic set up of MetaMorpheus as well as how to set up each individual task for the creation of custom data analysis workflows.

3.1 Starting MetaMorpheus

Open MetaMorpheus from the start menu or double click the MetaMorpheus desktop icon to open the GUI (*see* Fig. 1).

3.2 Loading Protein Databases

1. Select the Databases tab in the menu on the left side of the GUI to open the Protein Databases loading page (*see* Fig. 2).

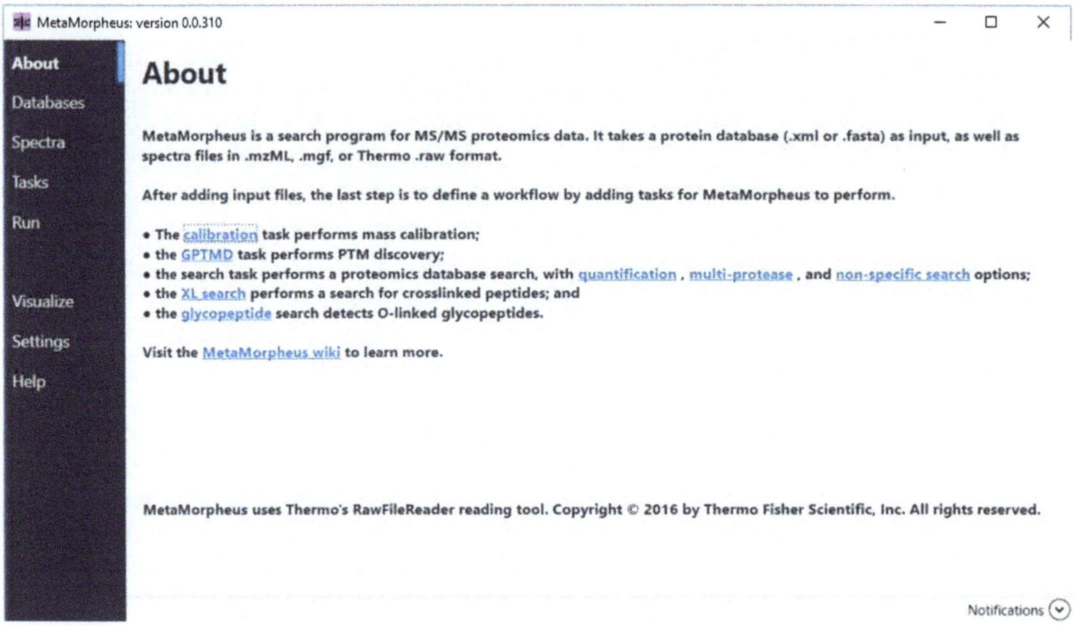

Fig. 1 MetaMorpheus application opens to the About window

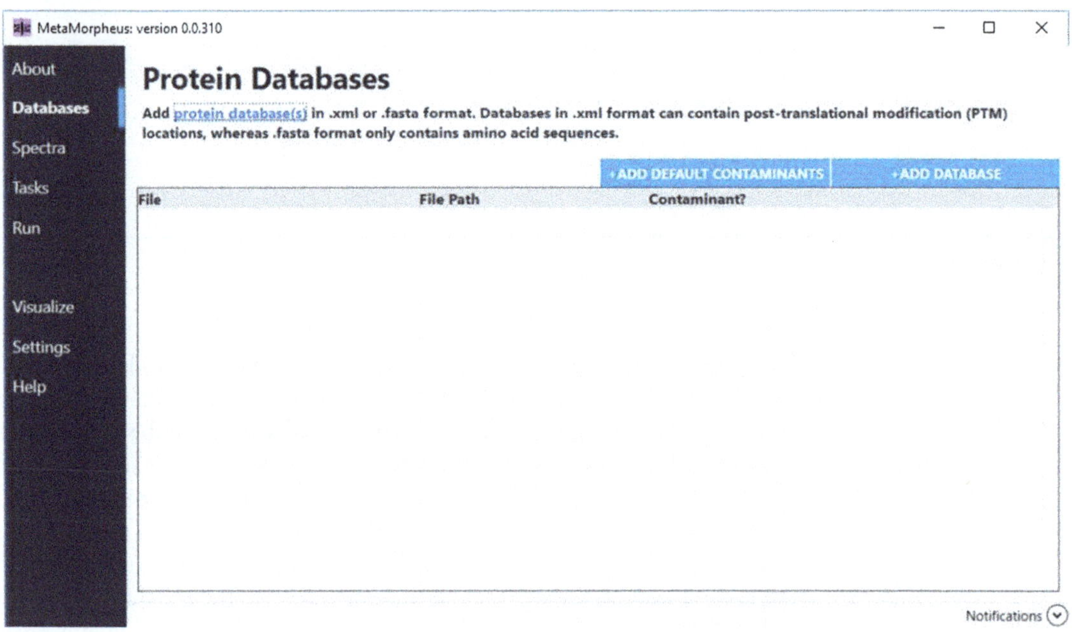

Fig. 2 Protein Databases window where users specify .FASTA or .XML databases for analysis

2. Database files can be added by two different approaches: (a) dragging and dropping database file(s) into the GUI (*see* **Note 7**), or (b) selecting the +ADD DATABASE button to open a file browsing window enabling navigation to the desired database file(s) for selection.

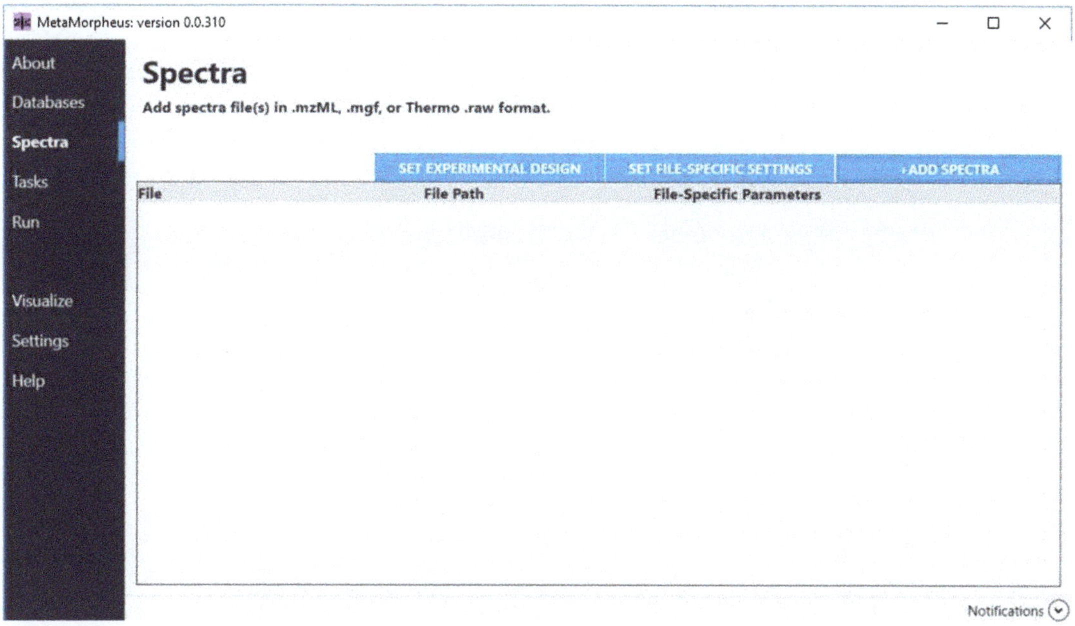

Fig. 3 Spectra window where users specify spectra files for analysis

3. MetaMorpheus accommodates the use of more than one protein database. This enables the use of a contaminant database as well as the ability to analyze multi-species samples.
4. To add a database of common contaminant protein sequences, click the ADD DEFAULT CONTAMINANTS button (*see* **Note 8**). Users can also supply their own contaminant databases. To have MetaMorpheus recognize a provided database as a contaminant database, right click the file and select "Set as contaminant database".

3.3 Loading Spectra Files

1. Select the Spectra tab in the menu on the left side of the GUI to open the Spectra loading page (*see* Fig. 3).
2. Mass spectra files can be added in a similar manner to databases either by (a) dragging and dropping the spectra file into the GUI (*see* **Note 7**) or by (b) selecting the +ADD SPECTRA button (*see* **Note 9**).

3.4 Set File-Specific Settings

1. The ability to specify file-specific search settings is an optional feature of MetaMorpheus that facilitates analysis of complex datasets containing samples generated by differing preparation methods; such as multi-protease experiments.
2. Select file(s) in the Spectra window requiring a particular file-specific setting.

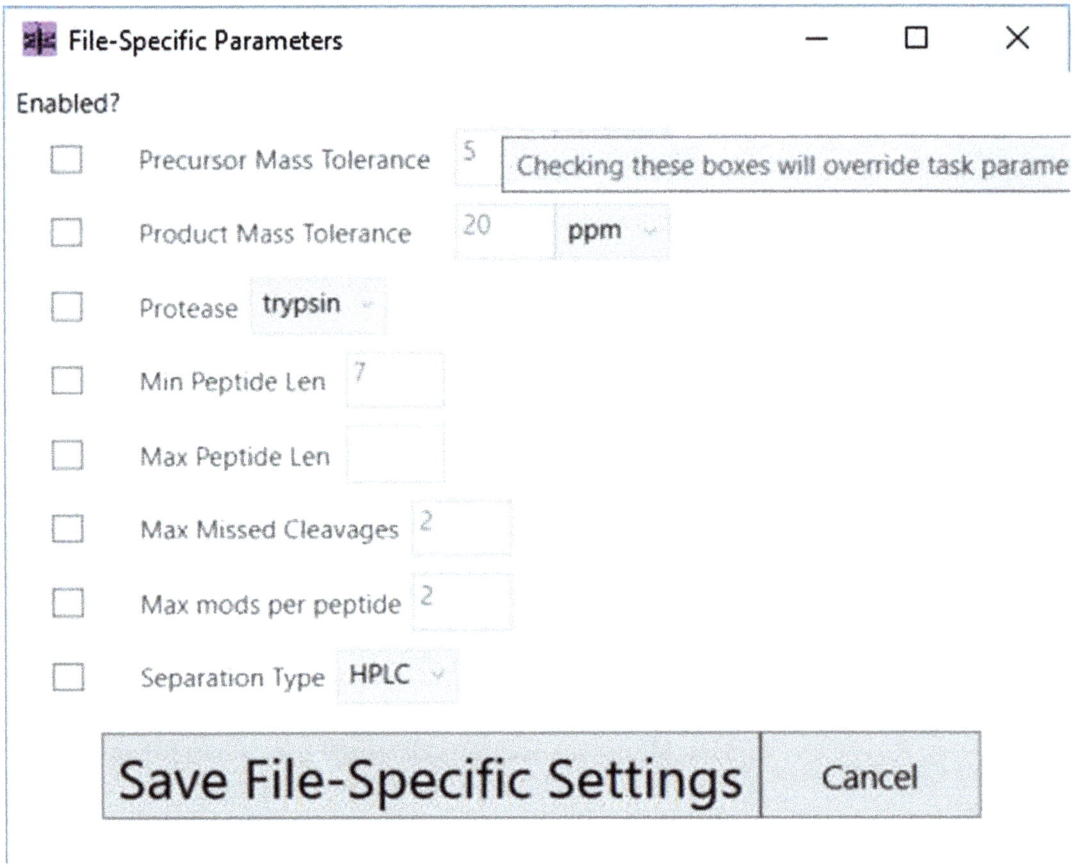

Fig. 4 File-specific parameters window

3. Click the SET FILE-SPECIFIC SETTINGS button to open a window displaying a list of parameters, which can be set on a file-specific basis. All parameters are disabled by default (*see* **Note 10**) (*see* Fig. 4).
4. Click the check box next to each parameter to enable editing (*see* **Note 11**).

3.5 Mass Calibration

The mass accuracy of MS1 and MS2 spectra can vary within a sample, or over multiple samples. Numerous factors can contribute to this variance such as systematic drift, random noise, changes in power supply voltage, vacuum system stability, and varying temperature or humidity. Spectral mass calibration corrects for this undesirable variance and almost always improves the number of confident peptide identifications made by MetaMorpheus [7]. Although it is not strictly necessary, the addition of a Calibration Task to any analysis workflow is recommended.

1. Select the Tasks tab in the menu on the left side of the GUI to open the Tasks page (*see* Fig. 5).

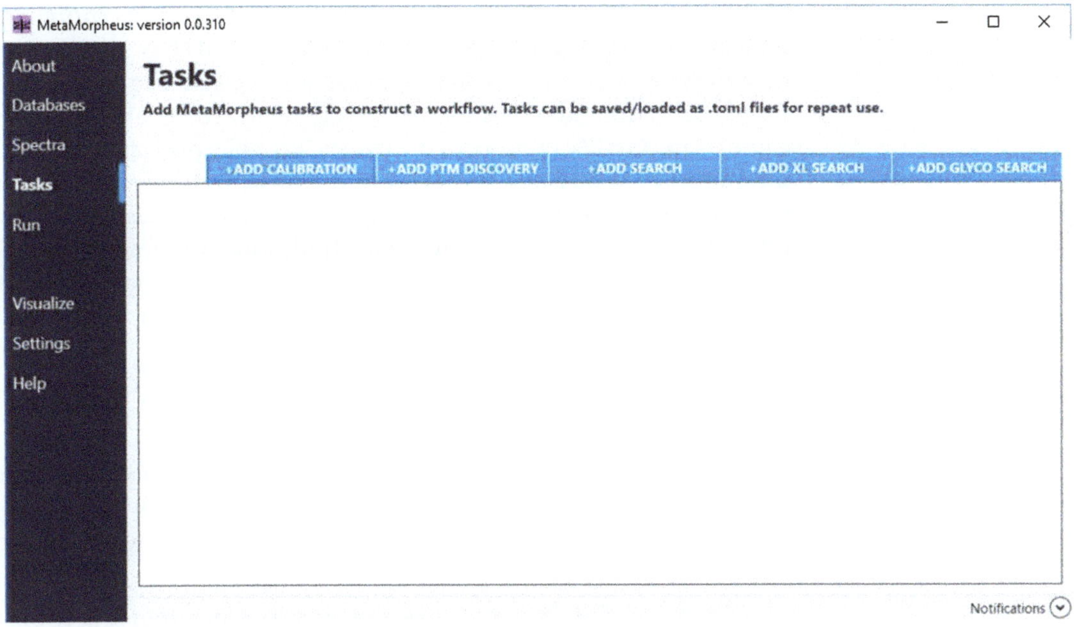

Fig. 5 Task window where users can add MetaMorpheus tasks to the data analysis workflow

2. Select the +ADD CALIBRATION button above the task panel to open the Calibrate Task window where all parameters for the Calibration Task can be adjusted as needed.
3. The Calibrate Task window has two parameter sections: (a) Search Parameters and (b) Modifications.
4. Expansion of the "Search Parameters" section exposes parameters affecting the search component of the mass calibration algorithm. A description of each of these parameters can be found below (*see* Subheading 3.15, **Steps 10**, **18**, **24**, **25**, **26**, **27**, **28**, **29**, **31**, **32**, **34**, **49**, **50** and **51**).
5. The selection of the drop-down arrow beneath the "Modifications" header displays lists of PTMs that can be selected as fixed (*see* Subheading 3.15, **Step 12**) or variable modifications (*see* Subheading 3.15, **Step 64**).
6. After parameters have been adjusted, select the "Add the Calibrate Task" button at the bottom of the window to add the Calibration Task to the MetaMorpheus analysis workflow (*see* **Notes 12**, **13** and **14**).

3.6 Global Post-Translational Modification Discovery

Global Post-Translational Modification Discovery (GPTMD) is a tool within MetaMorpheus that enables the identification of peptides containing PTMs, which are not annotated in the supplied database or provided as variable modifications [7]. GPTMD constructs a new reference database in .XML format by annotating the existing database with PTMs discovered by a first-pass search of the provided spectra. It utilizes a mass-tolerant search approach to identify peptide spectral matches (PSMs) with mass shifts (notches) indicative of PTMs selected by the user. GPTMD is superior to many other open-mass search approaches because the discovered PTMs are annotated, instead of being provided as a delta mass value. The GPTMD approach enables variable post-translational modification at targeted positions in the protein to improve search results without incurring the inflated search times or false discovery rates (FDR) typically associated with traditional variable modification searching. GPTMD illuminates larger, and potentially biologically interesting, portions of the proteome, which are not identified by other approaches.

1. Select the Tasks tab in the menu on the left side of the GUI to open the Tasks page (*see* Fig. 5).
2. Select the +ADD PTM DISCOVERY button above the task panel to open the GPTMD Task window where all parameters required for post-translational modification discovery can be adjusted.
3. The GPTMD Task window contains four parameter sections: (a) File Loading Parameters, (b) Search Parameters, (c) Fixed/Variable Modifications and (d) GPTMD Modifications.
4. Parameters for spectral processing can be found beneath the "File Loading Parameters" header. A detailed description of the involved parameters can be found below (*see* Subheading 3.15, **Steps 8**, **9**, **33**, **58**, **60**, **61** and **63**).
5. Parameters for the first-pass search used for the discovery of candidate post-translationally modified peptides are located under the "Search Parameters" header. A detailed description of the utilized parameters can be found below (*see* Subheading 3.15, **Steps 10**, **13**, **18**, **24**, **25**, **26**, **27**, **28**, **29**, **31**, **32**, **34**, **49**, **50** and **51**).
6. Selection of the drop-down arrow under the "Fixed/Variable Modifications" header displays lists of PTMs that can be selected as fixed (*see* Subheading 3.15, **Step 12**) or variable modifications (*see* Subheading 3.15, **Step 64**).
7. Expansion of the "GPTMD Modifications" section displays the same list of PTMs found in the "Fixed/Variable Modifications" panel. Selection of a PTM adds its mass shift to the list of notches investigated to identify previously unannotated PTMs (*see* **Notes 15** and **16**). Any modifications, except variable oxidation of methionine and fixed carbamidomethylation of cysteine, should be added here. Custom PTMs can be added to MetaMorpheus (*see* **Note 17**).
8. After parameters for the task have been finalized, select the "Add the GPTMD Task" button at the bottom of the window to add the GPTMD Task to the MetaMorpheus analysis workflow (*see* **Notes 12**, **13** and **14**).

3.7 Search

1. Select the Tasks tab in the menu on the left side of the GUI to open the Tasks page (*see* Fig. 5).
2. Select the +ADD SEARCH button above the task panel to open the Search Task window where all parameters for the final search of the spectra can be adjusted to fit the user's needs.
3. The Search Task window will open to display five expanded sections of parameters: (a) Search Parameters, (b) Modifications, (c) Protein Parsimony, (d) Quantification, and (e) Output Options. Additionally, there is a drop-down menu containing Advanced Options.
4. The parameters provided in the "Search Parameters" section are the most commonly adjusted. Detailed definitions of each of the parameters in this section can be found below (*see* Subheading 3.15, **Steps 10**, **18**, **25**, **27**, **28**, **31**, **49**, **50**, **51** and **56**.
5. Lists containing all PTM options for fixed (*see* Subheading 3.15, **Step 12**) and variable modifications (*see* Subheading 3.15, **Step 64**) are displayed beneath the "Modifications" header.
6. The parameters provided in the "Protein Parsimony" section dictate how peptide identifications undergo protein inference, the process by which peptides are mapped back to their potential protein(s) of origin. Detailed descriptions for each parameter can be found below (*see* Subheading 3.15, **Steps 1**, **54** and **59**).
7. Quantification parameters inform peptide and protein quantification. Parameters dealing with SILAC and label-free quantification are present within this section and are defined below (*see* Subheading 3.15, **Steps 20**, **22**, **39**, **42**, **48** and**57**).
8. The "Output Options" section contains parameters dictating what results are reported and how these results are exported. Detailed definitions of all the "Output Options" parameters can be found below (*see* Subheading 3.15, **Steps 3**, **11**, **65**, **67**, **68** and **69**).
9. Selection of the "Advanced Options" menu reveals three additional parameter sections: (a) File Loading Parameters, (b) Search Parameters and (c) Post-Search Analysis. Each of these sections contains parameters only adjusted, typically, by more advanced users of MetaMorpheus.
10. The advanced "File Loading Parameters" section contains parameters regarding spectra file loading and pre-processing. A detailed description of all the involved parameters can be found below (*see* Subheading 3.15, **Steps 8**, **9**, **33**, **40**, **41**, **45**, **58**, **60** and **61**

11. The options provided in the advanced "Search Parameters" section are not commonly changed. They provide the user the ability to alter miscellaneous details such as the search algorithm used by MetaMorpheus or how decoy proteins are generated. Detailed definitions of these advanced parameters can be found below (*see* Subheading 3.15, **Steps 7, 13, 14, 15, 17, 21, 23, 24, 26, 29, 32, 34, 43, 44, 53, 55** and **62**).
12. The final set of advanced parameters, the "Post-Search Analysis" section, dictates post-analysis processing. A detailed description of these parameters, and their uses can be found below (*see* Subheading 3.15, **Steps 4** and **70**).
13. After parameters for the task have been finalized, select the "Add the Search Task" button at the bottom of the window to add the Search Task to the MetaMorpheus analysis workflow (*see* **Notes 12, 13** and **14**).

3.8 Multi-Protease Protein Inference

1. To utilize MetaMorpheus' multi-protease protein inference capabilities, mass spectra files from more than one proteolytic digest must be loaded into the Spectra tab of the GUI.
2. Select a subset of files from the same proteolytic digest and use the SET FILE-SPECIFIC PARAMETERS button to set the appropriate protease (*see* **Note 11**). Repeat this process until all files have a file-specific protease assigned.
3. It is recommended to perform Calibration (*see* Subheading 3.5) and GPTMD (*see* Subheading 3.6) tasks prior to searching.
4. Add a Search Task to the MetaMorpheus workflow, ensuring the protein parsimony parameter (*see* Subheading 3.15, **Step 1**) is enabled (*see* **Note 18**).

3.9 Crosslink Search

1. Select the Tasks tab in the menu on the left side of the GUI to open the Task page (*see* Fig. 5).
2. Select the +ADD XL SEARCH button above the task panel to open the XL Search Task window where all parameters for the search of crosslinked peptides can be adjusted [8].
3. The XL Search Task window contains four sections of parameters: (a) Crosslink Search, (b) Search Parameters, (c) Modifications, and (d) Output Options.
4. Parameters within the "Crosslink Search" section are specific to the crosslink experiment performed, such as the chemical crosslinker used and how it was quenched. If the crosslinker used is not one of the provided options, custom crosslinkers can be added to MetaMorpheus (*see* **Note 19**). Definitions for each of these crosslink specific parameters can be found below (*see* Subheading 3.15, **Steps 5, 6, 19, 35, 36** and **37**).

5. Expansion of the "Search Parameters" sections reveals the parameters used to inform spectra pre-processing and the underlying search algorithm. Detailed definitions for each one of these parameters can be found below (*see* Subheading 3.15, **Steps 8**, **14**, **18**, **25**, **26**, **28**, **29**, **31**, **33**, **34**, **49**, **50**, **51**, **56**, **58**, **60**, **61** and **63**).
6. The "Modifications" section contains lists of PTMs for selection as fixed (*see* Subheading 3.15, **Step 12**) or variable modifications (*see* Subheading 3.15, **Step 64**).
7. MetaMorpheus exports its PSM results using a .psmtsv format. The final parameter section, "Output Options", provides the option to generate a results file in .pep.XML format in addition to the standard output. The results of the crosslink search task, in the .pep.XML format, can be visualized using ProXL (*see* **Note 20**). A more detailed description of this parameter can be found in Subheading 3.15 **Step 66**.
8. After parameters for the task have been finalized, select the "Add the XL Search Task" button at the bottom of the window to add the XL Search Task to the MetaMorpheus analysis workflow (*see* **Notes 12** and **13**).

3.10 Glycopeptide Search

1. Select the Tasks tab in the menu on the left side of the GUI to open the Task page (*see* Fig. 5).
2. Select the +ADD GLYCO SEARCH button above the task panel to open the Glyco Search Task window containing all necessary parameters for the identification of N or O-glycosylated peptides [9].
3. The Glyco Search Task window contains three parameter sections (a) Glyco Search, (b) Search Parameters, and (c) Modifications.
4. The "Glyco Search" section consists of parameters specific to the glycan searching capabilities of MetaMorpheus. Specific definitions for each of these parameters can be found below (*see* Subheading 3.15, **Steps 2**, **10**, **16**, **19**, **30**, **38**, **46** and **47**). A unique feature of MetaMorpheus' Glyco Search Task is the ability to add a custom glycan database for searching (*see* **Note 21**).
5. The parameters within the "Search Parameters" section inform spectra pre-processing and the underlying search algorithm. Detailed definitions for each one of these parameters can be found below (*see* Subheading 3.15, **Steps 8**, **14**, **18**, **25**, **26**, **27**, **28**, **29**, **31**, **33**, **34**, **49**, **50**, **51**, **58**, **60**, **61** and **63**).
6. The "Modifications" section contains lists of PTMs that can be selected as fixed (*see* Subheading 3.15, **Step 12**) or variable modifications (*see* Subheading 3.15, **Step 64**).

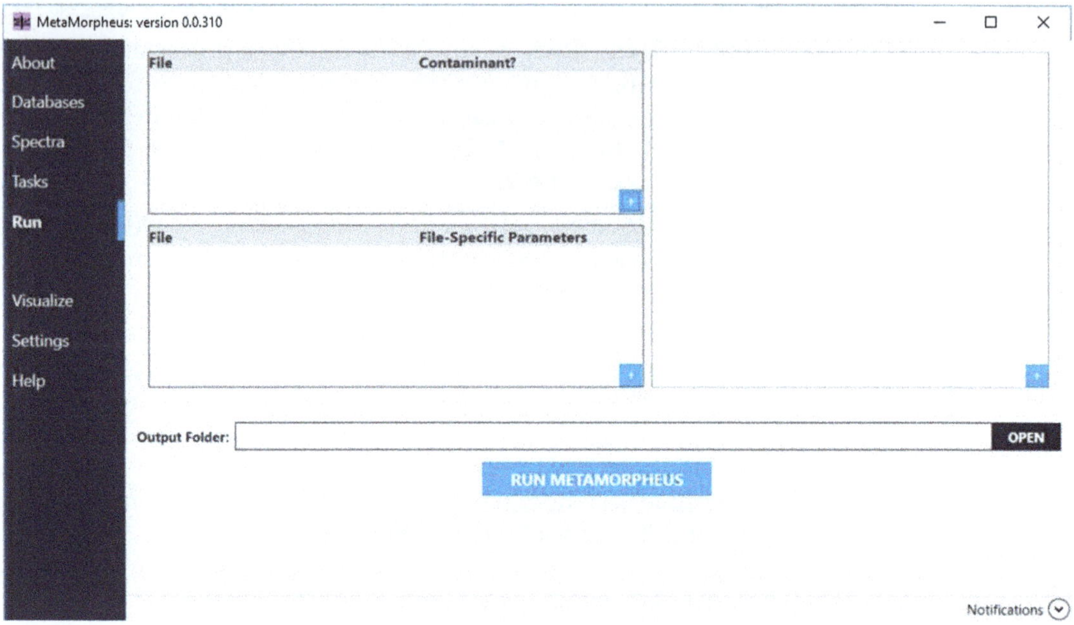

Fig. 6 The run summary page containing all spectra files, databases, and tasks necessary for analysis with MetaMorpheus

7. After parameters for the task have been finalized, select the "Add the GlycoSearch Task" button at the bottom of the window to add the Glyco Search Task to the analysis workflow (*see* **Notes 12** and **13**).

3.11 Starting Analysis in MetaMorpheus

1. After all desired tasks have been added to the workflow, select the Run tab in the menu on the left side of the GUI (*see* Fig. 6).
2. The Run window serves as a summary displaying all databases, spectra files and tasks for analysis (*see* **Note 22**). Any missing files or tasks can be added at this time using the small "+" button at the bottom right of the appropriate panel.
3. All the results from MetaMorpheus tasks will be placed in an output folder at the location noted within the output folder field. This is automatically set as a time-stamped directory within the directory of the first specified spectra file.
4. After all parameters and tasks are finalized, select the RUN METAMORPHEUS button to run the established workflow.
5. Analysis can be aborted at any time by clicking the CANCEL RUN button, which appears after the data analysis process has begun. If a run is canceled, the RESET TASKS button can be used to regenerate all tasks in the workflow for future analysis. Once tasks are reset, the RUN METAMORPHEUS button will re-appear, indicating the workflow can once again be run.

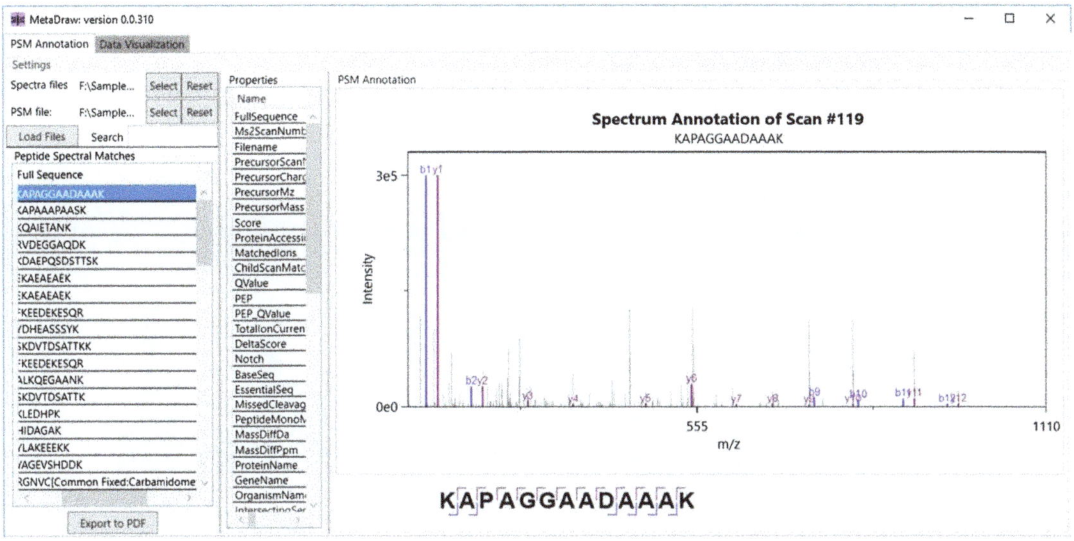

Fig. 7 PSM annotation window of MetaDraw within MetaMorpheus. A sample spectra annotation is displayed

3.12 Spectrum Annotation with MetaDraw

1. Select the Visualize tab in the menu on the left side of the GUI to open MetaDraw, MetaMorpheus' spectral annotator and data viewer.
2. A window will open to show the PSM Annotation tab where the spectra files can be uploaded for annotation (*see* Fig. 7). Files can be added either by dragging and dropping or by clicking the "Select" button to open a file explorer window.
3. The PSM or peptide results files (in .psmtsv format) must be also be added using either the drag and drop method or the Select button.
4. Once all the necessary files have been added, click the "Load Files" button to populate the Peptide Spectral Matches panel with all MS2 scans from the spectra file corresponding to PSMs in the MetaMorpheus results file.
5. To display an annotated spectrum, select the row of the desired PSM. The annotated spectra will be displayed in the PSM Annotation window and information about the PSM will be displayed in the Properties panel.
6. The annotated spectra can be exported as a PDF document using the "Export to PDF" button.

3.13 Data Visualization with MetaDraw

1. From the homepage, select the Visualize tab in the menu on the left side of the GUI to open MetaDraw.
2. Select the Data Visualization tab next to the PSM Annotation tab in the top left corner (*see* Fig. 8).

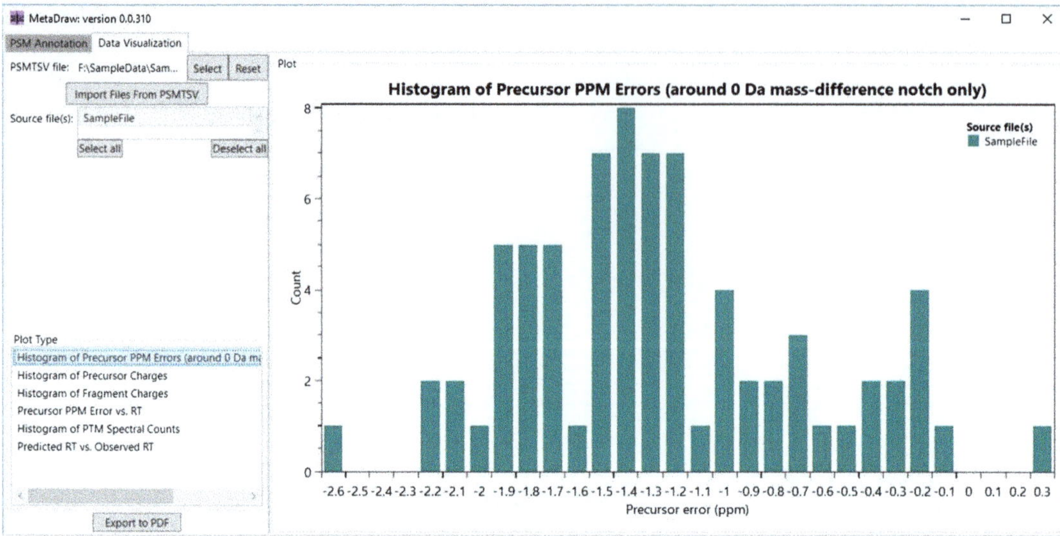

Fig. 8 Data Visualization window of MetaDraw within MetaMorpheus. A sample histogram of Precursor PPM errors is displayed

3. Provide the PSM or peptide results file of interest. The file can be added by dragging and dropping or by using the "Select" button to browse in file explorer.
4. Once the results file has been provided, click the "Import Files From PSMTSV" button to populate the Source file(s) panel with the spectra files containing peptide identifications.
5. Users can select one or more spectra files or can use the "Select all" button to generate plots containing peptide identifications above a 1% FDR from the selected spectra files.
6. Select the desired plot from the Plot Type panel in the bottom left-hand corner. Generated plots are displayed in the Plot panel on the right. A description for each plot type can be found in Table 1.
7. All plots can be exported as a PDF document using the "Export to PDF" button.

3.14 Command-Line Operation of MetaMorpheus

MetaMorpheus can be operated via the command-line interface in Windows, Linux, or MacOS operating systems. To run MetaMorpheus in Windows command line, run "`CMD.exe`" with the following arguments:

```
CMD.exe -d "C:\MyFolder\myDatabase.fasta"
        -t " C:\MyFolder\myTask1.toml"
           "C:\MyFolder\myTask2.toml"
        -s "C:\MyFolder\mySpectraFile.mzML"
```

Table 1
MetaDraw data visualization plots

Plot type	Description
Histogram of Precursor PPM error (around 0 Da mass-notch difference only)	Distribution of precursor mass errors for unmodified peptide identifications (1% FDR). The y-coordinate is the number of identifications in the specific precursor mass error bin (x-coordinate).
Histogram of Precursor charges	Distribution of precursor ion charge states for all PSM or peptide identifications (1% FDR)
Histogram of fragment charges	Distribution of fragment ion charge states for PSM or peptide identifications (1% FDR)
Precursor PPM error vs. RT	Scatter plot where the y-coordinate is the precursor error of each PSM or peptide identification (1% FDR) and the x-coordinate is the experimentally obtained retention time
Histogram of PTM spectral counts	This graph displays the number of spectra identified (1% FDR) to contain a specific PTM type. The y-coordinate is the count of PSMs containing a given PTM and the x-axis is the PTM
Predicted RT vs. observed RT	Scatter plot comparing the predicted hydrophobicity of a peptide based on its amino acid sequence (y-coordinate) to the observed experimental retention time of the species (x-coordinate)

To run the .NET Core version of MetaMorpheus in Linux or in MacOS, run `"dotnet CMD.dll"`with the following arguments:

```
dotnet CMD.dll -d "\home\myfolder\mydatabase.fasta"
               -t "\home\myfolder\myTask1.toml"
                  "\home\myfolder\myTask2.toml"
               -s "\home\mySpectraFile.mzML"
```

The command-line arguments for all environments are as follows:

1. `--help`: This argument prints a list of all MetaMorpheus command-line arguments and their definitions.
2. `-d`: This argument denotes what protein databases will be analyzed (.XML or .FASTA) and is required. Following the argument, provide the file paths for all databases being used, with a space delimiter between each file (*see* **Note 23**).
3. `-s`: This argument precedes all spectra files to be analyzed and is required. Provide the file paths for all spectra files to be analyzed after this argument with a space delimiter between each file (*see* **Note 23**).

4. `-t`: This argument indicates which tasks will be performed during the analysis and is required. Tasks are provided as .toml files, supplying all necessary parameters for each task (*see* Subheading 3.15 for descriptions of all parameters). There are different .toml file formats for each task type (Calibration, GPTMD, Search, XL Search, and Glyco Search). Provide the file paths for all .toml files in the order the tasks are to be performed following the argument, with a space delimiter between each file (*see* **Note 23**).
5. `-o`: This argument sets the output folder for all results of the MetaMorpheus workflow. This argument is usually optional. If no output folder is explicitly set, then a time-stamped folder will be automatically generated in the directory of the first provided spectra file.
6. `-g`: This argument generates .toml files for all tasks containing the default parameter settings. These .toml files can be modified so the parameters fit the experimental data being analyzed. When this argument is used, only the "`--o`" argument is required. The "`-o`" argument specifies the output where the default .toml files will be written.
7. `-v`: This optional argument deals with verbosity, determining the extent to which output and error messages are displayed. The default value for this argument is "normal" but can be set to "none" or "minimal."
8. `--test`: This argument runs a small test search using a yeast database and spectra file included with MetaMorpheus during installation. This command ensures proper installation of MetaMorpheus. When this argument is called, no other command arguments are required.
9. `--version`: This argument displays the version information for MetaMorpheus.

3.15 Parameters

This section provides definitions for all MetaMorpheus task parameters. Parameters are organized in alphabetical order, by their name as displayed in the GUI. Following the GUI name, the parameter name, as displayed in .toml setting files, is provided in parenthesis for command-line users. The default values provided for all parameters in MetaMorpheus are designed to facilitate the analysis of most high-resolution MS2 data without requiring alteration.

1. **Apply Protein Parsimony and Construct Protein Groups (*DoParsimony*):** This Search Task parameter indicates if protein parsimony will be performed on the identified peptides (`1%` PSM-level FDR). Selection of protein parsimony is required for match between runs, protein quantification and multi-protease protein inference.

2. **Child Scan Dissociation (*MS2ChildScanDissociationType*):** This Glyco Search Task parameter specifies the dissociation type used for generating MS3 scans or second dissociation MS2 scans.
3. **Compress Individual File Results (*CompressIndividualFiles*):** This Search Task parameter determines if MetaMorpheus' individual results files are compressed in order to minimize memory requirements.
4. **Construct Mass Difference Histogram (*DoHistogramAnalysis*):** This post-search analysis parameter within the Search Task allows for the creation of a histogram displaying the observed mass shifts for all peptide identifications (`1%` FDR). The mass shifts observed for PSMs are clustered into bins and analyzed for peaks corresponding to the molecular weight of a PTM or amino acid substitution. This analysis is primarily useful for interpreting open-mass search results.
5. **Crosslink at Cleavage Sites (*CrosslinkAtCleavageSite*):** This XL Search Task parameter dictates whether or not a crosslink can be identified at a proteolytic cleavage site.
6. **Crosslinker Type (all parameters under the *XlSearchParameters.Crosslinker* header):** This XL Search Task parameter specifies the crosslinker used in the experiment. The crosslinker type can be selected from a list of crosslinkers or a custom crosslinker can be added (*see* **Note 19**).
7. **C-Terminal Ions (*FragmentationTerminus*):** This Search Task parameter specifies the generation of fragment ions from the C-terminus (e.g., *x*-, *y*-, and *z*-ions) of all theoretical peptides.
8. **Deconvolute Precursor (*DoPrecursorDeconvolution*):** Present in the GPTMD, Search, XL Search, and Glyco Search Tasks, this parameter enables the identification of multiple peptides from a single MS2 scan. For each MS2 scan, the MS1 isolation window is investigated for precursors that could have been co-fragmented to yield the observed fragmentation pattern.
9. **Deconvolution Max Assumed Charge State (*DeconvolutionMaxAssumedChargeState*):** Present in the GPTMD and Search Tasks, this parameter dictates the maximum expected charge state for a peptide. Any isotopic envelopes with charge states larger than this value are discarded or are incorrectly identified as harmonics.
10. **Dissociation Type (*DissociationType*):** Present in all task types (Calibration, GPTMD, Search, XL Search, and Glyco Search), this parameter specifies the dissociation type used for the acquisition of MS2 spectra. MetaMorpheus was originally designed

for analysis of high-resolution MS2 data, because of this all dissociation types are assumed to be high-resolution, with the exception of the LowCID option (*see* **Note 3**).

11. **Filter Results to q-Value (*QValueOutputFilter*):** This Search Task parameter dictates the maximum q-value of peptide identifications in the output files. The filtering of identifications makes the exported result files more manageable for large datasets.

12. **Fixed Modifications (*ListOfModsFixed*):** Present in all tasks (Calibration, GPTMD, Search, XL Search, and Glyco Search), this parameter dictates which PTMS are "fixed" and should be applied to every possible location in the database specified. Typically, the only fixed modification necessary is carbamidomethylation of cysteine, which results when reduced samples have been alkylated with iodoacetamide. Other fixed modifications can be selected when appropriate such as TandemMassTag (TMT) labels.

13. **Generate Complementary Ions (*AddCompIons*):** Present in GPTMD and Search Tasks, this parameter adds artificial complementary masses to the experimental MS2 spectrum. Artificial fragment masses are inferred by subtracting the deconvoluted mass of each observed MS2 fragment ion from the observed precursor mass and adding a dissociation type-specific mass shift. This strategy can be helpful in identifying peptides with modifications near a terminus (e.g., C-terminal modifications of tryptic peptides) [12].

14. **Generate Decoy Proteins (*DecoyType*):** Present in all search tasks (Search, XL Search, and Glyco Search), this parameter indicates if MetaMorpheus automatically generates decoy protein sequences from the provided protein database(s). Decoy proteins provide known false-positive sequences which can be used to determine q-values. In the Search and Glyco Search Tasks, decoy proteins can be generated by using either the reversed or slide methods. Reversed decoys are generated by reversing the protein sequence provided in the target database. Slide decoys are generated by non-random shuffling of the amino acids within each provided protein sequence. If the protein database supplied already contains decoy protein sequences, uncheck this feature.

15. **Generate Target Proteins (*SearchTarget*):** This Search Task parameter indicates MetaMorpheus will search for target peptides generated by the in silico digestion of provided database (s). This parameter can be disabled for decoy-only searches, which are useful in analyses where target and decoy databases are searched separately.

16. **Glyco Search (*GlycanSearchType*):** This Glyco Search Task parameter determines whether O-glycopeptides or N-glycopeptides are discovered by the Glyco Search algorithm. Only one class of glycans can be investigated at a time using the Glyco Search Task.
17. **Handle Overlap Between Target and Contaminant Databases (*TCAmbiguity*):** This Search Task parameter specifies the classification of protein entries that are shared between target and contaminant databases as a contaminant entry, target entry or both.
18. **Initiator Methionine (*InitiatorMethionineBehavior*):** Present in all tasks (Calibration, GPTMD, Search, XL Search, and Glyco Search), this parameter specifies how MetaMorpheus addresses the potential cleavage of initiator methionine residues in the protein database. The initiator methionine for protein entries can always be cleaved, always be retained, or variable (both cleaved and retained versions are created) in the generation of theoretical peptides. It is recommended to treat the initiator methionine as variable.
19. **Keep Top N Candidates (*CrosslinkSearchTopNum* or *GlycoSearchTopNum*):** This parameter in the XL Search and Glyco Search Tasks specifies the maximum number of candidate peptides considered per MS2 scan to reduce computational complexity.
20. **LFQ: Quantify peptides/proteins with FlashLFQ (*DoQuantification*):** Selection of this quantification option within the Search Task establishes that FlashLFQ will be used to perform label-free peptide and protein level quantification. An experimental plan in MetaMorpheus is required for label-free quantification (*see* **Note 24**). Additional information on FlashLFQ can be found in Chapter 13.
21. **Mass Difference Acceptor Criterion (*MassDiffAcceptorType*):** This Search Task parameter determines the acceptable the mass notch(es) for the difference between a peptide's observed and theoretical precursor mass. Selections can be made from the provided options (`"Exact"`, `"1 Missed Monoisotopic Peak"`, `"1 or 2 Missed Monoisotopic Peaks"`, `"1,2 or 3 Missed Monoisotopic Peaks"`, `"+-3 Missed Monoisotopic peaks"`, `"-187 and Up"`, and `"Accept all"`). Additionally, MetaMorpheus supports the addition of a custom mass difference acceptor (*see* **Note 25**).
22. **Match Between Runs (*MatchBetweenRuns*):** This Search Task parameter indicates match between runs will be utilized as part of the quantification process. Match between runs allows peptides that were fragmented in at least one spectra file to be quantified across all other spectra files. Any peptide

identified in one spectra file is searched for in all other files within a small mass-to-charge and retention time window. To learn more about match between runs, refer to the FlashLFQ protocol (*see* Chapter 13).

23. **Max Fragment Mass (*MaxFragmentSize*):** This Search Task parameter imposes an upper limit for the mass of theoretical fragment ions.

24. **Max Heterozygous variants for Combinatorics (*MaxHeterozygousVariants*):** Present in the Calibration, GPTMD, and Search Tasks, this parameter is only relevant when one of the databases provided is generated via Spritz [13] and contains annotated sequence variants. It dictates the maximum number of variants that can be applied to a single protein sequence, thus determining the number of theoretical variant-containing proteins generated.

25. **Max Missed Cleavages (*MaxMissedCleavages*):** Present in all tasks (Calibration, GPTMD, Search, XL Search, and Glyco Search), this parameter specifies the maximum number of missed cleavages allowed during in silico digestion of the protein database(s). The protease utilized affects this parameter because certain proteases, such as Chymotrypsin, are more prone to missed cleavages [14].

26. **Max Modification Isoforms (*MaxModficationIsoforms*):**This parameter is present in all tasks (Calibration, GPTMD, Search, XL Search, and Glyco Search) and specifies the maximum number of different peptide forms (peptidoforms) possible for a single theoretical peptide sequence. A large number variable and/or annotated modifications for a peptide can drastically increase the number of peptidoforms present in the database, making this parameter crucial for controlling database size.

27. **Max Mods Per Peptide (*MaxModsForPeptide*):** This parameter, present in the Calibration, GPTMD, Search, and Glyco Search Tasks, defines the maximum number of PTMs allowed on an individual peptide. As this value increases, so does the number of PTM combinations, search space and computational time.

28. **Max Peptide Length (*MaxPeptideLength*):** Present in all task types (Calibration, GPTMD, Search, XL Search, and Glyco Search) this parameter establishes the maximum length of theoretical peptides generated by in silico database digestion. Any peptides present in the sample longer than the specified value will not be correctly identified.

29. **Max Threads (*MaxThreadsToUsePerFile*):** This parameter specifies the maximum number of threads MetaMorpheus can utilize. The default value is determined based on the CPU

running MetaMorpheus and is set to one less than the total number of threads.

30. **Maximum OGlycans Allowed (*MaximumOGlycanAllowed*):** This Glyco Search parameter specifies the maximum number of O-glycosylation sites possible for a single theoretical peptide. This parameter should be adjusted depending on prior knowledge of the sample being analyzed. For example, mucins are a class of proteins known for heavy O-glycosylation [15]. A sample of primarily mucin proteins should have a higher value set for this parameter than non-mucin samples.
31. **Min Peptide Length (*MinPeptideLength*):** Present in all task types (Calibration, GPTMD, Search, XL Search, and Glyco Search), this parameter establishes the minimum length of theoretical peptides generated by in silico database digestion. The default value for this parameter is seven, because peptides shorter than this length are difficult to confidently identify.
32. **Min Read Depth for Variants (*MinVariantDepth*):** Found in the Calibration, GPTMD, and Search Tasks, this parameter is only relevant when one or more of the databases provided are generated by Spritz [13] and contain annotated sequence variants. This parameter specifies the read depth, or coverage, that a specific variant must have in the RNA sequencing data in order to be included into theoretical protein sequences. This prevents variants without sufficient transcriptomic support from expanding the search space.
33. **Minimum Intensity Ratio (*MinimumAllowedIntensityRatioToBasePeak*):** This parameter, present in the GPTMD, Search, XL Search, and Glyco Search tasks, establishes the minimum intensity ratio required for each experimental fragment ion. The intensity ratio for each fragment ion is calculated by dividing its intensity by that of the highest intensity peak in the scan. If the minimum intensity threshold is not met, that fragment ion cannot be compared to the theoretical peptide spectra.
34. **Minimum Score Allowed (*ScoreCutoff*):** Present in all tasks (Calibration, GPTMD, Search, XL Search, and Glyco Search), this parameter defines the minimum score required to report a PSM. The score, for high-resolution MS2 data, is determined by summing the number of matched fragment ions with the fraction of the total ion current (TIC) accounted for by these matched ions.
35. **MS2 Child Scan Dissociation(*MS2ChildScanDissociationType*):** This XL Search Task parameter specifies the dissociation type used to generate MS3 scans or second dissociation MS2 scans. If this level of fragmentation is not relevant for the spectra being analyzed, the parameter can be set to Null.

36. **MS2 Scan Dissociation Type (*DissociationType2*):** This XL Search Task parameter specifies the dissociation type used to generate MS2 fragmentation spectra.
37. **MS3 Child Scan Dissociation (*MS3ChildScanDissociationType*):** This XL Search Task parameter specifies the dissociation type used to generate MS4 scans or second dissociation MS3 scans. If this level of fragmentation is not relevant for the spectra being analyzed, the parameter can be set to Null.
38. **N-Glycan Database (*NGlycanDatabasefile*):** This Glyco Search Task parameter determine which N-glycan database will be utilized for the identification of N-linked glycopeptides. Custom N-glycan databases can be added if necessary (*see* **Note 21**).
39. **No Quantification (*DoQuantification*):** Selection of this quantification option within the Search Task dictates that neither label-free or SILAC quantification will be performed.
40. **Nominal Window Width Thomsons (*WindowWidthThomsons*):** This Search Task parameter specifies the width of the MS1 and MS2 filtering windows in Thomsons (m/z units). Dividing MS1 and MS2 scans into windows helps prevent filtering bias that may result from prevalence of high intensity peaks in the center of the spectrum and lower intensity peaks at low and high m/z ranges.
41. **Normalize Peaks in Each Window (*NormalizePeaksAcrossAllWindows*):** This Search Task parameter enables the normalization of peak intensity values to the most intense peak within the defined window.
42. **Normalize Quantification Results (*Normalize*):** When label-free quantification with FlashLFQ is enabled, this Search Task parameter dictates the normalization of peptide intensity values. This normalization is based on the assumption that the majority of peptides do not change in abundance between conditions (*see* Chapter 13). The normalization algorithm requires the information provided in the experimental design (*see* **Note 24**).
43. **N-Terminal Ions (*FragmentationTerminus*):** This Search Task parameter specifies the generation of fragment ions from the N-terminus (e.g., *a*-, *b*-, and *c*-ions) of all theoretical peptides.
44. **Number of Database partitions (*TotalPartitions*):** The modern, semi-specific, and non-specific search algorithms generate an index of theoretical peptide spectra from the supplied database(s). This index can become prohibitively large and exceed the RAM capacity of the computer. This parameter allows for the search space to be divided into partitions, or

sections, before the search is performed to avoid complications. The theoretical peptides in each partition are searched separately and then aggregated to provide the same results as if the partitioning method was not applied.

45. **Number of Windows (*NumberOfWindows*):** This Search Task parameter defines the number of windows, or sections, the MS1 and MS2 scans are to be divided into for peak filtering. Often, peaks are most intense in the center of the spectrum and less intense on the edges. When filtering is applied to the entire spectrum there is a risk of removing quality peaks in the low and high m/z regions of the spectra and retaining noise peaks in the center. Division of the scans into filtering windows prevents this bias.
46. **O-Glycan Database (*OGlycanDatabasefile*):** This Glyco Search Task parameter determines which O-glycan database will be utilized for the identification of O-linked glycopeptides. Custom O-glycan databases can be added if necessary (*see* **Note 21**).
47. **OxoniumIonFilt (*OxoniumIonFilt*):** This Glyco Search Task parameter specifies that only MS2 scans containing an oxonium ion at 204 m/z will be investigated as potential glycopeptides.
48. **Peak-Finding Tolerance (*QuantifyPpmTol*):** This Search Task parameter defines the parent mass tolerance (in ppm) used for label-free quantification.
49. **Precursor Mass Tolerance (*PrecursorMassTolerance*):** This parameter, found in all tasks (Calibration, GPTMD, Search, XL Search, and Glyco Search), establishes the maximum mass difference between the observed and theoretical precursor masses permitted for a PSM. This value is typically specified in ppm but can also be represented in daltons.
50. **Product Mass Tolerance (*ProductMassTolerance*):** Found in all tasks (Calibration, GPTMD, Search, XL Search, and Glyco Search), this parameter establishes the maximum mass difference between theoretical and experimental fragment ion permitted for it to be considered a match. This value can also be set in either ppm or Daltons (*see* **Note 26**).
51. **Protease (*Protease*):** Present in all tasks (Calibration, GPTMD, Search, XL Search, and Glyco Search), this parameter establishes the protease used for in silico database digestion. This protease should be the same as was used experimentally to digest the sample. Selection can be made from a provided list of common proteases, or a custom protease can be specified (*see* **Note 27**).
52. **Quench Method (*XLQuench*):** This XL Search Task parameter specifies the method(s) utilized to quench the crosslinker.

53. **Report PSM Ambiguity (*ReportAllAmbiguity*):** This Search Task parameter defines how ambiguous peptide spectral matches are reported. When multiple theoretical peptide sequences match the same MS2 spectra, and these PSMs all have the same score, the identification is ambiguous. If this box is unchecked, a random peptide from the multiple ambiguous matches will be reported. Otherwise, all possible sequences are reported.
54. **Require at least Two Peptides to Identify Protein (*NoOneHitWonders*):** This Search Task parameter requires the identification of two unique peptides for the establishment of a protein group in protein parsimony. Historically, this parameter was developed to eliminate the presence of one-hit wonders but has since been considered to be overly stringent and be detrimental to protein parsimony overall [16].
55. **Search Mode (*SearchType*):** This Search Task parameter defines which search algorithm will be used for the task. MetaMorpheus includes 4 different search algorithms (or modes): (a) Classic Search (*see* **Note 28**), (b) Modern Search (*see* **Note 29**), (c) Semi-Specific Search (*see* **Note 30**) and (d) Non-Specific Search (*see* **Note 31**).
56. **Separation Type (*SeparationType*):** This parameter in the Search and XL Search Tasks specifies the online separation method utilized prior to mass spectrometric analysis. This determines whether predicted hydrophobicity or electrophoretic mobility values are calculated for the peptides.
57. **SILAC/SILAM: Quantify peptides/proteins with stable isotope labels (*DoQuantification* and *GenerateUnlabeledProteinsForSilac*):** Selection of this quantification option within the Search Task indicates a portion of the peptides and proteins within the sample have been isotopically labeled enabling relative quantification. Upon selection, additional parameters for SILAC-based quantification appear including a checkbox to quantify unlabeled peptides and a table in which to specify the amino acid labels being used (*see* **Note 32**).
58. **Top N Peaks per m/z window (*NumberOfPeaksToKeepPerWindow*):** Present in the GPTMD, Search, XL Search, and Glyco Search Tasks, this parameter indicates the maximum number of peaks allowed in a window with a specified m/z width. This parameter applies to the peak filtering process of MS1 and MS2 scans. The peaks within the window are ordered by intensity prior to the cutoff being applied.
59. **Treat Modified Peptides as Different Peptides (*ModPeptidesAreDifferent*):** This Search Task parameter requires the protein parsimony algorithm to consider modified peptides

distinct from their unmodified form. This can potentially disambiguate protein groups by the presence of annotated PTMs.

60. **Trim MS1 Peaks (*TrimMs1Peaks*):** This parameter is present in all search task types (Search, XL Search, and Glyco Search) and enables the filtering of MS1 peaks as part of spectra pre-processing.
61. **Trim MS2 Peaks (*TrimMsMsPeaks*):** This parameter is present in all search task types (Search, XL Search, and Glyco Search) and enables the filtering of MS2 peaks as part of spectra pre-processing.
62. **Use Delta Scores for FDR (*UseDeltaScore*):** This Search Task parameter specifies whether the Delta Score, instead of the Score, should be used for ranking PSMs prior to statistical analysis. The Delta Score is the difference between the scores of the two best matching peptides for the same MS2 spectrum. If the Delta Score produces fewer PSMs at a 1% FDR, then the Score will be automatically used instead.
63. **Use Provided Precursor (*UseProvidedPrecursorInfo*):** Present in the GPTMD, Search, XL Search, and Glyco Search Tasks, this parameter indicates that the precursor mass reported in the spectra should be used as the observed precursor mass for the search. This can be used in addition to deconvoluted precursor masses.
64. **Variable Modifications (*ListOfModsVariable*):** Present in all tasks (Calibration, GPTMD, Search, XL Search, and Glyco Search), this parameter dictates which PTMS are "variable" and that modified and unmodified forms of all peptides should be generated. Variable modifications should be used with caution because they massively increase the search space and typically lead to high false-positive rates. With the exception of variable oxidation of methionine, all other potentially present variable modifications should be searched for using the GPTMD approach (*see* Subheading 3.6).
65. **Write .mzID (*WriteMzId*):** This Search Task parameter requires additional search result files to be written in .mzID format. This is the output file type defined by the Human Proteome Organization (HUPO) and was designed to be a standardized format for reporting search results across different searching platforms.
66. **Write .pep.XML (*WritePepXml*):** This XL Search Task parameter requires additional result files to be written in .pep.XML format. This file format is widely accepted for the output of proteomics search engines. This result file format can be used as input for ProXL (*see* **Note 20**) for visualization of cross-linking results.

67. **Write Contaminants (*WriteContaminants*):** This Search Task parameter specifies the inclusion of contaminant peptide identifications in the result files. Contaminant identifications are clearly annotated as contaminants.
68. **Write Decoys (*WriteDecoys*):** This Search Task parameter specifying the inclusion of decoy peptide identifications in the result files. Decoy identifications are clearly annotated as decoys.
69. **Write Individual File Results (*WriteIndividualFiles*):** This output option within the Search Task specifies that result files for each individual spectra file be written in addition to the cumulative result files.
70. **Write Two Pruned Databases [Mod and Mod+Protein Pruned] (*WritePrunedDatabase*):** This post-search analysis parameter within the Search Task triggers the construction and export of two custom pruned .XML databases (modification pruned and modification + protein pruned). Modification pruning limits the PTMs annotated in the database to either those that were confidently identified at `1%` FDR, annotated in the original database, or both depending on the user's specifications. The process of protein pruning restricts the database to protein entries that have peptide-level support at `1%` FDR. These pruned databases can be beneficial for subsequent top-down and intact-mass proteoform analysis [17].

4 Notes

1. Versions of MetaMorpheus prior to 0.0.309 have a different GUI layout, but the operations remain the same, and instructions in this protocol can still be applied to older versions.
2. Conversion of spectra files to different formats can be achieved using the software program MSConvert. MSConvert is part of ProteoWizard and can be downloaded at http://proteowizard.sourceforge.net/download.html.
3. The traditional MetaMorpheus score was designed for the comparison of high-resolution MS2 spectra to theoretical spectra. The integer component of the score is the number of fragment ions that match between the observed and theoretical spectra. The decimal component is the fraction of the total ion current (TIC) of the experimental spectra that is represented by the matching fragment ions. This scoring approach is much less effective for low-resolution MS2 spectra. To adjust for this, MetaMorpheus has implemented an adapted XCorr [18] scoring algorithm for use with low-resolution CID fragmentation.

4. .XML-formatted protein databases contain more information than .FASTA format. One of the greatest advantages of the use of UniProt .XML databases is the presence of annotated PTMs. UniProt .XML databases for reference proteomes can be retrieved at https://www.uniprot.org/proteomes/.
5. The larger the proteomic dataset being analyzed, the more RAM the analysis will require.
6. Windows (https://dotnet.microsoft.com/download/dotnet-core/thank-you/runtime-desktop-3.1.3-windows-x64-installer). MacOS (https://dotnet.microsoft.com/download/dotnet-core/thank-you/runtime-3.1.3-macos-x64-installer). Linux (https://docs.microsoft.com/dotnet/core/install/linux-package-managers).
7. When files are added using the drag and drop approach, MetaMorpheus automatically detects the file type (spectra file, database file, etc.), based on the file extension, and places it in the proper location.
8. The inclusion of contaminant databases enables the correct identification of contaminant peptides, preventing their misidentification as false-positive target PSMs. MetaMorpheus has a custom contaminant database designed to include many common contaminants from basic sample preparation methods.
9. More than one file can be added at a time either by dragging and dropping a group of files or by selecting multiple files in the file explorer window.
10. If file-specific settings are not specified, parameters from the task's settings are used. When file-specific settings are specified, task parameters are ignored in favor of the file-specific parameters.
11. Select multiple spectra files to alter file-specific parameters for several spectra at once.
12. Once added, tasks appear in the Tasks window in the order they will be run.
13. To edit the parameters of any task after it has been added to the workflow, right click the task and select Edit task or double click the task and the parameters window will reopen for adjustment. Once necessary adjustments have been made, simply select the "Save [task name] task" button.
14. To alter the default parameters for a task, make the necessary parameter adjustments in the task window and select the "Save As Default" button at the bottom right of the window.
15. The default GPTMD modifications have been curated to contain modifications empirically determined to be common

biological modifications across multiple species, common metal adducts, and common chemical artifacts that can occur during sample preparation.

16. Modifications can be located and selected using the search bar in addition to the drop-down menus.
17. To create a custom PTM, select the Settings tab in the menu on the left side of the GUI window. Select Create new modification from the Settings window to open up a new window for PTM creation. Information in fields marked with a red asterisk are required for PTM generation. Provide a name for the modification, the category to place the modification under in the list (this can be an existing category or you can create a new one such as 'Custom'), what amino acid, or amino acid motifs the modification is present on, the chemical formula, and finally if the modification is located on a specific peptide/protein terminus. Additionally, neutral losses and diagnostic ions can be provided for the PTM based on the dissociation type. Once all of the required information has been provided, click the "Save Mod" button. The program must be restarted for the new modification to appear in the GUI.
18. When files from multiple proteolytic digests are loaded and protein inference is applied, the multi-protease protein inference algorithm [11] is automatically triggered providing improved protein inference results.
19. To add a custom crosslinker, exit the XL Search Task, and select the Settings tab in the menu on the left side of the GUI. Select Create new crosslinker from the Settings window to open a new window containing fields for the required information for crosslinker generation. Select "Save Crosslinker" once the required information has been provided. MetaMorpheus must be restarted before the custom crosslinker will show up as an option within the XL Search Task.
20. ProXL is a web based software tool for analyzing, visualizing, and sharing protein crosslinking mass spectrometry data and can be accessed at http://proxl-ms.org/.
21. Custom glycan databases can be constructed and added to MetaMorpheus for use in the Glyco Search Task. To add a database, navigate to the Settings tab in the menu on the left side of the GUI. Select Open mods/data folder to open up file explorer to MetaMorpheus' location. Select the Glycan_Mods folder followed by either the N- or O-glycan folder depending on what kind of database you are adding. Move the custom glycan database to this location and restart MetaMorpheus for its incorporation into the GUI. The format of the custom glycan databases should follow formatting of the existing glycan databases. Briefly, the database entries should contain a

composition based glycan description, a symbol based glycan description and the molecular weight of the glycan (e.g., HexNAc(1)Hex(1)Fuc(1) N1H1F1S0 511.1901). MetaMorpheus' Glyco Search currently supports the interpretation of the following monosaccharides and modifications from glycan databases: Hex (H), HexNAc (N), NeuAc (A), NeuGc (G), Fuc (F), Phospho (P), Sulfo (S), Na (Y), Ac (C), Xylose (X), SuccinylHex (U), and Formylation (M).

22. A typical MetaMorpheus run will consist of a Calibration, GPTMD and Search Task.
23. When specifying a file path following an argument, it is preferable to use the absolute file path inside of quotes. This is required if there are spaces in the path name.
24. The experimental design for label-free quantification can be added to MetaMorpheus by selecting the SET EXPERIMEN TAL DESIGN button to the left of the SET FILE-SPECIFC SETTINGS button in the Spectra window. Specify the file condition, biological replicate number, fraction number and technical replicate number for each spectra file. This information enables FlashLFQ [10] to perform normalization and protein quantification on the provided data (*see* Chapter 13).
25. Custom mass tolerances can be created using the following syntax: name dot # PPM, #, # where the first # is the ppm error and the subsequent #s are the missed monoisotopics (e.g., test dot 4 PPM, 1, 2). You can also perform an interval search using the syntax name interval [#,#], where each # is either a min or max in Da with both numbers relative to the precursor mass.
26. It is recommended that the product mass tolerance be set in ppm when an Orbitrap mass spectrometer is used for data acquisition.
27. To add a custom protease to MetaMorpheus, select the Settings tab in the menu on the left side of the GUI. Select Open mods/data folder to open a file explorer window at MetaMorpheus' location. Select the Proteolytic Digestion folder and open the proteases.tsv file. This file contains all the proteases, and their cleavage-specific information for MetaMorpheus. A new protease can be added by creating a new row in the file. For each new protease, provide the name and define its digestion motif using the following syntax: (a) specify the amino acids inducing cleavage by listing the single amino acid codes with the "—" character on the left for N-terminal cleavage and on the right for C-terminal cleavage, (b) provide any amino acids residue(s) that prevent proteolytic cleavage using "[ ]", (c) Use "X" to denote any amino acid within a cleavage motif that is a wildcard (could be any amino acid), (d) use "" to

define any exceptions to the wildcard character, and (e) Use "," to separate multiple cleavage motifs for a single protease (e.g., Chymotrypsin (do not cleave before proline) F[P]—,W[P]—,Y [P]—). Once all necessary protease information has been added, save the edited protease.tsv file and restart MetaMorpheus for the custom protease to appear in the GUI.

28. In the classic search algorithm, a MS2 spectrum is compared to every theoretical spectrum that has the same precursor mass within the task defined precursor mass tolerance. The highest scoring match is reported.

29. The modern search algorithm creates an index of all theoretical peptide spectra, which serves as a look-up table. During the comparison of theoretical and experimental spectra, experimental fragment ions can be rapidly compared to theoretical target and decoy peptide MS2 fragments in the look-up table. The theoretical target or decoy peptide with the most matching fragment ions is recorded in the results file. The modern search algorithm is much faster than the classic search when conducting an open-mass search.

30. The semi-specific search algorithm employs a novel digestion and search strategy to identify peptides where one of the termini may not follow the cleavage motif of the selected protease [19]. A separate q-value calculation prevents the semi-specific search from introducing significant false-positive identifications.

31. The non-specific search algorithm utilizes the same digestion and search strategy as the semi-specific approach [19], but differs because up to 2 of the peptide's termini may not conform to the protease's cleavage motif. If no protease was used for digestion of the sample (e.g., peptidomics), select the non-specific option for the protease parameter. A separate q-value calculation prevents the non-specific search from introducing significant false-positive identifications when a specific protease is selected.

32. Click the "Add Isotope Label" button that appears in the Search Task window when the SILAC/SILAM: Quantify peptides/protein with stable isotope labels parameter is selected. Specify the labeled amino acid in the top left corner. The remaining fields will be automatically filled with the information for the unlabeled amino acid. Alter the count of heavy isotope to match the composition of the label and the chemical formula and mass difference will be automatically updated. The final parameter specifies the type of labeling experiment performed (i.e., Multiplex or Turnover/Pulse). After completing the label design, select "Save Label(s)" to go back to the Search Task, or click "Add Additional Labels to This Condition" if

more than one heavy labeled amino acid was used. For the command-line version, include a [*SearchParameters.SilacLabels*] section containing the following parameters, *OriginalAminoAcid*, *AminoAcidLabel*, *LabelChemicalFormula*, and *MassDifference* for each of the labeled amino acids.

Acknowledgements

The development and maintenance of MetaMorpheus is supported by NIH-NIGMS grant R35GM126914. Rachel M. Miller was supported in part by the NIH Chemistry-Biology Interface Training Grant (T32GM008505). Robert J. Millikin was supported by the NIH Genomic Sciences Training Program (5T32HG002760). Zach Rolfs was supported by NIH-NCI grant U24CA199347.

References

1. Link AJ, Eng J, Schieltz DM, Carmack E, Mize GJ, Morris DR, Garvik BM, Yates JR (1999) Direct analysis of protein complexes using mass spectrometry. Nat Biotechnol 17(7):676–682. https://doi.org/10.1038/10890
2. Nesvizhskii AI (2010) A survey of computational methods and error rate estimation procedures for peptide and protein identification in shotgun proteomics. J Proteomics 73(11): 2092–2123. https://doi.org/10.1016/j.jprot.2010.08.009
3. Skinner OS, Kelleher NL (2015) Illuminating the dark matter of shotgun proteomics. Nat Biotechnol 33(7):717–718. https://doi.org/10.1038/nbt.3287
4. Chick JM, Kolippakkam D, Nusinow DP, Zhai B, Rad R, Huttlin EL, Gygi SP (2015) A mass-tolerant database search identifies a large proportion of unassigned spectra in shotgun proteomics as modified peptides. Nat Biotechnol 33(7):743–749. https://doi.org/10.1038/nbt.3267
5. Griss J, Perez-Riverol Y, Lewis S, Tabb DL, Dianes JA, Del-Toro N, Rurik M, Walzer M, Kohlbacher O, Hermjakob H, et al (2016) Recognizing millions of consistently unidentified spectra across hundreds of shotgun proteomics datasets. Nate Methods 13(8):651–656. https://doi.org/10.1038/nmeth.3902
6. Nesvizhskii AI, Roos FF, Grossmann J, Vogelzang M, Eddes JS, Gruissem W, Baginsky S, Aebersold R (2006) Dynamic spectrum quality assessment and iterative computational analysis of shotgun proteomic data: toward more efficient identification of post-translational modifications, sequence polymorphisms, and novel peptides. Mol Cell Proteom 5(4):652–670. https://doi.org/10.1074/mcp.M500319-MCP200
7. Solntsev SK, Shortreed MR, Frey BL, Smith LM (2018) Enhanced global post-translational modification discovery with metamorpheus. J Proteome Res 17(5):1844–1851. https://doi.org/10.1021/acs.jproteome.7b00873
8. Lu L, Millikin RJ, Solntsev SK, Rolfs Z, Scalf M, Shortreed MR, Smith LM (2018) Identification of MS-cleavable and noncleavable chemically cross-linked peptides with metamorpheus. J Proteome Res 17(7):2370–2376. https://doi.org/10.1021/acs.jproteome.8b00141
9. Lu L, Riley NM, Shortreed MR, Bertozzi CR, Smith LM (2020) O-pair search with metamorpheus for o-glycopeptide characterization. bioRxiv https://doi.org/10.1101/2020.05.18.102327
10. Millikin RJ, Solntsev SK, Shortreed MR, Smith LM (2018) Ultrafast peptide label-free quantification with FlashLFQ. J Proteome Res 17(1): 386–391. https://doi.org/10.1021/acs.jproteome.7b00608
11. Miller RM, Millikin RJ, Hoffmann CV, Solntsev SK, Sheynkman GM, Shortreed MR, Smith LM (2019) Improved protein inference from multiple protease bottom-up mass spectrometry data. J Proteome Res 18(9):3429–3438. https://doi.org/10.1021/acs.jproteome.9b00330
12. Kong AT, Leprevost FV, Avtonomov DM, Mellacheruvu D, Nesvizhskii AI (2017) MSFragger: ultrafast and comprehensive peptide identification in mass spectrometry–based

proteomics. Nat Methods 14(5):513–520. https://doi.org/10.1038/nmeth.4256

13. Cesnik AJ, Miller RM, Ibrahim K, Lu L, Millikin RJ, Shortreed MR, Frey BL, Smith LM (2020) Spritz: A proteogenomic database engine. bioRxiv https://doi.org/10.1101/2020.06.08.140681
14. Giansanti P, Tsiatsiani L, Low TY, Heck AJ (2016) Six alternative proteases for mass spectrometry–based proteomics beyond trypsin. Nat Protocols 11(5):993–1006. https://doi.org/10.1038/nprot.2016.057
15. Varki A (1993) Biological roles of oligosaccharides: all of the theories are correct. Glycobiology 3(2):97–130. https://doi.org/10.1093/glycob/3.2.97
16. Gupta N, Pevzner PA (2009) False discovery rates of protein identifications: a strike against the two-peptide rule. J Proteome Res 8(9): 4173–4181. https://doi.org/10.1021/pr9004794
17. Dai Y, Buxton KE, Schaffer LV, Miller RM, Millikin RJ, Scalf M, Frey BL, Shortreed MR, Smith LM (2019) Constructing human proteoform families using intact-mass and top-down proteomics with a multi-protease global post-translational modification discovery database. J Proteome Res 18(10): 3671–3680. https://doi.org/10.1021/acs.jproteome.9b00339
18. Howbert JJ, Noble WS (2014) Computing exact p-values for a cross-correlation shotgun proteomics score function. Mol Cell Proteom 13(9):2467–2479. https://doi.org/10.1074/mcp.O113.036327
19. Rolfs Z, Millikin RJ, Smith LM (2020) An algorithm to improve the speed of semi- and non-specific enzyme searches in proteomics. Curr Bioinf 15:1–9. https://doi.org/10.2174/1574893615999200429123334

Chapter 4

Validation of MS/MS Identifications and Label-Free Quantification Using Proline

Véronique Dupierris, Anne-Marie Hesse, Jean-Philippe Menetrey, David Bouyssié, Thomas Burger, Yohann Couté, and Christophe Bruley

Abstract

In the proteomics field, the production and publication of reliable mass spectrometry (MS)-based label-free quantitative results is a major concern. Due to the intrinsic complexity of bottom-up proteomics experiments (requiring aggregation of data relating to both precursor and fragment peptide ions into protein information, and matching this data across samples), inaccuracies and errors can occur throughout the data-processing pipeline. In a classical label-free quantification workflow, the validation of identification results is critical since errors made at this first stage of the workflow may have an impact on the following steps and therefore on the final result. Although false discovery rate (FDR) of the identification is usually controlled by using the popular target–decoy method, it has been demonstrated that this method can sometimes lead to inaccurate FDR estimates. This protocol shows how Proline can be used to validate identification results by using the method based on the Benjamini–Hochberg procedure and then quantify the identified ions and proteins in a single software environment providing data curation capabilities and computational efficiency.

Key words Mass spectrometry software, Discovery proteomics, Benjamini–Hochberg false discovery rate, Statistics, Label-free quantification, Data visualization, Software engineering

1 Introduction

High-throughput mass spectrometry-based proteomics is continuously evolving toward increased complexity through the analysis of larger sample cohorts, with more sophisticated experimental designs, and deeper proteome coverage. This complexity has dramatically increased the data volumes to be processed, so that the processing efficiency has become a major concern for many labs or core facilities. Another effect of this complexity is that the examination of the data and the ability to review and correct the data-processing steps require a solution allowing experimentalists to visualize and navigate a considerable amount of data in a very effective way. To address this, we developed a software suite called

Thomas Burger (ed.), *Statistical Analysis of Proteomic Data: Methods and Tools*,
Methods in Molecular Biology, vol. 2426, https://doi.org/10.1007/978-1-0716-1967-4_4,

Proline [1] which provides an efficient and integrated way to process, visualize, and publish proteomic datasets.

The validation process of tandem mass spectrometry (MS/MS) identification in Proline was originally based on predefined filters used to accept or reject a peptide-spectrum match (PSM) and on the widely used target–decoy competition (TDC) method [2] to control the false discovery rate (FDR). We recently proposed an alternative to TDC validation by a totally different method to control the FDR at the PSM, peptide, and protein levels, while benefiting from the theoretical guarantees of the Benjamini–Hochberg (BH) framework [3]. Since version 2.1.2, the BH FDR method is available in the official release cycle of Proline, in addition to the target–decoy competition based method. The BH method provides a simple and pragmatical way to validate identification results without requiring the use of decoy protein databases. Following validation, Proline can then be used to quantify peptides and proteins as shown in this protocol.

2 Material

2.1 Requirements

Proline is a client–server application that can either be installed on a machine running Linux, Windows, or MacOS. An all-in-one package called Proline-Zero, including both the client and the server parts as well as the required dependencies is available. This package does not required any complicated installation procedure (*see* Subheading 2.3). In that particular case, the amount of RAM available is the only requirement: A minimum of 8GB of RAM is recommended for the server especially to be able to run quantification processes; while 1GB of RAM is enough for the client (*see* **Note 1**). This protocol is based on Proline version 2.1.2 installed from the Proline-Zero distribution.

For more intensive usage, the server part of Proline can be installed on a centralized server allowing multiple clients to connect to the same Proline server. In this deployment scheme, Java SE 8 Runtime Environment is required for client and server parts and a PostgreSQL database server (versions above 9.4) must be installed and configured on server side.

2.2 Data Format

Identification results from different search engines can be imported into Proline in their native data format. This includes .dat files from Mascot, .omx files from OMSSA, .xml files from X!Tandem, and folders of .txt files from Maxquant. In addition, Proline supports mzIdentML files, enabling to import results from any search engines that support this standard format.

2.3 Software Install

To install Proline from the Proline-Zero distribution, select the correct archive file for your operating system from the Proline website (http://www.profiproteomics.fr/proline/#downloads) and unarchive it. On first launch, the different components are automatically initialized and configured, including the database schema that will be used by Proline.

2.4 Sample Data

This protocol focuses on validation of identification results based on the BH method to control the FDR. To do so, input data must be MS/MS identification searches performed by Mascot search engine on databases containing only target proteins: No decoy is needed using this strategy. Identification data from [3] are used throughout this protocol to illustrate the workflow. This dataset (.dat and .raw files) is available on the proteomeXchange repository [4] with the identifier PXD016669. For the sake of simplicity, the subset of data needed for this protocol as well as the Proline-Zero distribution version 2.1.2 are available at ftp://ftp.cea.fr/pub/edyp/Proline/MiMB/. This FTP folder contains:

1. Replicates 1 to 4 of the 10 replicates of samples analyzed with a Q-Exactive Plus instrument with HH settings and searched against a database containing only target proteins will be used (files `QEx_HH_no_decoy_R1.dat` to `QEx_HH_no_decoy_R4.dat`).
2. The Thermo .raw files corresponding to the MS analysis of these four replicates (`QEx2_020296.raw`, `QEx2_020300.raw`, `QEx2_020322.raw` and `QEx2_020419.raw`) that will be used in the second part of this protocol.
3. The protein sequence database (.fasta file) that has been used to perform the MS/MS search.

3 Methods

3.1 Starting and Configuration

1. To start Proline, double-click `Proline-Zero.exe` on Windows or type `Proline-Zero.sh` on Linux-like OS. The different components are started by the launcher. Afterwards, the graphical user interface (named ProlineStudio) starts.
2. At the first start, the default configuration is initialized, and a default user and a project are created.
3. The connection window appears. Fill the requested fields with the default authentication information (host: `localhost` and user/password: `proline/proline`).

The initial configuration creates a default user and a first project, but it also generates some predefined descriptions such as instrument configurations or software used to generate the peaklist submitted to the search engine. Once connected, multiple projects can be created and some of these definitions can be modified as described below:

4. To create a new project, click on the icon in the left panel. In the creation dialog, indicate the name and the description of the project. Click OK.
5. Click on the File » Admin menu. A dialog appears showing four tabs allowing to: create new users; create a new peaklist software definition; view all existing projects with their associated data; and add new fragmentation rule sets (*see* **Note 2**).

By default all data used and generated by Proline-Zero are located in a sub-folder of the installation folder named `data` For security reasons, the server side is only allowed to browse the content of a restricted list of folders that can however be manually modified by editing the server configuration. By default the folders added to this list are sub-folders of `<proline-install>/data` Files must be manually copied into the following folders to be accessible to the server:

6. Copy the Mascot .dat files into the `<proline-install>/data/mascot` sub-folder.
7. Copy the fasta file into `<proline-install>/data/fasta` sub-folder.
8. Copy the Thermo .raw files into `<proline-install>/data/mzdb` sub-folder.

3.2 Import Search Results

Importing a search engine result consists in parsing the search results file to extract information and meta-information and store them into the Proline databases. Neither filtering nor thresholding is applied at this stage: all submitted spectra, peptide-spectrum matches, and protein hits are stored in the database, enabling the subsequent validation of putative identifications. As Mascot is a prerequisite to validate identification results using the BH method, this protocol focuses on Mascot identifications results. However, Proline also supports other search engines (*see* **Note 3**) that can be validated through the classical target–decoy approach.

1. Right-click on the `Identifications` node in the upper left panel and select Import Search Result.
2. The import dialog appears (*see* Fig. 1).
3. Select the files to import by clicking on the file button.
4. Set the different options:
   (a) *Software engine* Mascot: The search engine used to generate the file to import.

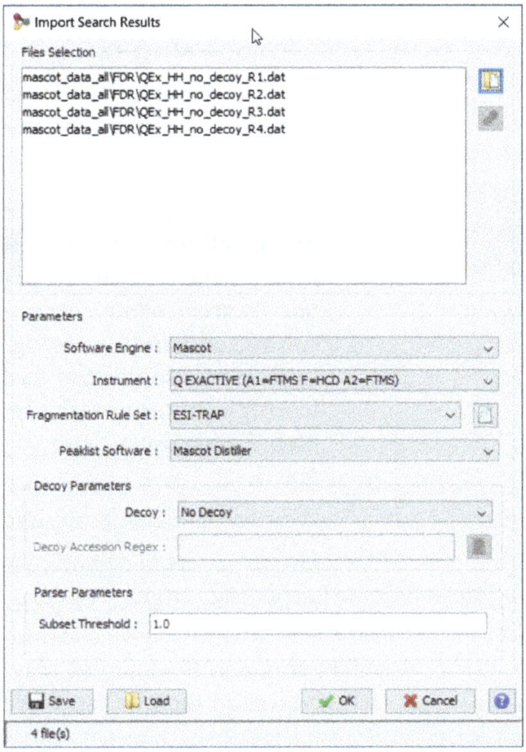

Fig. 1 Import dialog

(b) *Instrument* Q EXACTIVE (A1=FTMS F=HCD A2=FTMS): The mass spectrometer used to perform the MS/MS analysis.

(c) *Fragmentation Rule Set* ESI-TRAP: The fragmentation rules specified in the search engine. The button on the right can be used to visualize all rules of a specific rule set, and new rules can be created through the admin dialog (*see* Subheading 3.1, **Step 5**). This information is required to generate theoretical ion fragments and then annotated MS/MS spectra.

(d) *Peaklist software* Mascot Distiller: The software used to generate the peaklist submitted to the search engine. This information is used during the optional quantification step to extract scan number or retention time from the MS/MS spectrum titles. Additional software can be configured in the admin dialog (*see* Subheading 3.1, **Step 5**).

(e) *Decoy* No Decoy: the target–decoy strategy used during the search: `No Decoy` if the search was performed using a database containing only target protein sequences, `Concatenated Decoy` if the search was performed against a database containing both target and decoy proteins or

`Software Engine Decoy` if the decoy search is performed on the fly by the search engine.

If a wrong fragmentation rule set or a wrong peaklist software is selected at this step, they can be modified afterward.

5. Click OK.
6. Processes are all executed on the server side of Proline. The graphical user interface communicates with the server by submitting the tasks to be performed and modifies the user interface status while waiting for the tasks completion. The submitted and running tasks can be displayed in the `Logs` tab: Submitted tasks are represented by an hourglass icon (⌛), running tasks by a blue arrow (➔), successfully completed tasks by a green tick (✔) or a red cross (✖) when a task failed.
7. Wait for the task completion.

Once imported, the search result appears in the upper left panel under the `Identifications` node. User's actions on these results are reachable by right-clicking on the icon ◑ representing them in the left panel. By default a search result is designated by the name of the imported file but can be renamed.

3.3 Combine Identification Results

Identification results of different MS/MS spectrum searches, for example from replicates or sample fractionation, can be combined or merged to generate a non-redundant list of identified peptides and proteins. This merge can be performed before (on search results) or after validation (on identification summaries) and can be applied recursively, leading in a hierarchical organization of the datasets.

If applied before validation, all PSMs identified by the search engine are taken into account and are combined to create new PSMs into the aggregated dataset. This is useful to combine analytical replicates or fractions since peptides belonging to a single protein could be spread across different fractions. Then the created dataset can be validated as if it were a single search engine identification result.

1. To create a new dataset, Right-click on the `Identifications` node and select Add Dataset.
2. Name the new dataset.
3. Each imported file can be reused with no need to import it again. The `All Imported` node allows to retrieve the already imported files. Right-click on `All Imported` and then Display List (or double click on `All Imported`). Select the search results to be merged, drag and drop them in the newly created dataset. The search results are now hierarchically organized: The newly created dataset is the parent of the

hierarchy, and the dragged and dropped search results are the children.

4. To merge the search results, right-click on the parent dataset and select Merge Datasets » Union and wait for the completion of the merge task (*see* **Note 4**).
5. Once the merge is completed, the icon representing the dataset changes to give a feedback of the type of combination the dataset is made of. While represents an empty dataset, is used for an imported result set or a dataset just merged from non-validated result sets. The right part of the icon indicates properties before validation, whereas the left part indicates properties after validation. A letter is added in the blue half-circle to indicate the type of combination that has been made: `U` for `Union` and `A` for `Aggregate` (*see* **Note 4**). In our case, since the merge is done by union and before validation, the icon changes to .

3.4 Validate Identification Results

Validation of a search result can be performed at PSM, peptide, and protein levels independently. At each level, a set of predefined filters can be combined and applied to accept or reject PSMs or proteins based on a user specified threshold value applied to various properties such as rank, peptide length, minimum score for PSM, and minimum score or number of peptides for protein sets. In addition to these filters, a procedure can be applied to limit the false discovery proportion. This validation step can be done by using the popular target–decoy competition method [2] (*see* **Note 5**) or by using a method based on the Benjamini–Hochberg procedure, as proposed in [3].

1. Right-click on the node representing the merged search results and select Validate.
2. In the validation dialog (*see* Fig. 2), add filters to accept only PSM of rank 1 and length ≥7. Ensure that only one PSM is accepted per MS/MS spectrum (also named a query) by adding a `Single PSM per Query` filter (details of these filters are described in Table 1).
3. Select the BH procedure to control the PSM FDR and set the expected FDR value to 1%.
4. Select the BH procedure to control the Peptide FDR and set the expected FDR value to 1%.
5. Add a filter to invalidate protein set identified without at least one specific peptide (Specific peptides correspond to peptides identifying a unique protein set at the dataset level).
6. Select the BH procedure to control the Protein FDR and set the expected FDR value to 1%.

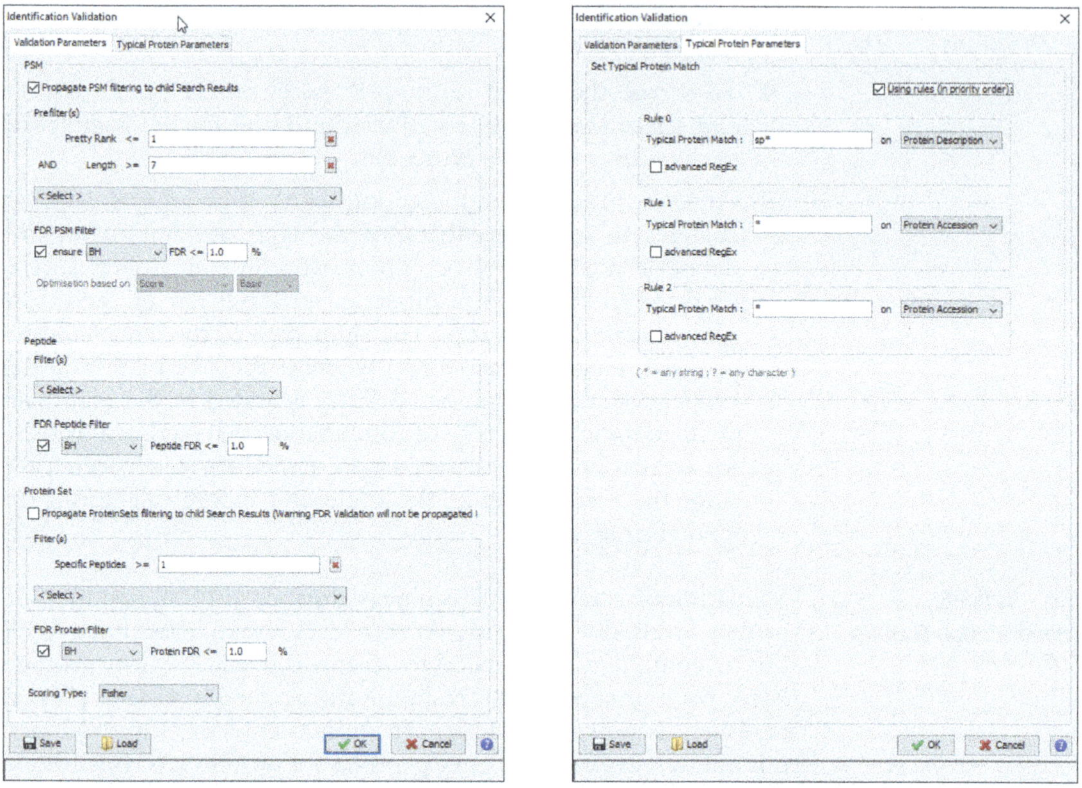

(a) (b)

Fig. 2 Validation dialog. (**a**) Validation parameters tab. (**b**) Change typical protein tab

Table 1
PSM filters

Filter	Description
Minimum peptide length	PSMs corresponding to peptide sequences shorter than the cut-off stipulated will be discarded when this parameter is applied.
Pretty rank	This filter is applied after having temporarily joined target and decoy PSMs corresponding to the same query. For each query, target/decoy PSMs are then sorted by score. As in Mascot, a pretty rank is computed for each PSM depending on their ranking: PSMs with almost equal score (difference < 0.1) are assigned the same rank. All PSMs with a pretty rank greater than the cut-off specified are discarded.
Single PSM per rank	This filter selects only one PSM per pretty rank, which is already the case when a given pretty rank is associated with a single PSM. When multiple PSMs have the same pretty rank, Proline retains the peptide associated with the protein that has the highest number of MS/MS events. Thus, if this filter is combined with the "Pretty rank" filter, the result obtained should be identical to the result of the "Single PSM per MS query" filter.

7. Choose Fisher as Scoring Type that relies on Fisher's test to calculate protein scores and *p*-values that will be used by the BH procedure (*see* **Note 6**).
8. Check the Propagate PSM filtering option. When validating a merged dataset, the applied filters as well as the threshold determined by the FDR control method can be propagated to the child datasets of the hierarchy that have been merged. Doing this, the global FDR at the top level of the hierarchy is controlled, but the user can still explore the individual validated search results, relying on the same validation criteria.
9. Move on to the Typical Protein Parameters tab. In the Rule 0 panel, fill the Typical protein match text field with sp* on Protein Description. A set of identified peptides can match to multiple proteins that cannot be distinguished [5]. This depends on the sample but also on the redundancy of the protein database used for the identification. The typical protein is the protein selected to be the representative of the identified group. The selection can be customized by specifying selection rules. In that particular example, the selection tries to select proteins from the SwissProt database instead of Trembl one if any (*see* **Note 7**).
10. Click OK and wait for the completion of the validation task. Once validated, the left half-circle is highlighted in orange (◐) (*see* **Note 8**).
11. Protein sequences are usually absent from the search engine identification result files. However, Proline can retrieve amino-acid sequences from the fasta file (*see* **Note 9**). Once validation is completed Right-click on the parent dataset and select Retrieve Protein Sequences (*see* **Note 10**).

3.5 Navigate Through Identification Summaries

MS/MS identification results before and after validation can be selected to display their content. Before validation, the navigation path through the data can start from submitted MS/MS spectrum, PSMs or proteins. After validation, the validated content can alternatively be displayed and the navigation then starts from PSMs, peptides, or protein sets. In any case, the selected data are displayed in views composed of one or more tables. The navigation principle is consistent across the graphical interface: The first table contains the data to start from, while the content of the following tables of the view is induced by the selected row in the preceding table. Each table can be searched, filtered, and sorted, and the column visibility can be modified to focus on the information of interest.

1. Right-click on the parent dataset node, and select Display » Identification Summary » Protein sets. The same view as represented in Fig. 3 appears in the right panel of the main

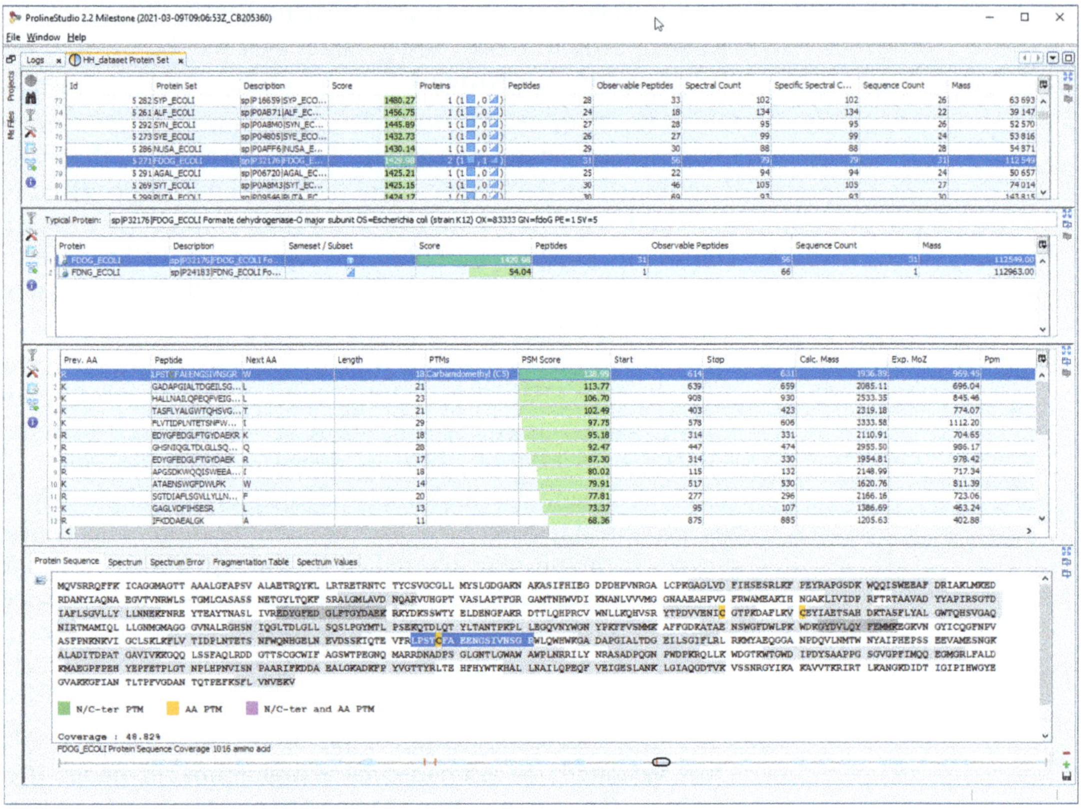

Fig. 3 Navigation through identification summary data

window. The first table shows the identified protein sets of the identification summary. Each row is a protein set represented by one of its protein. The menu bar on the left allows to search for a specific value in the table (🔍), to filter rows (filter icon), to modify the display (column visibility, order and width) (tools icon), or to export the content of the table to a CSV or XLS file (export icon) (*see* **Note 11**).

2. Select or search for the `FDOG_ECOLI` protein set in the list of identified protein sets in the first table: Click on the search button and indicate `FDOG*` in the search popup panel.

3. The second and third tables of the view show the content of the selected protein set: The different proteins identified by the same set of peptides (or by a subset of them) is represented in the second table, while the set of peptides identifying the proteins is displayed in the third table (*see* Fig. 3). If a protein is selected in the second table, for example a protein set marked as a subset by the icon (subset icon), the set of peptides displayed in the third table is restricted to peptides identifying the selected protein.

4. The third table shows all peptides identifying the selected protein. In addition to the sequence and the post-translational modifications associated with each peptide, information related to the highest scoring PSM identifying the peptide is shown. Click on a peptide in the third table to change the selection.
5. The protein sequence retrieved by Proline from the fasta file (*see* Subheading 3.4, **Step 11**) is displayed in the last panel at the bottom of the view. Amino acids matched by an identified peptide are represented on a gray background, and the selected peptide is highlighted in blue. Underneath the amino-acid sequence, a graphical display of the sequence laid out on a single line shows the position of identified peptides on the sequence.
6. Move on to the `Spectrum` tab representing the MS/MS spectrum submitted to the search engine. Click on the `Generate and Store Spectrum Match` icon ( ) on the left. In the dialog, select `ESI-TRAP` as the fragmentation rule set and Click Ok. The fragments matching the theoretical spectrum generated using the specified fragmentation rules are displayed over the MS/MS spectrum (*see* Fig. 4). Move on to the `Fragmentation table` tab to display the fragment matches in a tabular view (*see* **Note 12**).
7. Proline systematically stores and keeps track of metadata from processing steps, used parameters, and generated data. These metadata are available in different views and export outputs. Metadata from multiple dataset can be easily displayed, allowing to compare them (*see* Fig. 5). Select the four replicates in the left panel (left-click on the first dataset, hold the Shitf key, and then left-click on the last replicate) and then right-click and select Properties. In the `Properties` view, each row is a

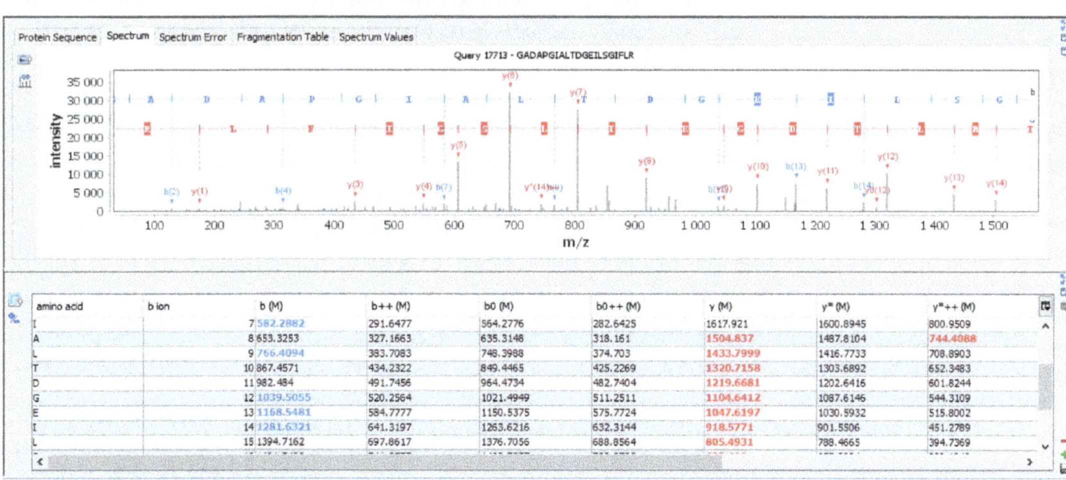

Fig. 4 Annotated MS/MS spectrum and fragmentation table

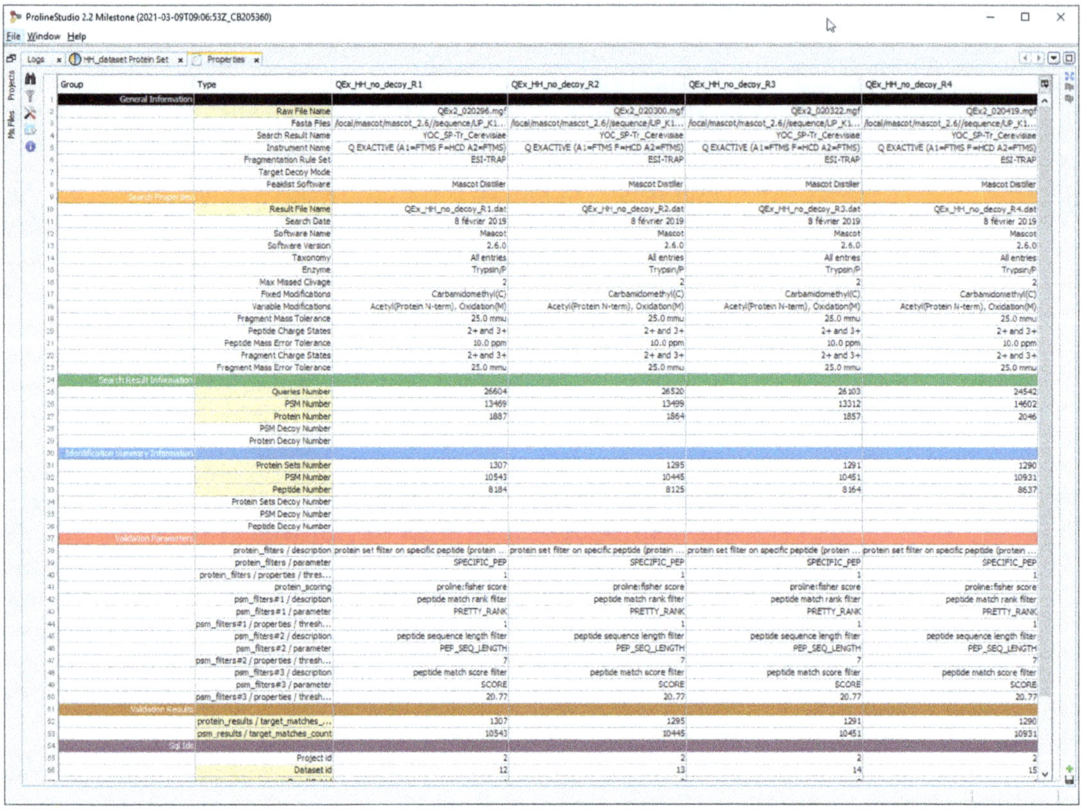

Fig. 5 Datasets properties

property or a metadata (grouped by theme) and each column represents a dataset. In the second column (referred to as `Type`) the property name is displayed on with an orange-light color if the values in the row are different. As an example, in the `Search properties` group, the `Result File Name` is different in each dataset but the other search parameters are all identical.

3.6 MS1 Label-Free Quantification

Once validated, identification summaries can be used for label-free quantification, either using spectral counting, or after detection of chromatographic peaks from MS1 signals. The different steps of the quantification process, the algorithms implemented in Proline, and their parameters have been extensively described in [1]. In this protocol, the focus is made on how to start a MS1 quantification from an identification summary.

1. The signal extraction is based on the mzDB format [6]. Raw files must be first converted by using the conversion tool which is embedded in Proline-Zero (*see* **Note 13**). In the left panel move on the `MS Files` tab. Browse the local

folders to the Proline installation path, then into the 🗀<proline-install>/data/mzdb sub-folder. Select the .raw files then Right-click and select Convert to mzDB.

2. In the conversion dialog, indicate the path to the converter: 🗀<proline-install>/raw2mzDB_0.9.10_build20170802/raw2mzDB

   Then choose as output path the 🗀<proline-install>/data/mzdb

3. Click OK and wait for the end of the conversion task (*see* Subheading 3.2, **Step 6**).

4. Move back to the Projects tab and select the parent dataset. Right-click on the dataset and select Quantify ≫ Label-Free (*see* **Note 14**).

5. In the dialog (*see* Fig. 6a), the first step is the experimental design definition. Groups and samples corresponding to conditions to compare and to biological samples can be created manually by right-clicking on the Quant node. However, a more automated solution is available: Drag and drop the parent dataset node from the right panel into the left panel, on the Quant node. An experimental design is automatically created and appears under the Quant node (in this case, it contains a single group, a single sample, and the four replicates). Right-Click on the Quant node and then select Rename. Modify the default dataset name and click OK. Click Next.

6. In **Step 2** (Associate MS files to sample analyses), mzdb files must be associated with their respective datasets (*see* Fig. 6b). In the right panel, browse the 🗀mzdb_files folder to display the four mzdb files. Select these files (Shift+left-click) and drag and drop them into the panel named Drop Zone. Proline takes advantage of the available metadata, especially the name of the peaklist associated with each dataset to match the peaklist name and the name of the mzdb files dropped in the Drop Zone. In this case, the names perfectly match so that there is nothing more to do. Click Next.

7. The last step consists in setting the parameters used during the quantification process (*see* Fig. 6c). By default the dialog shows a simplified set of parameters with default values suitable for high resolution mass spectrometry analysis: The mass over charge (moz) tolerance is set to 5.0 ppm, the alignment process is performed by using a maximum shift of retention time of 600.0 s, and the cross assignment is enabled between all runs with 60.0 s tolerance to match predicted retention time (*see* **Note 15**).

8. Click OK to start the quantification and wait for the completion of the quantification task.

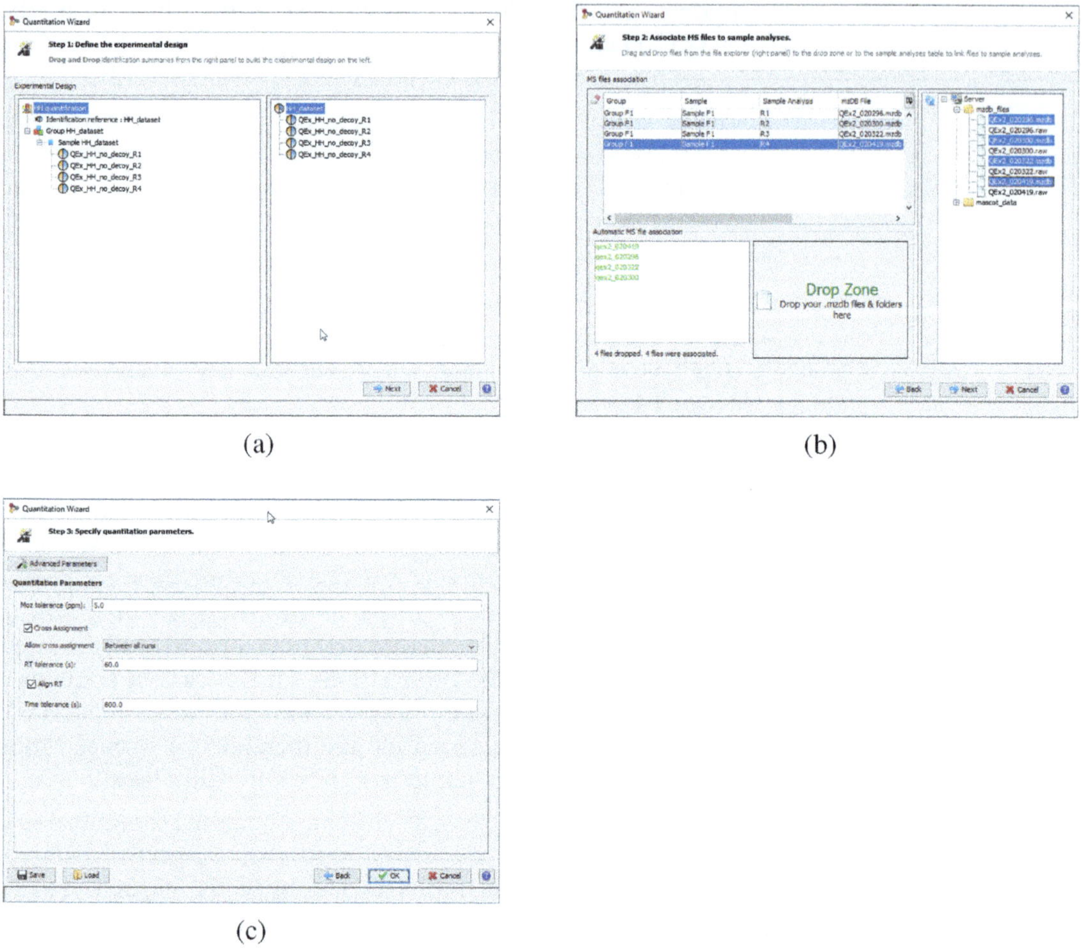

Fig. 6 Label-free quantification dialog. (**a**) Experimental design. (**b**) MzDB files association. (**c**) Label-free parameters

9. Once completed, the quantification dataset node icon changes to , the dataset is ready for browsing and navigation by right-clicking on this icon.

10. By default, peptide ion measurements are summarized as protein abundances using a simple sum operation. However, additional operations such as excluding peptides or ions based on their characteristics (missed cleavages, variable modifications, sequence specificity, etc.) or normalizing peptide and protein abundances between runs can be performed. These post-processing steps can be executed on-demand using different parameters or methods; there is no need to repeat the whole quantification process when changes are made. Right-click on the quantification dataset node and then select Compute Post Processing on Abundances. Choose how ions and peptides abundances are combined into peptide and

protein abundances in the parameter dialog and click OK. Wait for the task completion.

3.7 Navigate Through Quantification Datasets

1. Select the created quantification dataset and right-click and then select Display Abundances » Protein Sets. The predefined view opens (*see* Fig. 7). This view starts from the quantified list of protein sets shown in the first table. Below this table, the quantified peptides of the selected protein set are represented in a tabular display on the left hand side and in a graphical view on the right one. The graphical view represents the quantification profile of each peptide of the protein set (the abundance value in each MS analysis) as well as the protein set abundances (in yellow) calculated from these peptides on a second vertical axis. When a peptide is selected in the table, the graphical representation of the peptide is highlighted accordingly. The bottom panel shows the ions that have been quantified for the selected peptide (`Quanti. Peptide Ions` tab). The `XIC Features` tab shows the signal extracted from each MS analysis (called a feature) while the right hand side panel gives a graphical display of these features.
2. Click on the `Display all isotopes` icon () to display the signal extracted for each isotope of a currently selected feature. Click again to go back to the previous display showing all features.
3. The graphical display uses the identified features stored in the database. However a chromatogram can be extracted from the mzdb files, even if Proline was unable to assign a signal

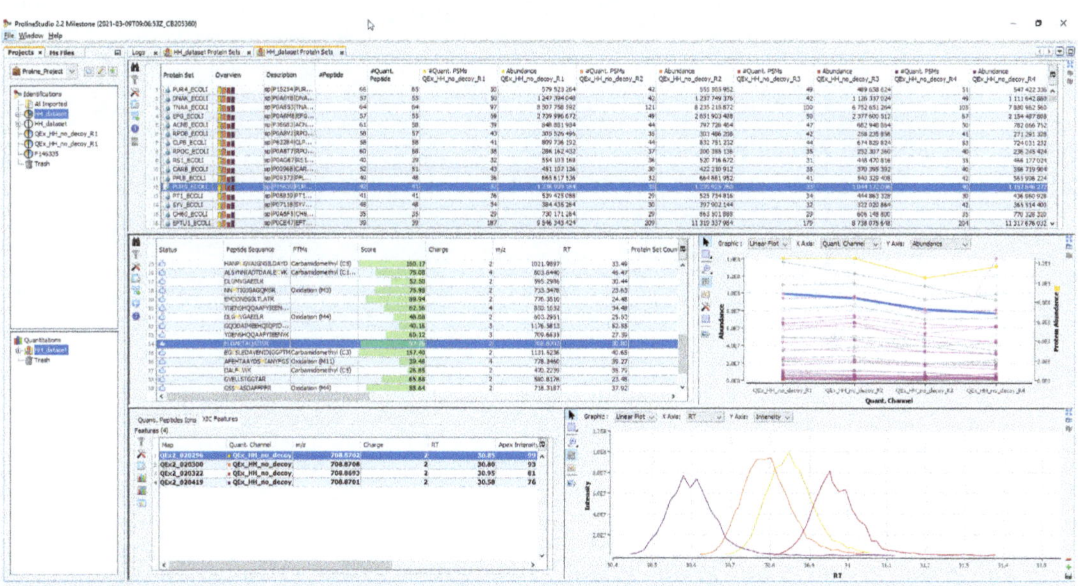

Fig. 7 Navigation through quantification dataset

abundance to the selected ion in some MS analyses. Select a peptide (for example, peptide `IFNADWVIDGEQQPK` of protein `PUR4_ECOLI`), then an ion for which an abundance value is missing, then move on to the `XIC Features`. The table shows the abundance values extracted for each MS analysis. If Proline fails to find an abundance value for an MS analysis, the corresponding row is filled with zero values. Right-click on any row and then select Extract All XIC. A task is created, requesting the extraction of the ion *m/z* chromatogram in all MS analysis. Once the response is received, the chromatograms are visible in the graphical display as dotted lines overlaid on the existing features.

4. In addition to abundance values, the alignment between MS analyses, which is a critical step of the label-free quantification process, can be visualized. Right-click on the quantification dataset (XIC) and then select Display Exp. Design » Map alignment. The upper panel of the view (*see* Fig. 8) shows the alignment curve between two selected MS analyses represented as the time difference (in seconds) between these two MS analysis over the overall analysis time (in minutes). The lower panel shows the alignment curves of all MS analysis with an analysis choose as the alignment reference.

5. In both panels the alignment curves give a general trend of the retention time shift. However in the upper panel the retention time difference of each individual ion can be represented as a

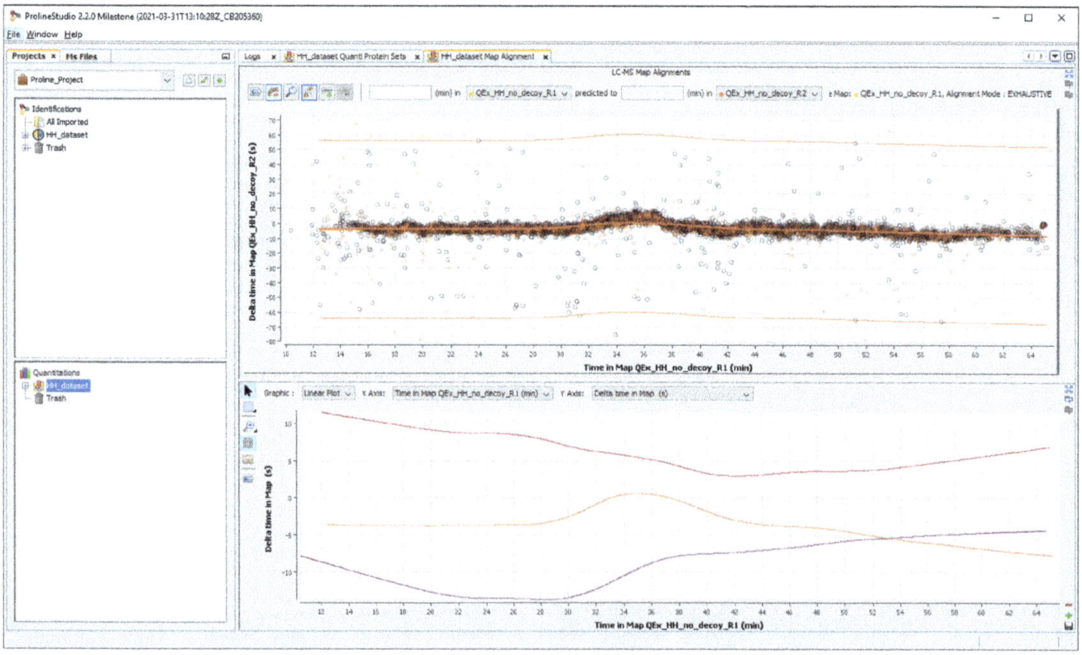

Fig. 8 View MS analyses alignments

scatter plot. Click on the icon. After a few seconds, the `Loading Data` tool-tip disappears and the points representing the ions are shown overlaid on the curve. Filled circles represent ions that have been identified in both MS analyses while black circles represent ions that have been cross-assigned in one of the two MS analyses (cross-assigned ions can be removed from the display by clicking on the icon). The upper and lower curves added to the alignment curve indicate the retention time tolerance used to perform the cross assignment. Visualizing this information enables the user to adjust the retention time tolerance used during the cross-assignment step to reduce the likelihood of errors or, conversely, to take into account the retention time discrepancy between the MS analyses. To modify only this single parameter, relaunch the quantification by right-clicking on the quantitation dataset and then select Clone & Extract Abundances: The quantification parameters dialog appears (*see* Subheading 3.6, **Step 5**) pre-filled with the experimental design and parameters of the selected quantification dataset. The parameter of interest can then be modified leaving all other parameters unchanged.

3.8 Customized Graphical Display

Any dataset in Proline can be browsed using different predefined views, starting from an initial list of objects: PSMs, peptides, protein sets, MS/MS queries, etc. Additionally, new views can be created dynamically and saved for future browsing.

1. Right-click on the quantification dataset and select Display Abundances > New User Window. The list of objects that can be used as starting point are shown. Select `Quanti Protein Sets` and click OK. A new view appears, composed of a single table showing the list of quantified protein sets (*see* **Note 16**).
2. Click on the icon in the right border of the panel to add a new panel into the view. A list of panels that are linked to the protein sets table is shown as well as an option to organize the panel into the view. Select the Customisable Graphical Display panel and add it Below the protein sets table.
3. In the graphical view, set the `Graphic` type to Scatter Plot, `X Axis` to `Abundance QEx_HH_no_decoy_R1` and `Y Axis` to `Abundance QEx_HH_no_decoy_R2`. As expected, the two replicates are similar so that abundances are linearly correlated. Right-click on the `X Axis` and select Log10 Axis. Do the same with the `Y Axis`. To select points in the scatter plot where the abundances are different between replicates, click on the selection tool icon ( ) to select a rectangular region (*see* **Note 17**)

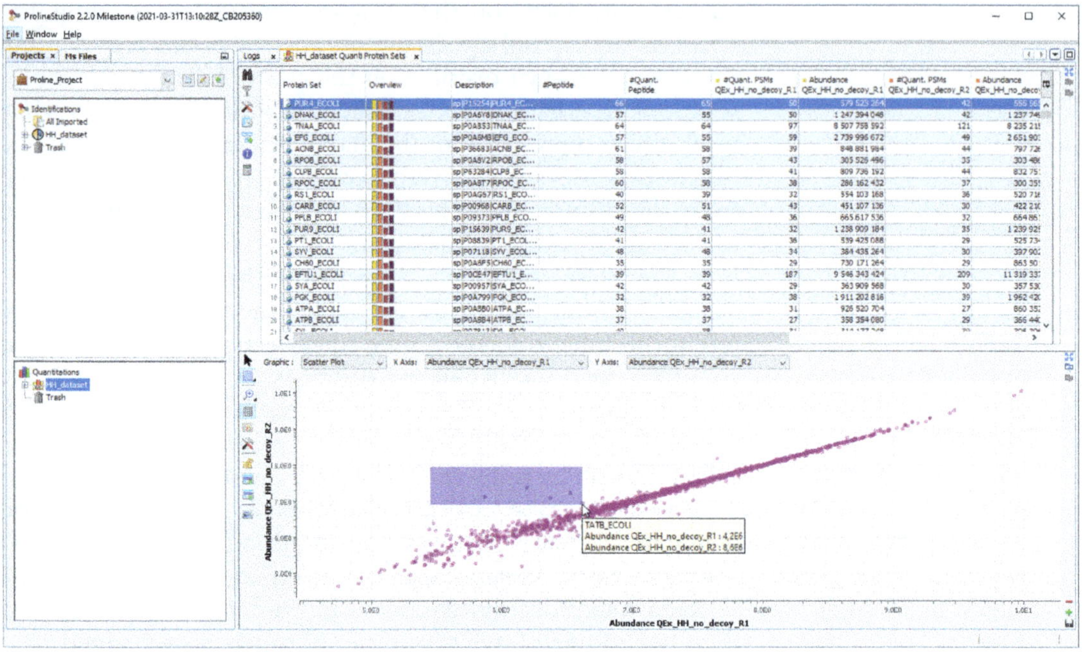

Fig. 9 Custom quantification view

containing these points (*see* Fig. 9). Selected points are then framed in black.

4. The selection can be transferred between tables and graphical display. To select the protein sets corresponding to the selected points in the scatter plot, click on the `Export selection` icon (). The corresponding rows are selected in the protein sets table, which can then be filtered to display only the content of the selection: Right-click on the first selected row and then select View Selected Data. To return to the original display right-click on a row and select View All Data.
5. Add a new table to the view by clicking on the + icon. Select the Quanti Peptides panel and add it Below. The content of this panel is now synchronized with the selected protein set allowing to display the peptide quantification underlying the protein calculated abundance.
6. Select a protein from the first table: The panel containing the peptides quantification information is updated accordingly, showing the quantified peptides belonging to the selected protein. Transfer this selection to the graphical display by clicking on the (): The corresponding point in the scatter plot is then selected.
7. This view can be saved to be reused with any other quantification dataset: Click on the icon in the left border of the panel to save the view. Choose a name for this view and click OK.

8. Right-click on the quantification dataset and select Display Abundances: The new view now appears in the sub-menu next to the existing predefined views.

3.9 Export Data

Identification and quantification results as well as metadata related to the different processing steps performed by Proline are stored in the Proline database and can be exported into various formats. MSExcel (.xlsx) and text (.csv) output formats contain data at PSMs, peptide ions, peptides, and protein sets levels that can be customized before export. In addition, results can be exported in standard-compliant (mzIdentML) formats for publication of data in public repositories such as PRIDE and ProteomeXchange.

1. Right-Click on a dataset and select Export » Excel.
2. Indicate the output file name.
3. Select Excel (.xlsx) as Export type.
4. Click on `Custom options` to choose the sheets, columns, and column names that must be exported (Fig. 10). Each exported sheet is represented by a tab in the customization dialog that can be unchecked to be removed from the exported file.
5. Click Export.
6. In addition to the identified PSM, peptides, and proteins, Proline offers more specialized exports. Export » Sequence Fasta exports a fasta file corresponding to the identified sequences.
7. Export » Spectra List exports a spectral library containing the list of identified peptides with their precursor mass, their observed fragment masses, and their retention time. This export can be achieved from identification or quantification datasets; however, MS/MS spectrum annotation must be generated beforehand (by Right-click on the dataset in the left panel and select Generate Spectrum Matches). Mainly used for Data Independent Acquisition (DIA), this output can be formatted to be compliant with PeakView or Spectronaut software (*see* **Note 18**).
8. Finally, the whole dataset can be exported in mzIdentML format for publication in ProteomeXchange. Right-click on the parent dataset icon and select Export » MzIdentML. Fill the different fields with the needed administrative data and click Next. Set the output file path and click OK.

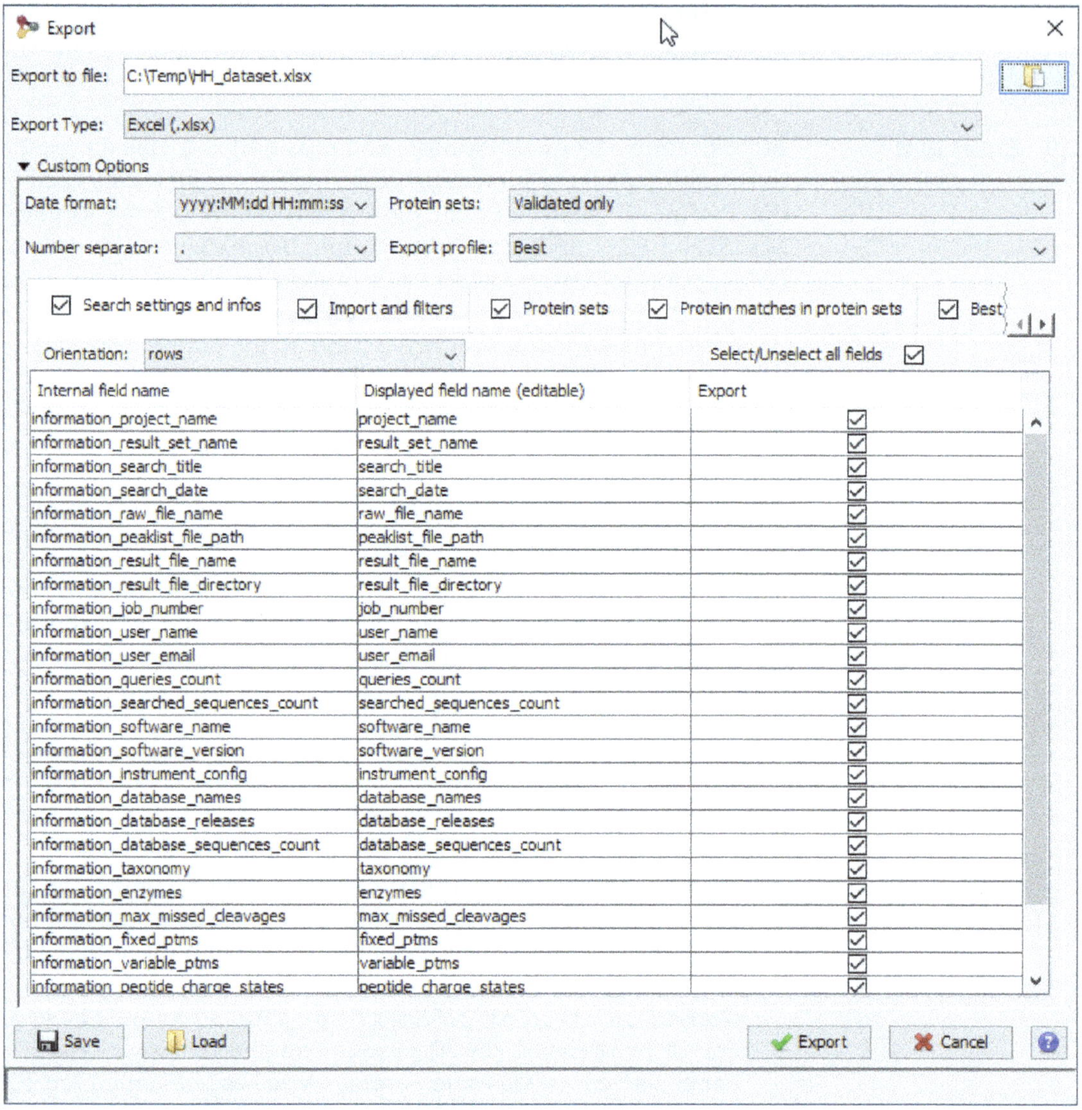

Fig. 10 Customize export dialog

4 Notes

1. These values obviously depend on the data size.
2. To modify these settings, use an administrator user profile (user: `admin` and password: `proline`).
3. Results files produced by Mascot, X! Tandem, OMSSA, and Andromeda search engines can be imported into Proline in their native format.

4. Combining datasets can be done in two different ways: `Union` duplicates all PSMs from the child datasets and creates a copy in the parent dataset; this is similar to what can be obtained by merging peak lists before submission to the search engine. `Aggregate` selects a representative PSM among PSMs matching the same peptide (same sequence and same post-translational modifications).
5. In this case, the search must have been performed against a target–decoy database (*see* Subheading 3.2) while in this protocol a classical search using a target database is sufficient. In the case of using a target–decoy strategy, any identification result performed against a target–decoy database from any search engine supported by Proline can be used. In addition, search results using the `decoy` option available in Mascot can also be imported and validated as a target–decoy search.
6. Proline applies the BH filters in the bottom-up order (PSM > peptides > proteins); however, it is not necessary to use them all: It is possible to validate the identification results using only one or any combination of two of these filters, depending on the user's need and objectives.
7. For Advanced users, a fully regular expression can be specified. In this case, check the corresponding option. A maximum of three rules can be specified. They are applied in priority order, i.e., if no protein of a protein set satisfies the first rule, the second one is tested and so on. If needed, the selection of the typical protein can be changed after the validation by right-clicking on the dataset and select `Change Typical Protein`. This operation can also be performed on multiple datasets at once by selecting multiple datasets (Control + left click).
8. In the same way as for non-validated results, a dataset combined or merged after validation is annotated with letters `U` or `A`, which are added in the orange half-circle.
9. The pre-requisites are: (a) The fasta file name must be the same as the one used during the MS/MS identification search; (b) a regular expression must be supplied to extract the protein accession from the fasta file to match the protein identification of the search result. Predefined regular expressions are configured into Proline so that uniprot accessions for example can be easily retrieved without modifying the configuration.
10. The default configuration can be modified by adding additional matching rules between fasta entries and protein accessions in the `parsing-rules.conf` file, located in the `SequenceRepository` sub-folder of Proline-Zero. The syntax of this file is explained in Proline Installation Guide, section "Sequence Repository Configuration," subsection "Protein description parsing rule."

11. When displaying a target/decoy identification summary, only target proteins are shown in this table. However, the decoy results are also available and can be displayed in a similar view by clicking on the icon in the menu bar. In this protocol, the dataset does not contain any decoy protein so the decoy dataset icon is disabled.
12. Instead of requesting the spectrum annotation PSM by PSM, this can be done systematically as follows: Right-click on a dataset in the left panel and select Generate Spectrum Matches. The same dialog appears and spectrum matches are generated and stored in the database for every validated PSM.
13. The converter is a standalone tool based on ProteoWizard (ensuring compatibility with a wide range of instrument vendors) available at https://github.com/mzdb/pwiz-mzdb. Note that the converter is only available on Windows platforms.
14. There are two ways to start a quantification process: from the identification dataset or from the `Quantitations` node in the lower left panel. Starting from the identification dataset ensures that the quantification process will start from a set of PSM/peptides and proteins that are controlled by the validation process and by the selected validation criteria.
15. The complete set of parameters, as described in [1], is accessible by clicking on the Advanced Parameters button.
16. Any predefined view can also be customized: as an example from the Protein sets view, in the right border of the bottom panel click on the minus icon (−) to remove the XIC feature graphical display. Click repeatedly until the protein sets table remains the only element in the view (the minus icon will then disappear).
17. Clicking on the arrow at the bottom right corner of this icon allows to modify the selection tool. The default selection tool selects the points within a rectangular region drawn by the user. The alternative selection tool allows the user to draw a free-hand selection region. Holding the Ctrl key allows points from disconnected regions to be added to the current selection.
18. Even if the spectra list can be exported from identification or quantification dataset, we recommend to start from a quantification node. Indeed, in such a case, the retention time of identified peptides corresponds to the apex of the extracted signal, which is more accurate than the retention time of the MS/MS spectrum (used for identification datasets).

Acknowledgements

This work was supported by grants from the "Investissement d'Avenir Infrastructures Nationales en Biologie et Santé" program (ProFI project, ANR-10-INBS-08), by the French National Research Agency (GRAL project, ANR-10-LABX-49-01) and by MIAI@Grenoble Alpes (ANR-19-P3IA-0003).

References

1. Bouyssié D, Hesse AM, Mouton-Barbosa E, Rompais M, Macron C, Carapito C, Gonzalez de Peredo A, Couté Y, Dupierris V, Burel A, Menetrey JP, Kalaitzakis A, Poisat J, Romdhani A, Burlet-Schiltz O, Cianférani S, Garin J, Bruley C (2020) Proline: an efficient and user-friendly software suite for large-scale proteomics. Bioinformatics 36(10):3148–3155. https://doi.org/10.1093/bioinformatics/btaa118
2. Elias JE, Gygi SP (2007) Target-decoy search strategy for increased confidence in large-scale protein identifications by mass spectrometry. Nat Methods 4(3):207–214. https://doi.org/10.1038/nmeth1019
3. Couté Y, Bruley C, Burger T (2020) Beyond target–decoy competition: stable validation of peptide and protein identifications in mass spectrometry-based discovery proteomics. Anal Chem 92(22):14898–14906. https://doi.org/10.1021/acs.analchem.0c00328
4. Vizcaíno JA, Deutsch EW, Wang R, Csordas A, Reisinger F, Ríos D, Dianes JA, Sun Z, Farrah T, Bandeira N, Binz PA, Xenarios I, Eisenacher M, Mayer G, Gatto L, Campos A, Chalkley RJ, Kraus HJ, Albar JP, Martinez-Bartolomé S, Apweiler R, Omenn GS, Martens L, Jones AR, Hermjakob H (2014) ProteomeXchange provides globally co-ordinated proteomics data submission and dissemination. Nat Biotechnol 32(3):223–226. https://doi.org/10.1038/nbt.2839
5. Nesvizhskii AI, Aebersold R (2005) Interpretation of shotgun proteomic data the protein inference problem. Mol Cell Proteomics 4(10): 1419–1440. https://doi.org/10.1074/mcp.R500012-MCP200
6. Bouyssié D, Dubois M, Nasso S, Peredo AGd, Burlet-Schiltz O, Aebersold R, Monsarrat B (2015) mzDB: a file format using multiple indexing strategies for the efficient analysis of large LC-MS/MS and SWATH-MS data sets. Mol Cell Proteomics 14(3):771–781. https://doi.org/10.1074/mcp.O114.039115

Chapter 5

Integrating Identification and Quantification Uncertainty for Differential Protein Abundance Analysis with Triqler

Matthew The and Lukas Käll

Abstract

Protein quantification for shotgun proteomics is a complicated process where errors can be introduced in each of the steps. Triqler is a Python package that estimates and integrates errors of the different parts of the label-free protein quantification pipeline into a single Bayesian model. Specifically, it weighs the quantitative values by the confidence we have in the correctness of the corresponding PSM. Furthermore, it treats missing values in a way that reflects their uncertainty relative to observed values. Finally, it combines these error estimates in a single differential abundance FDR that not only reflects the errors and uncertainties in quantification but also in identification. In this tutorial, we show how to (1) generate input data for Triqler from quantification packages such as MaxQuant and Quandenser, (2) run Triqler and what the different options are, (3) interpret the results, (4) investigate the posterior distributions of a protein of interest in detail, and (5) verify that the hyperparameter estimations are sensible.

Key words Shotgun proteomics, Label-free quantification, Protein quantification, Bayesian statistics, Probabilistic graphical models, Error propagation, Differential expression analysis

1 Introduction

Shotgun proteomics has proven to be a useful technique for identifying and quantifying proteins across multiple samples. The identification process of fragment mass spectra has received ample scrutiny in terms of statistical rigor, with target–decoy analysis [1] and false discovery rates (FDRs) [2] being widely adopted by the field. In contrast, as a result of its complexity, error estimation in the quantification process is often limited to a sequence of, frequently heuristic, thresholds. Application of these thresholds is often well-intended and does indeed eliminate false discoveries, but often fails to take into account biases and interactions these create with other thresholds. As a prime example, consider the common practice of applying a fold change threshold after obtaining a list of differentially abundant proteins by a *t*-test. Whereas the original list of differentially abundant proteins was controlled at, say, 5% FDR,

Thomas Burger (ed.), *Statistical Analysis of Proteomic Data: Methods and Tools*,
Methods in Molecular Biology, vol. 2426, https://doi.org/10.1007/978-1-0716-1967-4_5,

there are no guarantees that this FDR is valid for the list after application of the fold change threshold, as there could well be an enrichment of false positives at higher fold changes.

To make matters even more complicated, missing value imputation is an essential part of data analysis for label-free quantification, as often more than 50% of the data is missing due to the stochastic nature of data-dependent acquisition. These imputation methods can introduce many different types of errors [3, 4] which in turn can have a big impact on the statistical test employed for differential abundance testing.

To address these issues, we developed Triqler [5], a Python package that uses a Bayesian model to integrate errors at the different levels of protein quantification. Specifically, it combines identification and quantification errors into a single differential abundance FDR. The Bayesian model propagates and integrates the uncertainties from feature, PSM, peptide, protein, and treatment group level to a final posterior distribution of the fold change between two treatment groups. This is especially useful for missing values, for which we can specify a probability distribution over a range of likely values with a higher likelihood toward lower values. In this way, missing values are considered less reliable than observed values, as is intuitively clear.

The input format for Triqler is a simple tab-delimited file (*see* Subheading 2.6) and can most easily be obtained by converting output files from MaxQuant [6] (*see* Subheading 3.1) or Quandenser [7] (*see* Subheading 3.2). Alternatively, one can use any search engine of choice and add quantification information using Dinosaur [8] (*see* Subheading 3.3). For this tutorial, we explain how to run Triqler (*see* Subheading 3.4), what the individual steps inside Triqler are (*see* Subheading 3.5), and how to interpret the output (*see* Subheading 3.6). Finally, we dive deeper into the results by looking at posterior distributions for a protein of interest (*see* Subheading 3.7) and the hyperparameter estimation (*see* Subheading 3.8).

2 Material

2.1 Requirements

Triqler can be run on any system with Python 2 or 3 installed and has been tested on Windows, Mac OS X, and Linux. For typical datasets, only a modest amount of RAM is required (<2 GB) and can, thus, be run on any desktop computer. To accommodate large-scale dataset analysis, Triqler also supports multicore processing.

For users unfamiliar with the Python environment, we recommend installation of Python through the Anaconda environment, freely available for all major operating systems from https://www.anaconda.com/products/individual.

2.2 Software Install

Triqler is available through the pip command (*see* **Note 1**):

```
$ pip install triqler
```

Alternatively, you can clone the GitHub repository and build the package locally:

```
$ git clone https://github.com/statisticalbiotechnology/triqler.
    git
$ cd triqler
$ pip install .
```

These commands will download the latest release of Triqler (currently v0.6.1). To update the package, replace "pip install" by "pip install --upgrade" in either command.

2.3 Data Type

Triqler requires a list of PSMs, as well as corresponding information regarding intensity, sample, and experimental condition. Unfortunately, not all search engines provide intensity information with their PSMs and this information might, thus, have to be added in a separate step (*see* Subheading 3.3).

2.4 Data Size: Number of Samples

As Triqler aims to determine differential abundance, it requires at least 2 experimental conditions (e.g., case and control) and 3 biological or technical replicates per condition.

2.5 Data Size: Number of Proteins

The parameter estimation of the error model of Triqler depends on a background distribution of proteins that are not differentially abundant. Therefore, we recommend that the software should only be used in situations where the majority of the proteins (>70%) are expected not to be differentially abundant between the conditions and more than 100 proteins are present (*see* **Note 2**).

2.6 Input Format

To simplify the creation of input files, we provide a set of converters that take results from packages such as MaxQuant (*see* Subheading 3.1), Quandenser (*see* Subheading 3.2), Dinosaur (*see* Subheading 3.3), and Tide (*see* Subheading 3.3) and convert them to the Triqler input format.

Alternatively, it is not too complicated to generate a Triqler input file by yourself. An example input file is provided in the GitHub repository at https://github.com/statisticalbiotechnology/triqler/blob/master/example/iPRG2016.tsv. We will use

run	condition	charge	searchScore	intensity	peptide	proteins
A1.ms2	1:A+B	2	0.405342	267860228.2	K.KFIPFSR.V	HPRR1010001_poolA
A2.ms2	1:A+B	2	0.407504	251376674.6	K.KFIPFSR.V	HPRR1010001_poolA
A3.ms2	1:A+B	2	0.489888	292520251.7	K.KFIPFSR.V	HPRR1010001_poolA
B2.ms2	2:B	2	-1.05807	70019480.49	K.KFIPFSR.V	HPRR1010001_poolA
C1.ms2	3:A	2	0.372497	565660378.8	K.KFIPFSR.V	HPRR1010001_poolA
C2.ms2	3:A	2	0.40337	515010510.4	K.KFIPFSR.V	HPRR1010001_poolA
C3.ms2	3:A	2	0.291525	508428331.1	K.KFIPFSR.V	HPRR1010001_poolA
A1.ms2	1:A+B	3	-1.78404	9954876.791	K.YFPYRGSLLSLFIVG.-	decoy_HPRR4110123_poolB
A2.ms2	1:A+B	3	-1.80667	54937151.14	K.YFPYRGSLLSLFIVG.-	decoy_HPRR4110123_poolB
C1.ms2	3:A	3	-1.38149	228980354.7	K.YFPYRGSLLSLFIVG.-	decoy_HPRR4110123_poolB
C2.ms2	3:A	3	-1.43683	251341063.8	K.YFPYRGSLLSLFIVG.-	decoy_HPRR4110123_poolB
C3.ms2	3:A	3	-1.43252	237306632.5	K.YFPYRGSLLSLFIVG.-	decoy_HPRR4110123_poolB

Fig. 1 An example of the Triqler input format

this example file throughout this chapter to demonstrate several characteristics of Triqler.

The input format consists of 7 columns separated by tabs (*see* Fig. 1), indicated by the following headers: `run`, `condition`, `charge`, `searchScore`, `intensity`, `peptide`, `proteins`. Each row consists of a PSM result from a search engine of choice (`charge`, `searchScore`, `peptide`, `proteins`), together with the intensity from the corresponding MS1 feature (`intensity`) and meta-information regarding the sample the PSM corresponds to (`run`, `condition`). If a peptide is shared between multiple proteins, the protein names are separated by semicolons (*see* **Note 3**).

3 Methods

As the input to Triqler is unfortunately rather non-standard, we start with 3 different methods of generating an input file from existing pipelines. This will be followed by running Triqler itself and interpreting the results, as well as diving deeper into the results for particular proteins of interest.

3.1 Generating Triqler Input from MaxQuant

Triqler contains functionality to convert a MaxQuant `evidence.txt` output file into a Triqler input file [9] (an example is given below, *see* Subheading 3.1, **Step 3**):

```
$ python -m triqler.convert.maxquant --file_list_file <L>
--out_file <OUT_FILE> <IN_FILE>
```

Explanation of the command line parameters:

1. `<IN_FILE>`:The file called `evidence.txt` present in the `combined/txt` folder of the MaxQuant results.

A1.mzML	A+B
A2.mzML	A+B
A3.mzML	A+B
B1.mzML	B
B2.mzML	B
B3.mzML	B
C1.mzML	A
C2.mzML	A
C3.mzML	A

A1_frac1.mzML	A+B	A1	1
A1_frac2.mzML	A+B	A1	2
A1_frac3.mzML	A+B	A1	3
A2_frac1.mzML	A+B	A2	1
A2_frac2.mzML	A+B	A2	2
A2_frac3.mzML	A+B	A2	3
B1_frac1.mzML	B	B1	1
B1_frac2.mzML	B	B1	2
B1_frac3.mzML	B	B1	3

Fig. 2 An example of the Triqler file list input format. The simple format (left) only contains the file name and experimental condition. The extended format (right) also includes columns for the sample name and fraction number

2. --file_list_file <L>: A simple tab-separated text file with spectrum file names in the first column and treatment group in the second column (*see* Fig. 2, left). The spectrum file names should be the same as in the `Raw file` column name in `evidence.txt` without the preceding path. In the case of fractionated samples, the third and fourth columns should contain the sample name and fraction, respectively (*see* Fig. 2, right).
3. `--out_file` <OUT>: The output of this converter, a Triqler input file.

Below are the step-by-step instructions to process a MaxQuant `evidence.txt` file with Triqler:

1. For best performance, we recommended running MaxQuant by setting all FDR thresholds (PSM, peptide, and protein-level) to 100% (*see* **Note 4**) and turning off matches-between-runs (*see* **Note 5**).
2. Create a tab-delimited file with the file metadata as described above, e.g., `file_metainfo.txt`.
3. Run the MaxQuant converter:

```
$ python -m triqler.convert.maxquant --file_list_file
file_metainfo.txt --out_file triqler_input.tsv evidence.txt
```

4. MaxQuant uses the decoy prefix `REV__` by default, so remember to change this prefix accordingly:

```
$ python -m triqler --decoy_pattern REV__ triqler_input.tsv
```

3.2 Generating Triqler Input from Quandenser

Quandenser is a peptide quantification package that employs unsupervised clustering on both MS1 and MS2 data, without assigning peptide sequences. The benefit is that one can run the identification part as often as desired, without having to redo the quantification.

The current Quandenser converter relies on that the search results are post-processed by Percolator [10, 11].

The interface for this converter is as follows (an example is given below, *see* Subheading 3.2, **Step 4**):

```
$ python -m triqler.convert.quandenser --file_list_file <L>
--psm_files <TARGET>,<DECOY> --out_file <OUT> <IN_FILE>
```

Explanation of the command line parameters:

1. `<IN_FILE>`: The file called `Quandenser.feature_groups.tsv` in the Quandenser results folder.
2. `--file_list_file <L>`: A simple tab-separated text file with spectrum file names in the first column and condition in the second column. In the case of fractionated samples, the third and fourth columns should contain the sample name and fraction, respectively (*see* Fig. 2).
3. `--psm_files <TARGET>,<DECOY>`: The target/decoy PSM output files from Percolator, separated by commas. Both output files from stand-alone Percolator as well as from `crux percolator` are supported (*see* **Note 6**).
4. `--out_file <OUT>`: the output of this converter, a Triqler input file.

Below are the step-by-step instructions to process the Quandenser output files with Triqler (*see* **Note 7**).

1. Run Quandenser on your input files. This will produce a file called `Quandenser.feature_groups.tsv` as well as one or more consensus spectra files in the `consensus_spectra` folder.
2. Search all consensus spectra files with a search engine of choice, preferably one supported by Percolator (*see* **Note 8**).
3. Run Percolator on the search results files, and make sure to use both the `--results-psms` and `--decoy-results-psms` to obtain results on PSM level with both targets and decoys reported.
4. Convert the search results:

```
$ python -m triqler.convert.quandenser --file_list_file
file_metainfo.txt --out_file triqler_input.tsv
--psm_files percolator.target_psms.txt,percolator.decoy_psms.txt
Quandenser.feature_groups.tsv
```

5. Run Triqler:

```
$ python -m triqler triqler_input.tsv
```

3.3 Generating Triqler Input from Dinosaur

Triqler also provides functionality to do quantitative analysis on search results from a search engine of choice, post-processed by Percolator (*see* **Note 8**), by using Dinosaur to add MS1 quantification information to the MS2 spectra. The interface for this converter is as follows (an example is given below, *see* Subheading 3.3, **Step 4**):

```
$ python -m triqler.convert.dinosaur --file_list_file <L>
--psm_files <TARGET>,<DECOY> --out_file <OUT> <IN_FILES>
```

The command line parameters are the same as in Subheading 3.2, with the exception that the input files now are feature-to-spectrum mapping files, most easily produced by using our Dinosaur adapter for Python (*see* **Note 9**). Below are the step-by-step instructions to obtain quantification information with Triqler using a search engine of choice (*see* **Note 10**).

1. Run the Dinosaur adapter for Python on your input mzML files. This will produce several files named `dinosaur/<file_name>.feature_map.tsv`, and, optionally, spectrum files named `dinosaur/<file_name>.recalibrated.<spectrum_format>`.
2. Search your input mzML files or the recalibrated MS2 spectrum files (*see* **Note 11**) with a search engine of choice, preferably one supported by Percolator (*see* **Note 8**).
3. Run Percolator on the search results files, make sure to use both the `--results-psms` and `--decoy-results-psms` to obtain results on PSM level with both targets and decoys reported.
4. Convert the search results:

```
$ python -m triqler.convert.dinosaur --file_list_file
file_metainfo.txt --out_file triqler_input.tsv
--psm_files percolator.target_psms.txt,percolator.decoy_psms.txt
dinosaur/*.feature_map.tsv
```

5. Run Triqler:

```
$ python -m triqler triqler_input.tsv
```

3.4 Triqler Interface

To verify that Triqler was installed correctly and to get an overview of the command line parameters, run the following command:

```
$ python -m triqler --help
```

This should produce the following help text. The individual command line arguments are explained further below:

```
Triqler version 0.6.1
Copyright (c) 2018-2020 Matthew The. All rights reserved.
Written by Matthew The (matthew.the@scilifelab.se) in the
School of Engineering Sciences in Chemistry, Biotechnology and Health at the
Royal Institute of Technology in Stockholm.
Issued command: triqler.py --help
usage: _ _main_ _.py [-h] [--out_file OUT] [--fold_change_eval F]
                     [-decoy_pattern P] [--min_samples N]
                     [-num_threads N]
                     [--ttest] [--write_spectrum_quants]
                     [--write_protein_posteriors P_OUT]
                     [--write_group_posteriors G_OUT]
                     [--write_fold_change_posteriors F_OUT]
                     IN_FILE

positional arguments:
  IN_FILE            List of PSMs with abundances (not log transformed!)
                     and search engine score.
                     See README for a detailed description of the columns.

optional arguments:
  -h, --help           show this help message and exit
  --out_file OUT       Path to output file (writing in TSV format).
                       N.B. if more than 2 treatment
                       groups are present, suffixes will be
                       added before the file extension.
                       (default: proteins.tsv)
  --fold_change_eval F log2 fold change evaluation threshold.
                       (default: 1.0)
  --decoy_pattern P    Prefix for decoy proteins. (default: decoy_)
  --min_samples N      Minimum number of samples a peptide
                       needed to be quantified in. (default: 2)
  --num_threads N      Number of threads, by default this is equal to
                       the number of CPU cores available
                       on the device. (default: 6)
  --ttest              Use t-test for evaluating differential expression
                       instead of posterior
                       probabilities. (default: False)
  --write_spectrum_quants
                       Write quantifications for consensus
                       spectra. Only works if consensus spectrum
                       index are given in input.
                       (default: False)
```

```
--write_protein_posteriors P_OUT
                    Write raw data of protein posteriors to
                    the specified file in TSV format. (default: )
--write_group_posteriors G_OUT
                    Write raw data of treatment group
                    posteriors to the specified file in
                    TSV format. (default: )
--write_fold_change_posteriors F_OUT
                    Write raw data of fold change posteriors
                    to the specified file in TSV format.
                    (default: )
```

A detailed description of the command line arguments follows below:

1. `<IN_FILE>`: Tab-separated input file with the format described previously (*see* Subheading 2.6).
2. `--out_file <OUT>`: Tab-separated results file on protein-level. If more than 2 treatment groups are specified, multiple output files are generated, where the comparison is inserted before the `.tsv` extension. For example, if the output file is specified as `proteins.tsv`, then the output files will be named `proteins.1vs2.tsv`, `proteins.1vs3.tsv`, etc.
3. `--fold-change-eval <F>`: The $\log_2$ fold change used for evaluation of differential abundance. Specifically, we integrate the probability of the fold change distribution outside the region $[-F, F]$ as the probability that the protein is differentially abundant (*see* Subheading 3.7, **Step 9**). Note that this test is quite different from the *t*-test, and setting $F = 0$ to supposedly obtain all differentially abundant proteins regardless of the fold change will result in nonsensical results (*see* Subheading 3.7, **Step 10**).
4. `--decoy_pattern <P>`: Prefix for decoy proteins used by the search engine. For MaxQuant searches, this is typically `REV__`. This pattern is used to recognize which of the PSMs are targets and decoys.
5. `--min_samples <N>`: This flag controls which peptides are discarded because they have too many missing values. For example, if $N = 3$ and a total of 10 samples are provided, all peptides with more than 7 missing values are discarded (*see* **Note 12**).
6. `--num_threads <N>`: This flag controls how many threads are used to calculate the posterior distributions (*see* **Note 13**).
7. `--ttest`: Specifying this flag computes a *t*-test based on the expected values of the protein abundances instead of using the Bayesian posterior calculation. This is only meant for

comparison purposes and we do not claim or support any validity of these results.

8. `--write_spectrum_quants`: Specifying this flag produces an extra intermediate output very similar to the peptides output specifically for Quandenser input. Instead of outputting the peptide abundances across all runs, it outputs the abundances for an MS1 feature group across all runs.
9. `--write_protein_posteriors` <P_OUT>: Writes the raw results of the posterior distributions of the relative protein abundance for each protein. How to visualize these results is described later (*see* Subheading 3.7, **Step 7**).
10. `--write_group_posteriors` <G_OUT>: Writes the raw results of the posterior distributions of the treatment group mean abundance for each protein. How to visualize these results is described later (*see* Subheading 3.7, **Step 8**).
11. `--write_fold_change_posteriors` <F_OUT>: Writes the raw results of the posterior distributions of the fold change for each protein. How to visualize these results is described later (*see* Subheading 3.7, **Step 9**).

3.5 Running Triqler

1. Download the example file from our GitHub repository: https://github.com/statisticalbiotechnology/triqler/blob/master/example/iPRG2016.tsv (*see* **Note 14**).
2. Run Triqler, keeping all parameters at their default values

```
$ python -m triqler iPRG2016.tsv
```

Below, the internal steps of Triqler and the produced outputs are explained section by section.

3. Boilerplate welcome message, including the specified command line parameters for easy future reference:

```
Triqler version 0.6.1
Copyright (c) 2018-2020 Matthew The. All rights reserved.
Written by Matthew The (matthew.the@scilifelab.se) in the School of Engineering
Sciences in Chemistry, Biotechnology and Health at the Royal Institute
of Technology in Stockholm.
Issued command: triqler.py iPRG2016.tsv
```

4. Triqler parses the input file and recalculates q-values and posterior error probabilities (PEPs) using a Python re-implementation of `qvality` [12].

```
Parsing triqler input file
  Reading row 0
Calculating identification PEPs
  Identified 12113 PSMs at 1% FDR
```

5. Some peptides may have been assigned to multiple MS1 features in a single run, each with a different XIC intensity (*see* **Note 15**). In this step, Triqler selects the best MS1 feature based on the search engine score and, if available, the feature-match error probability.

```
Selecting best feature per run and spectrum
  featureGroupIdx: 0
```

6. Triqler divides the XIC intensities by a power of 10 while preserving 2 significant digits for the smallest observed intensity. This does not affect subsequent analysis or the content of the output files but serves to increase the readability of the intermediate files (*see* **Note 16**):

```
Dividing intensities by 100000 for increased readability
```

7. Peptide-intensity pairs from the different runs are grouped based on sequence and charge state. Subsequently, the FDR and PEPs of the unique peptides are calculated [13]. The grouped peptide-intensity pairs are then written to an intermediate file, in this case `iPRG2016.tsv.pqr.tsv` (*see* **Note 17**):

```
Calculating peptide-level identification PEPs
  Identified 1988 peptides at 1% FDR
Writing peptide quant rows to file: iPRG2016.tsv.pqr.tsv
```

8. Protein inference is executed using the best peptide score as the protein's score, together with the picked protein approach [14]:

```
Calculating protein-level identification PEPs
  Identified 349 proteins at 1% FDR
```

9. Triqler makes use of the empirical Bayes method to estimate hyperparameters from the input data. We fit distributions to naive estimates of peptide and protein abundance values (*see* Subheading 3.8):

```
Fitting hyperparameters
  params["muDetect"], params["sigmaDetect"] = 1.056334,
0.372395
  params["muXIC"], params["sigmaXIC"] = 3.276315, 0.953023
  params["muProtein"], params["sigmaProtein"] = 0.066437,
0.239524
  params["muFeatureDiff"], params["sigmaFeatureDiff"] =
-0.013907, 0.149265
  params["shapeInGroupStdevs"],params["scaleInGroupStdevs"]
= 1.027176, 0.089433
```

10. Based on the hyperparameter estimation of the standard deviation of the protein prior distribution, Triqler estimates a lower bound for the `--fold_change_eval` parameter (default: 1.0). Setting this threshold below this lower bound estimate will trigger a warning, as it can lead to false positives unaccounted for by the reported FDR [9]. Usually, this lower bound estimate will be around 0.5 but for this dataset a much higher lower bound of 1.62 was estimated. This is a result of the unusual study design of the iPRG2016 dataset in which half the proteins are intentionally absent in a number of samples. In this case, we can safely ignore this warning for this reason:

```
Minimum advisable --fold_change_eval: 1.62
  WARNING: --fold_change_eval is set below recommended lower
bound, increased
 risk of false positives.
```

11. The hyperparameters are now used in the probabilistic graphical model to estimate posterior distributions for the relative protein abundances, as well as for the fold change between the treatment groups:

```
Calculating protein posteriors
  50 / 422 11.85%
  100 / 422 23.70%
  150 / 422 35.55%
  200 / 422 47.39%
  250 / 422 59.24%
  300 / 422 71.09%
  350 / 422 82.94%
  400 / 422 94.79%
```

12. Finally, the protein identification PEP and the probability that the fold change exceeds the fold change threshold specified by the `--fold_change_eval` parameter are combined. This

combined PEP can then be used to compute the differential abundance FDR (*see* **Note 18**). For each comparison, Triqler then reports the number of differentially abundant proteins at 5% FDR and outputs the final results to one file per comparison (e.g., `proteins.1vs2.tsv`):

```
Comparing 1:A+B to 2:B
  output file: proteins.1vs2.tsv
  Found 204 target proteins as differentially abundant at 5%
      FDR
Comparing 1:A+B to 3:A
  output file: proteins.1vs3.tsv
  Found 216 target proteins as differentially abundant at 5%
      FDR
Comparing 2:B to 3:A
  output file: proteins.2vs3.tsv
  Found 352 target proteins as differentially abundant at 5%
      FDR
Triqler execution took 22.869995618006214 seconds wall
clock time
```

In this particular dataset, there are a total of 383 truly differentially abundant (spiked-in) proteins. However, in the $A + B$ vs A and $A + B$ vs B comparisons, the expected $\log_2$ fold change coincides with the chosen `--fold_change_eval` parameter of 1.0, which makes it hard for about half of these proteins to be called differentially abundant (*see* **Note 19**).

3.6 Interpreting the Triqler Output

The protein output files can be opened in Excel, or another spreadsheet package of choice. For an explanation of the columns, *see* Table 1.

To help understand the output better, we will look at a specific example in the `proteins.1vs2.tsv` file.

1. Search for the protein `HPRR3730445_poolB`, which should be around the 174th line.
2. In the 1st column, the *q*-value for this protein is shown to be 0.006985 (the exact value may vary in newer versions), meaning that the list of proteins up to this position has an FDR of 0.7%, meaning that we expect approximately $173 \times 0.006985 = 1.2$ false positives in this list of 173 proteins. This *q*-value combines the error probabilities of the identification and the quantification process, which improves on the typical practice of separately reporting identification and quantification FDRs.

Table 1
Triqler protein output format

Column name	Description
q_value	Differential abundance FDR. Includes identification and quantification error probabilities. Is the cumulative average of the sorted column posterior_error_prob
posterior_error_prob	Differential abundance PEP. Combines identification and quantification error probabilities from protein_id_posterior_error_prob and diff_exp_prob_1.0
protein	Protein identifier, as specified in the input file
num_peptides	Number of unique peptides for this protein (without applying an FDR filter)
protein_id_posterior_error_prob	Identification PEP, based on the best scoring PSM for this protein
log2_fold_change	Expected value of the posterior distribution of the $\log_2$ fold change
diff_exp_prob_1.0	The integrated probability of the fold change distribution outside the region $[-F, F]$ specified by the fold_change_eval parameter (*see* Subheading 3.7, **Step 9**). The 1.0 refers to the default value of the fold_change_eval parameter and will change accordingly
1:A+B:A1.ms2	Expected value of the posterior distribution of the relative protein abundance relative for each sample. Here, relative means relative to the mean across all samples (*see* Subheading 3.7, **Step 3**) (*see* **Note 20**).
1:A+B:A2.ms2 1:A+B:A3.ms2 2:B:B1.ms2 2:B:B2.ms2 2:B:B3.ms2 3:A:C1.ms2 3:A:C2.ms2 3:A:C3.ms2	
peptides	All peptides mapped to this protein, disregards shared peptides

3. In the 2nd column, we can see that the posterior error probability is 0.1302. This value, again, combines both identification and quantification error probabilities. Practically speaking this value means that this particular protein has a probability of 87% to be both correctly identified and differentially abundant.
4. In the 4th column, we see that we have found 3 unique peptides for this protein. However, this does not tell us how many were identified at 1% peptide-level FDR.
5. In the 5th column, the identification posterior error probability is listed as 0.0001162. Meaning that the protein has a probability of only 0.01% to be incorrectly identified.
6. In the 6th column, we see that the expected $\log_2$ fold change between the groups $A + B$ and B for this protein is -1.522. This means that we expect that this protein is present in group $A + B$ at a concentration of $2^{-1.522} = 0.34$ relative to group B. According to the spike-in ratios, this value should have been 0.5.
7. In the 7th column, the probability that the $\log_2$ fold change between the $A + B$ and B treatment group is larger than 1.0 (the default value for the `fold_change_eval` parameter) is given as 0.1301, meaning that there is a 13% chance that the absolute fold change was in fact below 0.5. If we would have specified a lower value for the `fold_change_eval` parameter, this probability would have increased, as we would be integrating a larger part of the posterior distribution of the fold change. This in turn will affect the combined posterior error probability and FDR in the first and second columns.
8. In the 8–16th column, we can see the individual expected protein abundance value per sample, as obtained from the posterior distributions for the relative protein abundance (relative to the mean over all samples, which would itself be represented by the value 1.0, also *see* Subheading 3.7, **Step 3**). For example, we see that in `A3.ms2` the expected value for the relative protein abundance is 5.475, meaning that it is 5.5 times higher than the average protein abundance across all samples. In `B2.ms2` this value is 13.64, meaning that we expect the abundance in this sample to be $\frac{13.64}{5.475} = 2.5$ times higher than in `A3.ms2`.
9. In the 17th column and beyond, the 3 peptides that were used to identify and quantify this protein are listed.

3.7 Visualizing and Interpreting Posterior Distributions

In certain cases, one would like to examine how Triqler arrived at the conclusion of differential abundance for a particular protein. For this, Triqler provides functionality for extracting the relevant data for proteins of interest and plotting the posterior distributions at different levels (protein, treatment group, fold change). Triqler also provides functionality to plot a heatmap of the fold change posterior distributions of a set of proteins, such as proteins in a pathway of interest [9] (*see* **Note 21**).

1. Generate the posterior distribution plots for our protein of interest (we need to increase `--plot_max_fold_change` from its default value of 2.0, since we have extreme fold changes in this particular sample):

```
$ python -m triqler.distribution.plot_posteriors --protein_id
HPRR3730445_poolB --plot_max_fold_change 10.0 iPRG2016.tsv
```

   Three plot windows will be opened, which we will examine shortly. First, we will look at the command line output.

2. After the boilerplate text and hyperparameter estimation, the peptides for this protein are listed in descending order of confidence. First, the "raw" intensity values are listed in the same order as in the regular protein output (*see* Table 1); note that these have been divided by the power of 10 as mentioned previously (*see* Subheading 3.5, **Step 6**) compared to the original Triqler input. Missing values are indicated by `nan` values. These are followed by the `combinedPEP`, which is a combination of the identification and feature-match error probability. In this case, since the latter is not included, `combinedPEP` just reflects the identification PEP:

```
Peptide absolute abundances
760.43 509.03 1028.25 842.80 1610.55 1289.44 nan nan nan
combinedPEP=3.4e-06 peptide=R.WTAQGHANHGFVVEVAHLEEK.Q
99.93 166.59 3184.98 1868.59 6260.46 5909.35 nan nan 59.71
combinedPEP=2.2e-05 peptide=R.LVNQNASRWESFDVTPAVMR.W
10064.52 12531.44 nan 27429.83 26226.20 23061.53 nan 242.17 19.53
combinedPEP=0.0023 peptide=R.WESFDVTPAVMR.W
```

   Here, we can see that all 3 peptides are identified with high confidence. There are many missing values in the group *A* samples (the last three columns of each row), and one missing value in the third $A + B$ sample (third column).

3. The values from above are repeated, but now divided by the geometric mean across all samples, resulting in "relative" peptide abundances:

```
Peptide relative abundances
0.81 0.54 1.09 0.90 1.71 1.37 nan nan nan
combinedPEP=3.4e-06 peptide=R.WTAQGHANHGFVVEVAHLEEK.Q
0.12 0.21 3.96 2.32 7.78 7.34 nan nan 0.07
combinedPEP=2.2e-05 peptide=R.LVNQNASRWESFDVTPAVMR.W
```

```
2.70 3.37 nan 7.37 7.05 6.20 nan 0.07 0.01
combinedPEP=0.0023 peptide=R.WESFDVTPAVMR.W
```

4. The above information (peptide intensities and identification PEPs) are then processed by Triqler's probabilistic graphical model, resulting in several posterior distributions. First, the protein-level results are printed as expected values of the protein abundance posterior distribution. For comparison purposes, also the results of a regular *t*-test (2 treatment groups) or ANOVA test (3 or more treatment groups) on these expected values are shown:

```
Protein abundance (expected value) and p-value
3.27 2.97 5.48 7.66 13.64 11.55 0.01 0.07 0.02
p-value: 3.282253815003453e-05
```

5. Next, Triqler calculates the probability that the $\log_2$ fold change is below the specified `--fold_change_eval` parameter. By default, this is 1.0, but we will explore how these probabilities change for different values (*see* Subheading 3.7, **Step 10**). We can see that for the $A + B$ vs B comparison, there is a 13% probability that the $\log_2$ fold change is below 1.0, whereas for the other two comparisons, this probability is practically 0. This also makes sense, as in group A, this protein is in fact not present:

```
Posterior probability |log2 fold change| < 1.00
  Group A+B vs Group B: 0.130072
  Group A+B vs Group A: 0.000000
  Group B vs Group A: 0.000000
```

6. Finally, we also summarize the posterior distributions of the group means by fitting normal distributions to them. These values are, again, relative to the mean across all group abundances. Note that these values are based on the $\log_{10}$ transformed values, rather than $\log_2$ as is the case for the values reported for fold changes. The standard deviations reflect the uncertainty in the estimates. In this case, the standard deviation for group A is the largest, which reflects the fact that the missing values cause a higher degree of uncertainty:

```
Normal distribution fits for posterior distributions of
treatment
group relative abundances:
  Group A+B: mu, sigma = 0.472918, 0.114328
  Group B: mu, sigma = 0.918317, 0.071100
  Group A: mu, sigma = -1.782820, 0.269253
```

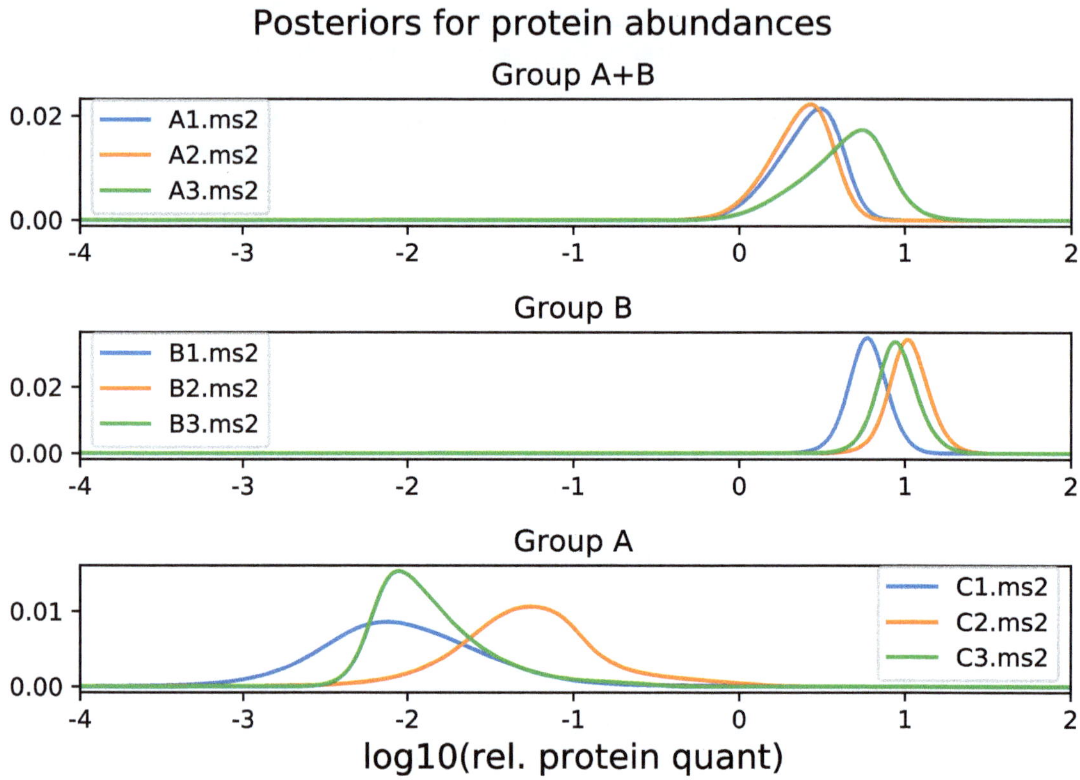

Fig. 3 Posterior distributions for the relative protein abundance of HPRR3730445_poolB in each of the 9 samples

7. Now, we can examine the 3 posterior distribution plots, starting with the relative protein abundance distributions in Fig. 3. Note that the abundance values are, again, $\log_{10}$ transformed. The distributions for the samples in the group $A + B$ and B are relatively narrow, due to the high number of observed values, in contrast to the samples in group A, which have a large number of missing values combined with very low intensities.
8. For the posterior distributions of the group means, we can clearly see the effect of the individual samples from the previous step (*see* Fig. 4). Here, we can also verify that the distributions do resemble normal distributions, as was assumed earlier (*see* Subheading 3.7, **Step 6**).
9. Finally, for each pair of groups, we "subtract" one group mean posterior distribution from the other to obtain a fold change posterior distribution (*see* Fig. 5). Again, note the logarithm change from $\log_{10}$ to $\log_2$ compared to the previous step. In these violin plots, we show the distribution on the *y*-axes. The green distributions represent the Triqler estimations and, for

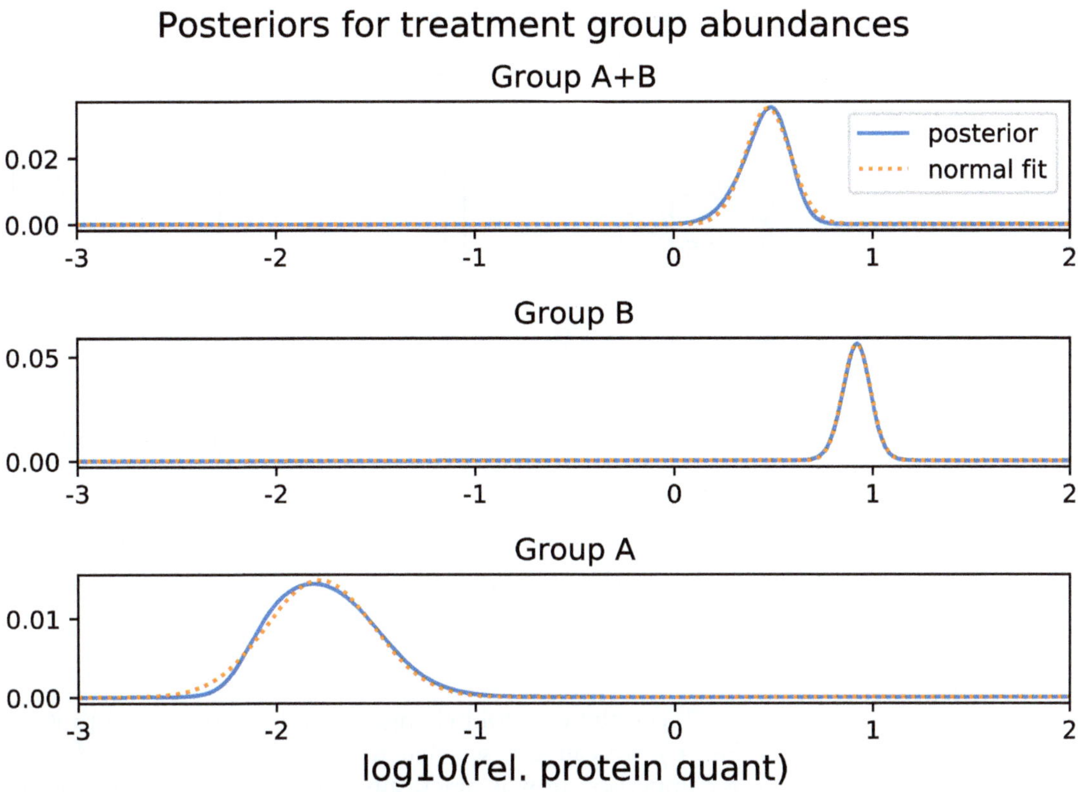

Fig. 4 Posterior distributions for the relative mean group abundance of HPRR3730445_poolB in each of the 3 groups

comparison, we also display the estimate that would have been given by a naive method (Top 3, mean row imputation) in blue.

10. To show the effect of the `--fold_change_eval` parameter, we now change this to 0.5 instead of the default value of 1.0:

```
$ python -m triqler.distribution.plot_posteriors --protein_id
HPRR3730445_poolB --plot_max_fold_change 10.0
--fold_change_eval 0.5 iPRG2016.tsv
```

All of the command line output stays the same, except for the second to last block, where the $A + B$ vs B comparison now also shows a very low probability of being below the lowered `--fold_change_eval`:

```
Posterior probability |log2 fold change| < 0.50
  Group A+B vs Group B: 0.010441
  Group A+B vs Group A: 0.000000
  Group B vs Group A: 0.000000
```

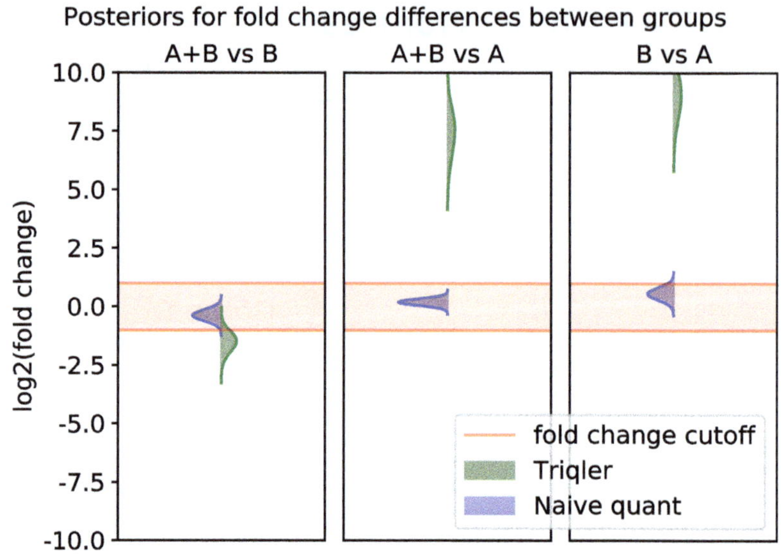

Fig. 5 Posterior distributions for the $\log_2$ fold changes of HPRR3730445_poolB for each of the 3 group comparisons

From the new fold change posterior distribution plot (*see* Fig. 6), we can also see that the distribution for the $A + B$ vs B comparison now is practically completely outside of the red fold change region. At the same time, this demonstrates the danger of using a too low value for `--fold_change_eval`, as the full width of the green distribution is now larger than the red region. Even if the green distribution would be centered around 0, there would still be some probability outside of the red region, which would then be reported as a non-negligible probability of the protein being differentially abundant even though no actual difference can be seen.

3.8 Visualizing and Interpreting Hyperparameter Estimation

A common concern for using Bayesian methods is the dependence on the prior distributions. Triqler employs the Empirical Bayes method to estimate the hyperparameters for the prior distributions (*see* **Note 22**) for the probabilistic graphical model. To check if these hyperparameter estimations are reasonable, Triqler provides functions to investigate these through graphical inspection (*see* **Note 23**).

1. Generate the fits for the hyperparameter estimations:

```
$ python -m triqler.distribution.plot_hyperparameter_fits
  iPRG2016.tsv
```

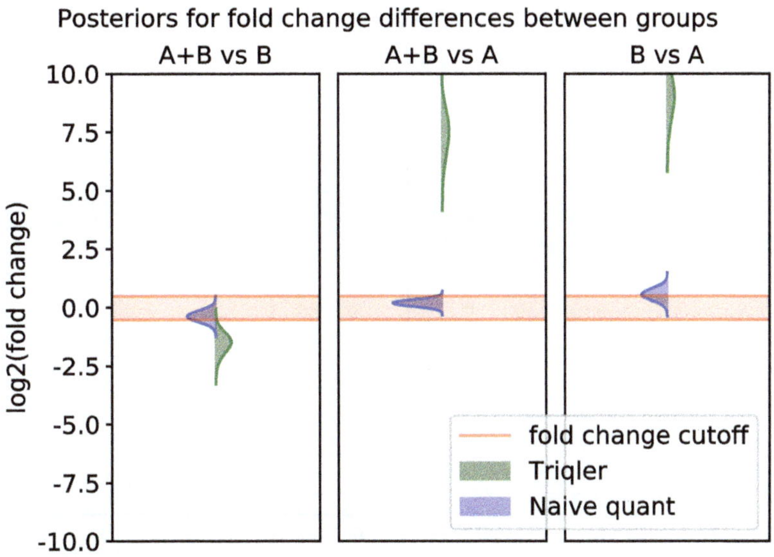

Fig. 6 Same as Fig. 5 but with the $\log_2$ fold change evaluation threshold changed to 0.5 instead of the default value of 1.0

2. To estimate the probability of a missing value as a function of the XIC, Triqler postulates that the $\log_{10}$(XIC) of all peptides across all samples can be modeled as a left-censored normal distribution (*see* Fig. 7). Here, the left-censored normal distribution (green) is a normal distribution (cyan) with some "mass" missing for low XICs due to a sigmoidal censoring function (purple). Note that the XIC values have been divided by a power of 10 (*see* Subheading 3.5, **Step 6**). Note that, in this particular example, the influence of the sigmoidal censoring function is not very apparent, likely due to the design of the study with many actually missing values.
3. To estimate the prior distribution for the relative protein abundances, we fit a hyperbolic secant function (*see* **Note 24**) to naive estimations (simple average of relative peptide abundances) of the protein abundances (*see* Fig. 8).
4. To estimate the distribution of the difference between the true and observed XIC, we fit a hyperbolic tangent distribution to the difference between the observed XICs and the expected XIC based on the relative protein abundance estimated in the previous step and the estimated ionization efficiency (*see* **Note 25**). (*see* Fig. 9). This can be seen as an estimate for the measurement uncertainty which will then propagate through the graphical model.

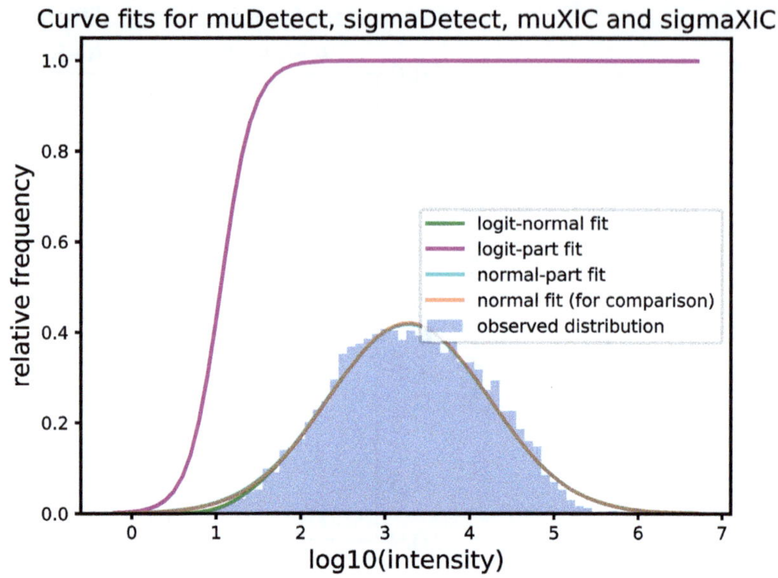

Fig. 7 Estimating the hyperparameters for the missing value probability as a function of XIC by a left-censored normal distribution

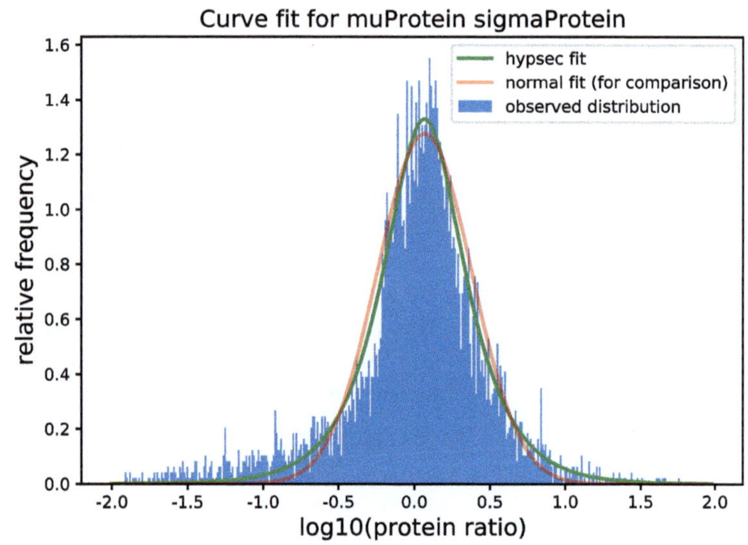

Fig. 8 The prior for the mean of the relative protein abundance is estimated with a hyperbolic secant distribution

5. Finally, we estimate the distribution of the standard deviation between protein abundances of samples within the same treatment group using a Gamma distribution (*see* Fig. 10). This serves to capture the biological and/or technical variance within a group.

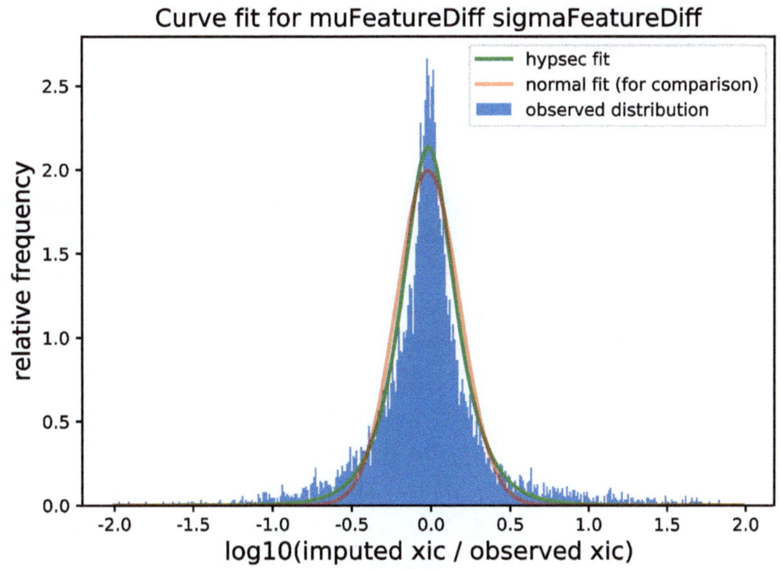

Fig. 9 The measurement uncertainty distribution is estimated with a hyperbolic secant distribution

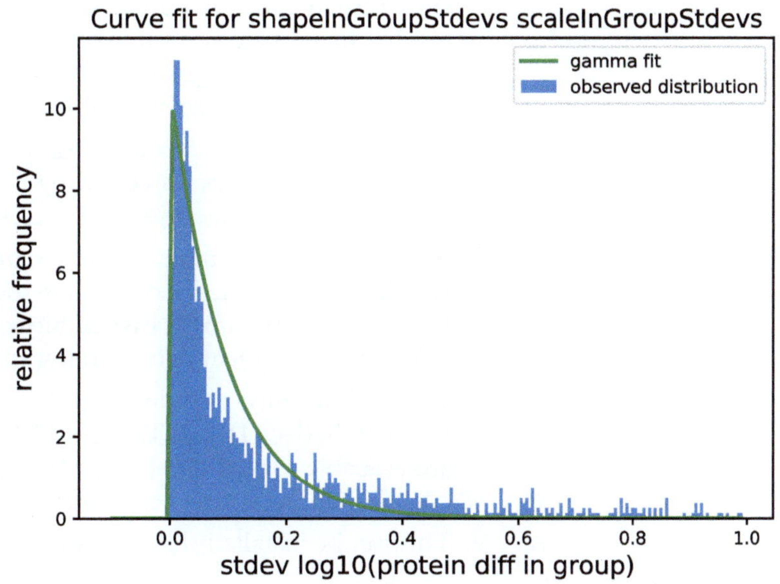

Fig. 10 The within-group standard deviation distribution is estimated by a Gamma distribution

4 Notes

1. Commands on the command line are prepended by the `$` character (e.g., `$ python --version`), which should be omitted when actually running the command (e.g., `python --version`). Command line inputs are written between the < and > symbols, e.g., <X>, which should be replaced by the user by the relevant value or file path.
2. We have, however, observed that Triqler still is able to correctly distinguish between changing and unchanging proteins in engineered datasets if the former requirement is violated.
3. Currently, shared peptides are not considered for identification and quantification. However, we expect to include functionality to include shared peptides in the near future.
4. For some test datasets, the results look reasonable for the default thresholds of 1% FDR as well, but as the error estimates are likely unreliable we do not recommend running Triqler in this way.
5. We have verified the validity of the results without MBR on several test datasets. Triqler will work with match-between-runs turned on, but we cannot guarantee the validity of the results. In a future version, we will try to incorporate the error estimates from MaxQuant's MBR step, similar to what we have done with Quandenser (*see* Subheading 3.2).
6. If the output from `crux percolator` is used, make sure that the order of the files processed by `crux tide-search` is the same as in the file specified at `--file_list_file`.
7. A more detailed example of how to run Quandenser and obtain search results from the consensus spectra can be found here: https://github.com/statisticalbiotechnology/quandenser/wiki/Example:-Quandenser-followed-by-Tide-and-Triqler.
8. Currently, this list includes SEQUEST [15] (Comet [16], Tide [17]), MSGF+ [18], X!Tandem [19]. Custom Python scripts are available for MODa [20], MSFragger [21], and Andromeda [22] upon request to the authors.
9. This can be installed through the following command:

```
$ pip install simsalabim
```

10. A more detailed example of how to run the dinosaur adapter and obtain search results can be found here: https://github.com/MatthewThe/simsalabim/wiki/Example:-Dinosaur-followed-by-Tide-and-Triqler.
11. We strongly recommend searching the recalibrated MS2 spectrum files, which now have accurate MS1 precursors assigned.

This generally improves the identification rate and allows for multiple identifications per spectrum for chimeric spectra. In the dataset we used as an example in this manuscript (iPRG2016), we could increase the number of identified peptides by 34%.

12. We have observed in [7] that even for low values of N, e.g., $N = 3$, the differential abundance FDR remains under some form of control, but advise users to be careful with setting this value so low, as the number of false positives does increase.
13. On some systems, the python multiprocessing module causes issues. Setting $N = 1$ will bypass this multiprocessing module at the cost of longer runtimes.
14. This dataset is described in [23]. Briefly, this dataset was designed to test how well protein identification and quantification pipelines can deal with shared peptides, by using two pools of synthesized proteins which are similar to human proteins (PrESTs). Each of the pools contained one out of a pair of proteins that share a number of peptides. In the first 3 samples, both pools were mixed in equal parts into a background of E. Coli lysate, in the second 3 samples, only the *B* pool proteins were mixed in, and in the last 3 samples, only the *A* pool proteins were added.
15. This can, for example, be a result of problems with MS1 feature detection or due to false positives in the MS2 spectrum identification.
16. Specifically `example/iPRG2016.tsv.pqr.tsv`, the output of the `--write-spectrum-quants` file, `example/iPRG2016.tsv.sqr.tsv` and the command line output of the `plot_posteriors` commands.
17. This file summarizes the Triqler input file by grouping the rows by combinations of peptide and charge. The format consists of a column called `combinedPEP`, which combines the peptide-level identification PEP and the feature-match PEP. In the absence of the latter, it is just the identification PEP. After the `charge` column and two columns specific to Quandenser (feature group index and consensus spectrum scan number), there are three groups of N columns, where N is the number of samples. The first group contains the feature-match PEPs, the second group the intensity values (divided by the power of 10 as mentioned earlier), and the last group contains the identification PEP. The last two columns contain the peptide sequence and protein identifier, respectively.
18. This is done by sorting the PEPs in ascending order and taking the cumulative average [24]. This works because the PEP is the derivative of the FDR, and is therefore sometimes called the local-FDR.

19. The 192 poolA proteins and 191 poolB proteins are present at half the concentration in the $A+B$ samples compared to the A and B samples, respectively. The posterior distribution therefore, ideally, centers around a $\log_2$ fold change of 1.0. Here, half of the probability distribution would indicate a $\log_2$ fold change below 1.0 resulting in a posterior error probability of 0.5. In practice, the $\log_2$ fold change posterior distributions should include the true $\log_2$ fold change of 1.0, but do not necessarily center around 1.0. Therefore, some of these proteins will have a fold change posterior distributions centered above 1.0 and a posterior error probability <0.5, which might result into them being called significantly differentially abundant at 5% FDR.
20. Another way to consider these values is as the regular protein abundance values (e.g., summed intensity) divided by the mean over all samples.
21. To plot a heatmap of posterior fold change distributions of a set of proteins, use the following command:

```
$ python -m triqler.distribution.plot_posteriors --protein_id_list
   protein_list.txt
```

where `protein_list.txt` is a simple text file with one protein identifier per line.

22. For a short introduction for non-statisticians into the basics of Bayesian statistics, including priors, hyperparameters, posteriors, and some simple examples, refer to [25].
23. See the Supplemental Data of [5] for more examples of fitted hyperparameter distributions for a range of datasets.
24. The hyperbolic secant distribution (green) has heavier tails than the normal distribution (red) and is, thus, better able to deal with outliers that are common in real-world data.
25. The ionization efficiency is estimated as the average across all samples of the XIC divided by the relative peptide abundance, where missing values are not considered.

References

1. Elias JE, Gygi SP (2007) Target-decoy search strategy for increased confidence in large-scale protein identifications by mass spectrometry. Nat Methods 4(3):207–214
2. Storey JD, Tibshirani R (2003) Statistical significance for genomewide studies. Proc Nat Acad Sci 100(16):9440–9445
3. Webb-Robertson BJM, Wiberg HK, Matzke MM, Brown JN, Wang J, McDermott JE, Smith RD, Rodland KD, Metz TO, Pounds JG, et al (2015) Review, evaluation, and discussion of the challenges of missing value imputation for mass spectrometry-based label-free global proteomics. J Proteome Res 14(5):1993–2001
4. Lazar C, Gatto L, Ferro M, Bruley C, Burger T (2016) Accounting for the multiple natures of missing values in label-free quantitative proteomics data sets to compare imputation strategies. J Proteome Res 15(4):1116–1125

5. The M, Käll L (2019) Integrated identification and quantification error probabilities for shotgun proteomics. Mol Cell Proteom 18(3): 561–570
6. Cox J, Hein MY, Luber CA, Paron I, Nagaraj N, Mann M (2014) Accurate proteome-wide label-free quantification by delayed normalization and maximal peptide ratio extraction, termed MaxLFQ. Mol Cell Proteom 13(9):2513–2526
7. The M, Käll L (2020) Focus on the spectra that matter by clustering of quantification data in shotgun proteomics. Nat Commun 11(1): 1–12
8. Teleman J, Chawade A, Sandin M, Levander F, Malmström J (2016) Dinosaur: a refined open-source peptide MS feature detector. J Proteome Res 15(7):2143–2151
9. The M, Kaäll L (2021) Triqler for maxquant: Enhancing results from MaxQuant by Bayesian error propagation and integration. J Proteome Res 20(4):2062–2068
10. Käll L, Canterbury JD, Weston J, Noble WS, MacCoss MJ (2007) Semi-supervised learning for peptide identification from shotgun proteomics datasets. Nat Methods 4(11):923–925
11. The M, MacCoss MJ, Noble WS, Käll L (2016) Fast and accurate protein false discovery rates on large-scale proteomics data sets with Percolator 3.0. J Amer Soc Mass Spectrom 27(11): 1719–1727
12. Käll L, Storey JD, Noble WS (2009) QVALITY: non-parametric estimation of q-values and posterior error probabilities. Bioinformatics 25(7):964–966
13. Granholm V, Navarro JF, Noble WS, Käll L (2013) Determining the calibration of confidence estimation procedures for unique peptides in shotgun proteomics. J Proteomics 80: 123–131
14. Savitski MM, Wilhelm M, Hahne H, Kuster B, Bantscheff M (2015) A scalable approach for protein false discovery rate estimation in large proteomic data sets. Mol Cell Proteom 14(9): 2394–2404
15. Eng JK, McCormack AL, Yates JR (1994) An approach to correlate tandem mass spectral data of peptides with amino acid sequences in a protein database. J Amer Soc Mass Spectr 5(11):976–989
16. Eng JK, Jahan TA, Hoopmann MR (2013) Comet: an open-source MS/MS sequence database search tool. Proteomics 13(1):22–24
17. Diament BJ, Noble WS (2011) Faster SEQUEST searching for peptide identification from tandem mass spectra. J Proteome Res 10(9):3871–3879
18. Kim S, Pevzner PA (2014) MS-GF+ makes progress towards a universal database search tool for proteomics. Nat Commun 5:5277
19. Craig R, Beavis RC (2003) A method for reducing the time required to match protein sequences with tandem mass spectra. Rapid Commun Mass Spectr 17(20):2310–2316
20. Na S, Bandeira N, Paek E (2012) Fast multi-blind modification search through tandem mass spectrometry. Mol Cell Proteom 11(4): M111.010199
21. Kong AT, Leprevost FV, Avtonomov DM, Mellacheruvu D, Nesvizhskii AI (2017) MSFragger: ultrafast and comprehensive peptide identification in mass spectrometry–based proteomics. Nat Methods 14(5):513–520
22. Cox J, Neuhauser N, Michalski A, Scheltema RA, Olsen JV, Mann M (2011) Andromeda: a peptide search engine integrated into the MaxQuant environment. J Proteome Res 10(4): 1794–1805
23. The M, Edfors F, Perez-Riverol Y, Payne SH, Hoopmann MR, Palmblad M, Forsström B, Käll L (2018) A protein standard that emulates homology for the characterization of protein inference algorithms. J Proteom Res 17(5): 1879–1886
24. Käll L, Storey JD, Noble WS (2008) Non-parametric estimation of posterior error probabilities associated with peptides identified by tandem mass spectrometry. Bioinformatics 24(16):i42–i48
25. Van de Schoot R, Kaplan D, Denissen J, Asendorpf JB, Neyer FJ, Van Aken MA (2014) A gentle introduction to Bayesian analysis: applications to developmental research. Child Develop 85(3):842–860

Chapter 6

Left-Censored Missing Value Imputation Approach for MS-Based Proteomics Data with GSimp

Runmin Wei and Jingye Wang

Abstract

Missing values caused by the limit of detection or quantification (LOD/LOQ) were widely observed in mass spectrometry (MS)-based omics studies and could be recognized as missing not at random (MNAR). MNAR leads to biased statistical estimations and jeopardizes downstream analyses. Although a wide range of missing value imputation methods was developed for omics studies, a limited number of methods were designed appropriately for the situation of MNAR. To facilitate MS-based omics studies, we introduce GSimp, a Gibbs sampler-based missing value imputation approach, to deal with left-censor missing values in MS-proteomics datasets. In this book, we explain the MNAR and elucidate the usage of GSimp for MNAR in detail.

Key words Mass spectrometry, Proteomics, Missing not at random, Left censor, Imputation, Gibbs sampler

1 Introduction

Missing values have commonly existed in mass spectrometry (MS)-based omics datasets. Most statistical methods require a complete dataset, which makes missing data an unavoidable problem for subsequent data analysis. Missing values can be divided into three categories from the data distribution [1, 2]: missing completely at random (MCAR; this type of missingness is unrelated to any other reasons or any study variables), missing at random (MAR; the missingness is related to other variables but not related to the value of the variable itself), and missing not at random (MNAR; the missingness is related to the value of the variable itself, e.g., the lowest detection limits of the instrument or the content of a tested substance below the detection limits will cause a nonrandom missing).

Thomas Burger (ed.), *Statistical Analysis of Proteomic Data: Methods and Tools*,
Methods in Molecular Biology, vol. 2426, https://doi.org/10.1007/978-1-0716-1967-4_6,

In MS-based analysis, according to other researches, the missing values account for 10–50% of the data [3]. Protein with high abundance is more detectable than those with low abundance. Due to the limit of compound quantification (LOQ), missing values are usually caused by signal intensities lower than LOQ, also known as left-censored MNAR (as the data distribution is truncated on its left side).

The simplest way is to remove the samples with missing values directly. However, if the dataset contains missing variables across different samples, this approach will drop useful nonmissing information at the same time. Thus, the "80% rule" was proposed [4]: when the nonmissing part of a substance is less than 80% of the total sample size, it is recommended to remove this substance for downstream analysis. A "modified 80% rule" suggests that when the nonmissing part of a substance is less than 80% of all biological subgroups, it is recommended to delete this substance. It is helpful to apply any of these rules to filter variables first before any samplewise deletion or imputation. Another simple missing value processing approach is to substitute missing values with deterministic values, e.g., mean or median values for MAR/MCAR and minimum or zero for left-censored data.

At present, most of the commonly used imputation methods are developed for the MCAR/MAR situation, e.g., missForest [5], k-nearest neighbor (kNN) [6], singular value decomposition (SVD) [7, 8], etc. However, imputation algorithms for MNAR are very limited, while this type of missingness causes even more difficulties for statistical analyses. Few examples of the literature are quantile regression imputation of left-censored data (QRILC) [9] and Gibbs sampler-based left-censored missing value imputation approach (GSimp) [10]. GSimp utilizes predictive information of nonmissing variables and holds a truncated normal distribution for each missing element simultaneously, via embedding a prediction model into the Gibbs sampler framework. Therefore, it can efficiently impute left-censored missing data in MS-based omics datasets.

2 Material

2.1 Hardware and Package Requirements

The environment for GSimp depends on the installation of R and RStudio. Currently, R runs on both 64- and 32-bit operating systems and most 64-bit OSs (e.g., Linux, Solaris, Windows, and macOS). The memory limits depend mainly on the build, but for a 32-bit build of R on Windows, they also depend on the underlying OS version.

The list of packages that must be installed prior to GSimp is provided in Table 1.

Table 1
Package dependencies along with their version

Package	Version
Amelia	1.7.4
abind	1.4–5
doParallel	1.0.11
FNN	1.1
foreach	1.4.4
ggplot2	2.2.1.9
glmnet	2.0–13
impute	1.50.1
imputeLCMD	2
knitr	1.17
magrittr	1.5
markdown	0.8
missForest	1.4
pheatmap	1.0.8
randomForest	4.6–12
reshape2	1.4.3
ropls	1.8.0
vegan	2.4–5

2.2 Data Format and Source Code

The default input dataset should be an $n \times p$ table with columns corresponding to the features or variables (proteins) and rows corresponding to samples. Row names and column names must be unique. Other cells will need to be numeric, and missingness should be shown as NA (*see* Fig. 1). The R code for GSimp, evaluation pipeline, tutorial, real-world, and simulated MS datasets are available at: https://github.com/WandeRum/GSimp. The online web tool MetImp is available at: https://metabolomics.cc.hawaii.edu/software/MetImp/.

2.3 Software Installation

To install the downloaded code, use the following commands:

```
## Set working directory to GSimp unzipped folder
## and source the functions
setwd('/GSimp-master/')
source('GSimp.R')
```

Column 1 : row identifiers. Typically sample IDs. The first cell should keep blank and the rest cells must be unique.

Row 1 : variable names. The first cell should keep blank. Variable names either starts at the second cell.

Other cells : numeric values and missingness should be NA.

	cmpd_1	cmpd_2	cmpd_3	cmpd_4	cmpd_5	cmpd_6	cmpd_7	cmpd_8	cmpd_9	cmpd_10
S_1	5.174	6.02	4.748	39.37	3.458	89.591	0.318	NA	0.298	8.198
S_2	39.718	0.944	1.419	34.127	4.326	67.739	1.534	NA	0.365	10.988
S_3	20.301	0.329	0.684	49.892	2.428	58.887	1.611	0.699	0.127	9.456
S_4	20.837	0.541	0.765	35.853	2.44	17.473	0.263	1.047	NA	1.261
S_5	45.223	0.72	0.653	9.704	0.799	19.411	0.032	0.171	0.085	2.187
S_6	62.464	0.258	0.327	44.7	2.04	37.057	1.8	0.724	0.194	6.179
S_7	13.494	2.612	1.139	55.735	2.232	182.772	0.502	0.14	0.11	15.805
S_8	11.554	1.799	0.989	28.458	1.512	176.7	0.927	0.199	NA	11.424
S_9	47.865	0.292	1.575	65.121	5.485	60.606	1.993	0.812	1.188	6.173
S_10	13.312	0.232	0.46	15.75	1.459	10.616	0.002	NA	NA	1.505
S_11	1.298	5.851	6.133	252.956	9.987	316.321	11.606	NA	1.265	39.731
S_12	5.754	3.338	2.469	274.048	8.583	165.402	17.405	0.467	0.762	22.156
S_13	0.92	5.006	4.022	194.742	6.19	84.355	5.52	0.144	0.529	10.317
S_14	0.687	7.835	5.241	105.538	1.426	24.812	2.586	NA	0.098	3.051

Fig. 1 Data format example

3 Methods

3.1 Data Processing

In GSimp, we recommend the following data processing steps:

1. Log transformation (for non-normal data),
2. Initialization for missing values (e.g., QRILC [9]),
3. Centralization and scaling (for elastic net prediction),
4. Imputation using GSimp,
5. Scaling recovery,
6. Exponential recovery,
7. Imputed data output.

All the above steps have been wrapped into the `pre_processing_GS_wrapper` function for a one-step processing and imputation. The following function will give the final imputed dataset:

```
# One-step wrapper function with data pre-processing
pre_processing_GS_wrapper <- function(data) {
  data_raw <- data
  # Log transformation #
  data_raw_log <- data_raw
  # Initialization #
  data_raw_log_qrilc <- impute.QRILC(data_raw_log)
  extract2(1)
  # Centralization and scaling #
  data_raw_log_qrilc_sc <- scale_recover(
                              data_raw_log_qrilc,
                              method = 'scale'
                                         )
  # Data after centralization and scaling #
  data_raw_log_qrilc_sc_df <- data_raw_log_qrilc_sc[[1]]
  # Parameters for centralization and scaling
  # (for scaling recovery)
  data_raw_log_qrilc_sc_df_param <- data_raw_log_qrilc_sc[[2]]
  # NA position #
  NA_pos <- which(is.na(data_raw), arr.ind = T)
  # NA introduced to log-scaled-initialized data #
  data_raw_log_sc <- data_raw_log_qrilc_sc_df
  data_raw_log_sc[NA_pos] <- NA
  # Feed initialized and missing data into GSimp imputation #
  result <- data_raw_log_sc
    GS_impute(.,
              iters_each=50, iters_all=10,
              initial = data_raw_log_qrilc_sc_df,
              lo=-Inf, hi= 'min', n_cores=2,
              imp_model='glmnet_pred'
              )
  data_imp_log_sc <- result$data_imp
  # Data recovery #
  data_imp <- data_imp_log_sc
    scale_recover(.,
                  method = 'recover',
                  param_df =
                     data_raw_log_qrilc_sc_df_param
                  )
    extract2(1)
  return(data_imp)
}
```

3.2 GSimp Input Arguments

The function `GS_impute` is the core function for the imputation of missing data and tracing the Gibbs sampler with certain missing positions (*see* Fig. 2). Here follows the main arguments of `GS_impute`:

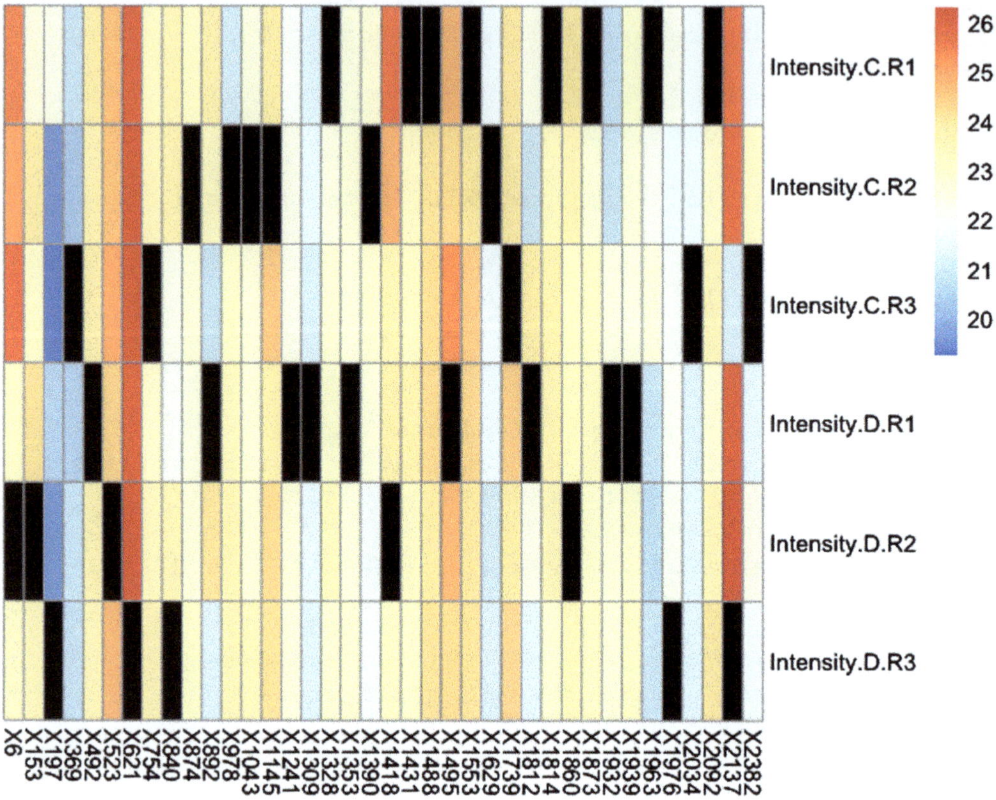

Fig. 2 Heat map of the data before imputation

1. **iters_each** is the number of iterations for imputing each missing variable (`default=100`).
2. **iters_all** is the number of iterations for imputing the whole data matrix (*see* **Note 1**). The default value is 20.
3. **initial** is the initialization method for missing values (`default=qrilc`). We provided three options: `lsym`, `qrilc`, and `rsym`. `lsym` will draw samples from the right tail of the distribution and symmetrically transform to the left tail. `rsym` will draw samples from the left tail of the distribution and symmetrically transform to the right tail; this is for the right-censored missing. `qrilc` will use QRILC-imputed values as initial. A preinitialized data frame is also acceptable for this argument.
4. **lo** is the lower limit (`default=-Inf`) and **hi** (`default=min`) is the upper limit for missing values. These two arguments can be defined as `-Inf`, `Inf`, `min`, `max`, `median`, `mean` or any single determined value or a vector of values (same length with number of variables; the values of nonmissing variables will not affect the results). Here, `lo=-Inf` and `hi=min` are default setting for left-censored missing values, where the upper bound is set to the minimum value of nonmissing part (*see* **Note 2**).

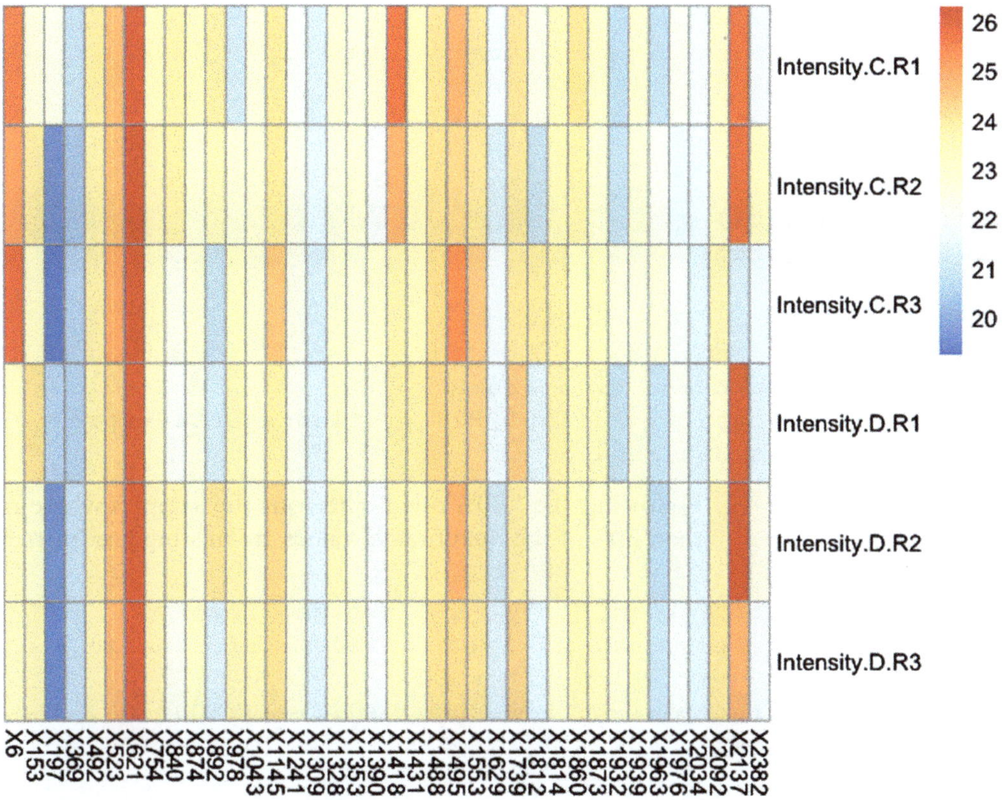

Fig. 3 Heat map of the data after imputation

5. **n_cores** is the number of cores for computing (*see* **Note 3**).
6. **gibbs** is the missing elements you want to trace across the whole Monte Carlo Markov chain (MCMC; `default=data.frame(row=integer(), col=integer())`). This argument must be set as the positions of missing elements (*see* **Note 4**).

3.3 GSimp Outputs

Once the imputation is over, the following elements are outputted by `GS_impute` (*see* Fig. 3):

1. **data_imp** is the imputed data frame.
2. **gibbs_res** is a three-dimensional array that records the whole process of specified missing elements across MCMC iterations. The first dimension represents `std`, `yhat`, and `yres`, which respectively stands for the standard deviation, predicted value, and sampling value. The second dimension represents the specified missing elements, and the third dimension, the iterations.

3.4 GSimp in Practice

1. We use a protein-level proteomics data [11] available in the DAPARdata package [12] to exemplify our algorithm. The following commands load the dataset:

```
# Package installation
if (!requireNamespace("BiocManager", quietly = TRUE))
    install.packages("BiocManager")
BiocManager::install("DAPARdata")

# Load data
require(DAPARdata)
data(Exp1_R25_prot)



# Change data to n x p data frame
pro_dat_raw <- data.frame(Exp1_R25_prot@assayData$exprs)
pro_dat <- t(pro_dat_raw)
```

2. Following the "80% rule," filter out the samples where more than 20% of the quantitative values are missing (*see* **Note 5**):

```
## Missing proportion filtering and subsampling ##
na_prop <- apply(pro_dat, 2, function(x) mean(is.na(x)))
# Use random 500 variables after 80
idx_fil <- sort(sample(which(na_prop < .2))[1:500])
pro_dat <- pro_dat[, idx_fil]
```

3. QRILC was used for initialization:

```
## Initialization ##
# We used QRILC as the methods for initialization #
pro_dat_qrilc <- impute.QRILC(pro_dat)
```

4. Since the expression data was already log2 transformed, conduct scaling (and centering) on the initialized data with empirical mean and standard deviation as parameters:

```
## Scaling ##
# scale_recover function can be used for scaling before
# missing value imputation and can also be used for
# scaling recovery after the missing value imputation
pro_dat_qrilc_sc <- scale_recover(pro_dat_qrilc,


                                  method = 'scale'
                                  )
pro_dat_qrilc_sc_df <- data.frame(pro_dat_qrilc_sc[[1]])
# Save parameters for centralization and scaling
# (for scaling recovery)
pro_dat_qrilc_sc_df_param <- pro_dat_qrilc_sc[[2]]
```

5. Finally, apply GSimp with 10×50 iterations while setting 20% quantile as upper limit (less stringent limit). Note that Glmnet was used for the prediction model construction:

```
# Extract NA positions
NA_pos <- which(is.na(pro_dat), arr.ind = T)

# NA introduced to log-scaled-initialized data
pro_dat_sc <- pro_dat_qrilc_sc_df
pro_dat_sc[NA_pos] <- NA

## Imputation ##
# Feed initialized and missing data into GSimp imputation
# We set the parameter hi to the 20
result <- pro_dat_sc
GS_impute(.,
          iters_each=50, iters_all=10,
          initial = pro_dat_qrilc_sc_df,
          lo=-Inf,
          hi= apply(pro_dat_sc,
                    2,
                    function(x) quantile(x, .2, na.rm=T)
                    ),
          n_cores=4,
          imp_model='glmnet_pred'
          )
data_imp_sc <- result$data_imp

## Scaling recovery ##
data_imp <- data_imp_sc
  scale_recover(.,
                method = 'recover',


                param_df = pro_dat_qrilc_sc_df_param
                )
  extract2(1)
# data_imp is the final dataset after imputation
```

4 Notes

1. Although a large number of iterations (e.g., `iters_all=20` and `iters_each=100`) are recommended for the convergence of Monte Carlo Markov chain (MCMC), a smaller number of iterations (`iters_all=10, iters_each=50`) will not severely affect the imputation accuracy, based on our tests on simulated data.

2. Notably, quantile values can be applied if minimum is too strict. For example, `hi=apply(data, 2, function(x) quantile (x, .1, na.rm=T))` amounts to the following: the 10% quantile of each variable is set to the upper bound. When noninformative bounds for both upper and lower limits (e.g., `Inf`, `-Inf`) were applied, GSimp could be extended to the situation of MCAR/MAR.
3. Parallel computing will impute all missing variables simultaneously, while nonparallel computing will impute missing variables sequentially from the least number of missing variables to the most.
4. For example, `gibbs=data.frame(row=c(1, 3), col=c (2, 5))` represents that you want to trace the missing elements in row 1, column 2 and row 3, column 5.
5. For data with more than 1000 dimensions (variables), considering the computational efficiency, we suggest reducing the number of variables through either dimension reduction on nonmissing data or splitting data into several blocks for imputation.

Acknowledgments

We would like to express our thanks of gratitude to the editor of this book, Dr. Thomas Burger, who accepted to proof read and perform the necessary formatting of this book.

References

1. Gelman A, Hill J (2006) Data analysis using regression and multilevel/hierarchical models. Analytical methods for social research. Cambridge University Press. https://doi.org/10.1017/CBO9780511790942
2. Little RJ, Rubin DB (2019) Statistical analysis with missing data, vol 793, 3rd edn. Wiley. https://doi.org/10.1002/9781119482260
3. Albrecht D, Kniemeyer O, Brakhage AA, Guthke R (2010) Missing values in gel-based proteomics. Proteomics 10(6):1202–1211. https://doi.org/10.1002/pmic.200800576
4. Bijlsma S, Bobeldijk I, Verheij ER, Ramaker R, Kochhar S, Macdonald IA, Van Ommen B, Smilde AK (2006) Large-scale human metabolomics studies: a strategy for data (pre-) processing and validation. Analytical Chemistry 78(2):567–574. https://doi.org/10.1021/ac051495j
5. Stekhoven DJ, Bühlmann P (2012) MissForest—non-parametric missing value imputation for mixed-type data. Bioinformatics 28(1): 112–118. https://doi.org/10.1093/bioinformatics/btr597
6. Troyanskaya O, Cantor M, Sherlock G, Brown P, Hastie T, Tibshirani R, Botstein D, Altman RB (2001) Missing value estimation methods for DNA microarrays. Bioinformatics 17(6):520–525. https://doi.org/10.1093/bioinformatics/17.6.520
7. Hastie T, Tibshirani R, Sherlock G, Eisen M, Brown P, Botstein D (1999) Imputing missing data for gene expression arrays. Stanford University Statistics Department Technical report. http://ww.web.stanford.edu/~hastie/Papers/missing.pdf
8. Stacklies W, Redestig H, Scholz M, Walther D, Selbig J (2007) pcaMethods—a bioconductor package providing PCA methods for incomplete data. Bioinformatics 23(9):1164–1167. https://doi.org/10.1093/bioinformatics/btm069

9. Lazar C, Gatto L, Ferro M, Bruley C, Burger T (2016) Accounting for the multiple natures of missing values in label-free quantitative proteomics data sets to compare imputation strategies. J Proteome Res 15(4):1116–1125. https://doi.org/10.1021/acs.jproteome.5b00981
10. Wei R, Wang J, Su M, Jia E, Chen S, Chen T, Ni Y (2018) Missing value imputation approach for mass spectrometry-based metabolomics data. Scientific Reports 8(1):1–10. https://doi.org/10.1038/s41598-017-19120-0
11. Giai Gianetto Q, Combes F, Ramus C, Bruley C, Couté Y, Burger T (2016) Calibration plot for proteomics: A graphical tool to visually check the assumptions underlying FDR control in quantitative experiments. Proteomics 16(1):29–32. https://doi.org/10.1002/pmic.201500189
12. Wieczorek S, Combes F, Lazar C, Giai Gianetto Q, Gatto L, Dorffer A, Hesse AM, Couté Y, Ferro M, Bruley C, et al (2017) DAPAR & ProStaR: software to perform statistical analyses in quantitative discovery proteomics. Bioinformatics 33(1):135–136. https://doi.org/10.1093/bioinformatics/btw580

Chapter 7

Towards a More Accurate Differential Analysis of Multiple Imputed Proteomics Data with mi4limma

Marie Chion, Christine Carapito, and Frédéric Bertrand

Abstract

Imputing missing values is a common practice in label-free quantitative proteomics. Imputation replaces a missing value by a user-defined one. However, the imputation itself is not optimally considered downstream of the imputation process. In particular, imputed datasets are considered as if they had always been complete. The uncertainty due to the imputation is not properly taken into account. Hence, the `mi4p` package provides a more accurate statistical analysis of multiple-imputed datasets. A rigorous multiple imputation methodology is implemented, leading to a less biased estimation of parameters and their variability, thanks to Rubin's rules. The imputation-based peptide's intensities' variance estimator is then moderated using Bayesian hierarchical models. This estimator is finally included in moderated *t*-test statistics to provide differential analyses results.

Key words Label-free quantitative proteomics, Differential analysis, Missing values, Multiple imputation, Moderated *t*-testing

1 Introduction

Current statistical methods used in label-free quantitative proteomics rely on peptides' intensities, measured by liquid chromatography coupled with tandem mass spectrometry (LC-MS/MS). While major instrumental improvements over the past years have allowed great progress in terms of sensitivity, dynamic range, and acquisition speed, the generated datasets remain incomplete and contain a variable proportion of missing values. Usual statistical tools do not optimally consider peptides, whose intensities are missing in some conditions, although they might be particularly interesting in differential analyses [1]. Imputation methods have been described and are currently applied in the state-of-the-art quantification software tools [2–6]. Imputation consists of replacing a missing value with a value derived using a user-defined formula (such as the mean, the median, or a value provided by an expert, thus taking into account

Thomas Burger (ed.), *Statistical Analysis of Proteomic Data: Methods and Tools*,
Methods in Molecular Biology, vol. 2426, https://doi.org/10.1007/978-1-0716-1967-4_7,

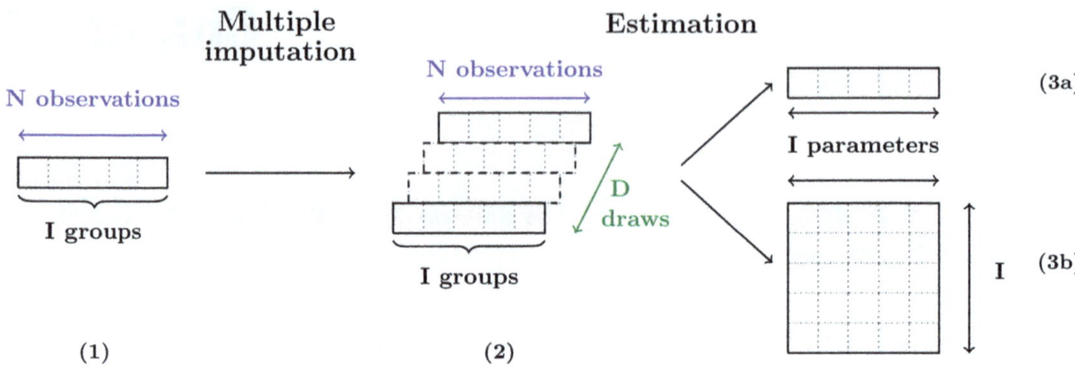

Fig. 1 Multiple imputation strategy. (1) Initial dataset with missing values. It is supposed to have **N** observations that are split into **I** groups. **(2)** Multiple imputation provides **D** estimators for the vector of parameters β of interest. **(3a)** The **D** estimators are combined using the first Rubin's rule to get the combined estimator. **(3b)** The estimator of the variance–covariance matrix of the combined estimator is provided by the second Rubin's rule

the knowledge of the user). However, the uncertainty due to the imputation is currently not properly taken into account, as the imputed dataset is considered as if it has always been complete in further statistical analysis.

Multiple imputation [7] partially addresses this issue by generating several imputed datasets. A recommendation takes the number of imputed datasets as the percentage of missing values in the original dataset [8]. These datasets are then used to build a combined estimator of the vector of parameters of interest, by usually using the mean of the estimators among all the imputed datasets (*see* Fig. 1). This combined estimator is known as the first Rubin's rule. The second Rubin's rule states a formula to estimate the variance–covariance matrix of the combined estimator, decomposing it as the sum of the intra-imputation variance component and the between-imputation component.

Common statistical methods generally conclude with a study of differences in protein abundances between different conditions either using Student or Welch tests or using more sophisticated approaches such as the moderated-*t* testing techniques, which are based on empirical Bayesian approaches [9].

The `mi4p` package suggests an enhanced version of this approach, which accounts for the variability arising from the missing data and the imputation process. The protocol is comprised of four main steps: (1) the missing values of the quantitative dataset are imputed, thanks to multiple imputation algorithms, (2) leading to an estimation of the parameters of interest and their variance–covariance matrix for each peptide, (3) the variance–covariance matrices are projected to get a univariate parameter of dispersion for each peptide, (4) which are used for a more accurate testing procedure through moderated-*t* testing procedure.

2 Material

2.1 Requirements

The workflow presented in this protocol is implemented under the R package `mi4p`. To use it, the R environment is required [10]. For a better user experience, the `R Studio` integrated development environment is recommended [11].

2.2 Data Format: Quantitative Data

The quantitative data should be provided as a data frame or a matrix. Rows should describe the peptides and columns the biological samples. Thus, each cell of the matrix contains the measured (or missing) abundance of the peptide in the considered sample. Although statistical analysis at the peptide level is recommended, the methodology described in this chapter can be used at protein level. A schematic view of the quantitative dataset is pictured in Fig. 2.

2.3 Data Format: Experimental Data

The experimental data should be provided as a two-column data frame or matrix. The first column should contain the names of the biological samples and should be named "Sample.Name," and the second column should contain the names of the corresponding considered condition and should be named "Condition."

2.4 Data Format: Imputed Data

The multiple imputed data should be provided as an array of as many matrices as the **D** draws used for multiple imputation. Each imputed matrix should be of the same size as the quantitative data. A schematic view of imputed data is pictured in Fig. 3.

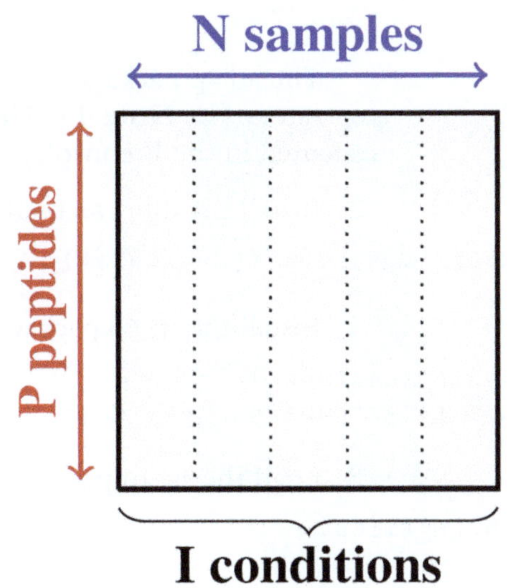

Fig. 2 Schematic representation of a quantitative dataset to be provided. There should be **P** rows corresponding to **P** peptides and **N** columns corresponding to **N** samples, which are spread between **I** conditions

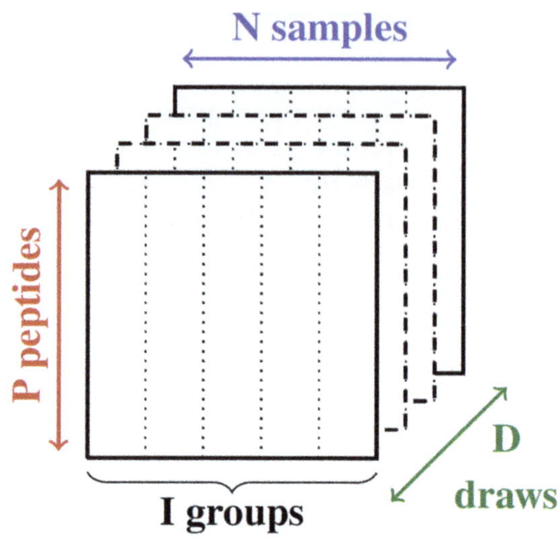

Fig. 3 Schematic representation of the imputed datasets to be provided. An array of **D** matrices corresponding to the **D** draws in the multiple imputation algorithm should be yielded. Each matrix should have **P** rows corresponding to **P** peptides and **N** columns corresponding to **N** samples, which are spread between **I** conditions

2.5 Package Installation and Loading

The `mi4p` package requires the following packages to be successfully installed: `BiocManager`, `DAPAR`, `emmeans`, `imp4p`, `limma`, and `mice` [5, 12–16]:

```
install.packages(c("BiocManager","emmeans","imp4p","mice"))
BiocManager::install(c("DAPAR","limma"))
```

The `mi4p` package itself can be downloaded from the GitHub repository (*see* **Note 1**). The following command lines should be executed in the R console:

1. Install the `devtools` R package:

```
install.packages("devtools")
```

2. Install the `mi4p` package from GitHub:

```
library(devtools)
install_github("mariechion/mi4p")
```

3. Load the `mi4p` package:

```
library(mi4p)
```

3 Methods

3.1 Multiple Imputation

Multiple imputation consists of imputing $\mathbf{D}$ times the missing values in the original quantitative dataset. This results in $\mathbf{D}$ imputed datasets. Multiple imputation is provided in `mi4p`, using the `multi.impute` function:

```
multi.impute(data, metadata, imp.meth, nb.imp)
```

The `data` argument refers to the original quantitative dataset that contains missing values. The `metadata` argument refers to the experimental dataset. The `imp.meth` argument denotes the chosen multiple imputation algorithm. The `mi4p` package is for now compatible with algorithms from the `imp4p` and `mice` packages [16, 17] (*see* **Note 2**). The default algorithm is set to `imp4p`. The `nb.imp` argument describes the number of draws to be done. By default, it is equal to the percentage of missing values in the original quantitative dataset. The `multi.impute` function returns an array of as many imputed matrices as `nb.imp`.

3.2 Estimation

The objective of multiple imputation is to estimate from $\mathbf{D}$ drawn datasets the vector of parameters of interest and its variance–covariance matrix. Notably, accounting for multiple imputation-based variability is possible, thanks to Rubin's rules, which provide an accurate estimation of these parameters. In `mi4p`, the vectors of parameters of interest are the vectors of the peptides' intensity mean in each condition considered. There are as many vectors to be estimated (and as many corresponding variance–covariance matrices) as the number of peptides in the quantitative dataset:

1. The first Rubin's rule leads to a combined estimator of the vector of intensity means (*see* **Note 3**). To compute the estimators for all peptides in the quantification dataset, the `rubin1.all` function should be used:

```
rubin1.all(imp.data, metadata, funcmean)
```

The `imp.data` argument refers to the array of imputed matrices and the `metadata` argument to the experimental dataset. The `funcmean` argument specifies the method for mean estimation. Here the default `funcmean` function is `meanImp_emmeans` and relies on the estimated marginal means algorithm (*see* **Note 4**). The `rubin1.all` function returns a list of estimated vector of intensity means in each condition for all peptides in the quantitative dataset (i.e., the length of the returned list equals the number of rows of `imp`.

data). To return only the combined estimator for a specific peptide, the rubin1.one function should be used:

```
rubin1.one(peptide, imp.data, metadata, funcmean)
```

The peptide argument denotes the row index of the considered peptide in the quantitative dataset.

2. The second Rubin's rule leads to a combined estimator of the variance–covariance matrix for each estimated vector of parameters of interest (*see* **Note 5**). The idea behind this rule is to decompose the variability into two components: the within-imputation variability and the between-imputation variability. To compute the estimators for all peptides in the quantification dataset, the rubin2.all function should be used:

```
rubin2.all(imp.data, metadata, funcmean, funcvar)
```

The imp.data argument refers to the array of imputed matrices and the metadata argument to the experimental dataset. The funcmean and funcvar arguments specify the method for mean and variance–covariance estimation, respectively. Here the default funcmean and funcvar functions are meanImp_emmeans and within_variance_comp_emmeans, both relying on the estimated marginal means algorithm (*see* **Note 6**). To return the within-imputation component only (respectively, the between-imputation component) for all peptides, the rubin2wt.all function (respectively, the rubin2bt.all function) should be used:

```
rubin2wt.all(imp.data, metadata, funcvar)
rubin2bt.all(imp.data, metadata, funcmean)
```

The rubin2.all, rubin2wt.all, and rubin2bt.all functions return lists of square matrices. The length of the list equals to the number of peptides considered, i.e., to the number of rows in imp.data. The size of the matrices is equal to the number of conditions considered, i.e. to the number of levels of the "Condition" factor in the metadata dataset. To return only the combined estimator for a specific peptide, the rubin2.one function should be used:

```
rubin2.one(peptide, imp.data, metadata, funcmean, funcvar)
```

The peptide argument denotes the row index of the considered peptide in the quantitative dataset. Likewise, to return the within-imputation component and/or the between-imputation component for a specific peptide, the rubin2wt.all and rubin2bt.all functions should be used:

```
rubin2wt.one(peptide, imp.data, metadata, funcvar)
rubin2bt.one(peptide, imp.data, metadata, funcmean)
```

The rubin2.one, rubin2wt.one, and rubin2bt.one functions return a square matrix. The size of the matrix is equal to the number of conditions considered, i.e., to the number of levels of the "Condition" factor in the metadata dataset.

3.3 Projection

State-of-the-art tests, including Student's *t*-test, Welch's *t*-test, and moderated *t*-test, rely on the variance estimation. Here, the variability induced by multiple imputation is described by a variance–covariance matrix. Therefore, a projection step is required to get a univariate variance parameter (*see* **Note 7**). This step is performed using the proj_matrix function:

```
proj_matrix(VarRubin.mat, metadata)
```

The VarRubin.mat denotes a variance–covariance matrix, as computed with rubin2.one, or a list of variance–covariance matrices, as computed with rubin2.all (*see* Subheading 3.2). The metadata argument refers to the experimental dataset. The proj_matrix function returns either a variance estimator for a given peptide or a list of variance estimators for all the peptides considered.

3.4 Moderated t-Test

Several testing methods can be used. For gene expression data, the recommended method is moderated *t*-testing [18]. In mi4p, the projected variance from multiple imputation serves as an input to the usual moderated *t*-test. This step is performed using the mi4limma function:

```
mi4limma(imp.data, metadata, VarRubin.S2)
```

The imp.data argument refers to the array of imputed datasets. The metadata refers to the experimental dataset. The VarRubin.S2 corresponds to the list of projected variance estimator for each peptide, as computed with the proj_matrix function (*see* Subheading 3.3). The mi4limma function returns a list of *p*-values and a list of log-transformed fold change for all peptides.

3.5 Complete Workflow

As an alternative to the step-by-step workflow described above, the complete mi4p workflow can be run with a single command:

```
mi4p.uni(data, metadata, imp.meth)
```

The data argument refers to the quantitative dataset, the metadata argument refers to the experimental dataset, and the imp.meth specifies the imputation method to be used (*see* Subheading 3.1). The mi4p.uni function returns a list of *p*-values and a list of log-transformed fold change for all peptides. The mi4p.uni function includes the four steps described above:

Table 1
Overview of the functions included in mi4p package

	Imputation	Estimation	Projection	Test
For one specific peptide		rubin1.one rubin2.one rubin2wt.one rubin2bt.one	proj_matrix	
For all peptides	multi.impute	rubin1.all rubin2.all rubin2wt.all rubin2bt.all	proj_matrix	mi4limma
	mi4p.uni			

multiple imputation (*see* Subheading 3.1), estimation (*see* Subheading 3.2), projection (*see* Subheading 3.3), and moderated *t*-testing (*see* Subheading 3.4). A synoptic view of the functions that can be used in each step is provided in Table 1.

3.6 Example Use Case

A detailed example use case of the workflow presented above can be found in the vignette of the `mi4p` package. It can be accessed using the following command:

```
vignette("mi4p")
```

The vignette will be updated along with the package.

4 Notes

1. The `mi4p` package is being regularly updated. It is, therefore, recommended to reinstall the package to use the most recent version.
2. While the only suggested algorithms for multiple imputation are taken from `imp4p` and `mice` packages [16, 17], the user can choose any other algorithm and recall the imputed matrices in the next steps under the aforementioned constraints (*see* Subheading 2.4).
3. Let $\hat{\beta}_{p,d}$ be the estimated vector of parameters for peptide p in the d-th imputed dataset. The first Rubin's rule gives the combined estimator for peptide p through the $\mathbf{D}$ imputed datasets such as:

$$\hat{\beta}_p = \frac{1}{\mathbf{D}} \sum_{d=1}^{\mathbf{D}} \hat{\beta}_{p,d} \quad (1)$$

4. The `meanImp_emmeans` function computes the estimated marginal means for specified factors or factor combinations in a linear model for a given imputed dataset. Estimated marginal means are also known as least-squares means or predicted marginal means and are predictions from a linear model over a reference grid.
5. The second Rubin's rule gives the combined estimator of the variance–covariance matrix for each estimated vector of parameters of interest for peptide p through the D imputed datasets such as:

$$\hat{\Sigma}_p = \frac{1}{\mathrm{D}}\sum_{d=1}^{\mathrm{D}} W_d + \frac{\mathrm{D}+1}{\mathrm{D}(\mathrm{D}-1)}\sum_{d=1}^{\mathrm{D}}(\hat{\beta}_{p,d} - \hat{\beta}_p)^T(\hat{\beta}_{p,d} - \hat{\beta}_p) \quad (2)$$

where W_d denotes the variance–covariance matrix of $\hat{\beta}_{p,d}$, i.e., the variability of the vector of parameters of interest as estimated in the d-th imputed dataset.
6. The `within_variance_comp_emmeans` function computes the symmetric variance–covariance matrix of the marginal means estimator for a given imputed dataset.
7. To keep all the pieces of information contained in the variance–covariance matrix, an extended version of the workflow presented in this chapter to the multivariate case is currently being implemented in `mi4p`. This multivariate extension will make it possible to fully take into account the effect of the imputation process, and thus the presence of missing values, on the precision of the estimate.

Acknowledgment

This work was funded through a PhD grant (2018–2021) from the Agence Nationale de la Recherche (ANR) [ANR-11-LABX-0055_IRMIA].

References

1. OB'rien JJ, Gunawardena HP, Paulo JA, Chen X, Ibrahim JG, Gygi SP, Qaqish BF (2018) The effects of nonignorable missing data on label-free mass spectrometry proteomics experiments. Ann Appl Stat 12(4): 2075–2095. https://doi.org/10.1214/18-AOAS1144
2. Chang C, Xu K, Guo C, Wang J, Yan Q, Zhang J, He F, Zhu Y (2018) PANDA-view: an easy-to-use tool for statistical analysis and visualization of quantitative proteomics data. Bioinformatics 34(20):3594–3596. https://doi.org/10.1093/bioinformatics/bty408
3. Choi M, Chang CY, Clough T, Broudy D, Killeen T, MacLean B, Vitek O (2014) MSstats: An R package for statistical analysis of quantitative mass spectrometry-based proteomic experiments. Bioinformatics 30(17): 2524–2526. https://doi.org/10.1093/bioinformatics/btu305
4. Tyanova S, Cox J (2018) Perseus: A bioinformatics platform for integrative analysis of

proteomics data in cancer research. In: Methods in molecular biology, vol 1711. Humana Press, pp 133–148. https://doi.org/10.1007/978-1-4939-7493-1_7

5. Wieczorek S, Combes F, Lazar C, Giai Gianetto Q, Gatto L, Dorffer A, Hesse AM, Coute Y, Ferro M, Bruley C, Burger T (2017) DAPAR & ProStaR: software to perform statistical analyses in quantitative discovery proteomics. Bioinformatics 33(1):135–136. https://doi.org/10.1093/bioinformatics/btw580
6. Webb-Robertson BJM, Wiberg HK, Matzke MM, Brown JN, Wang J, McDermott JE, Smith RD, Rodland KD, Metz TO, Pounds JG, Waters KM (2015) Review, evaluation, and discussion of the challenges of missing value imputation for mass spectrometry-based label-free global proteomics. J Proteome Res 14(5):1993–2001. pMID: 25855118. https://doi.org/10.1021/pr501138h
7. Little R, Rubin D (2019) Statistical analysis with missing data, vol 26, 3rd edn. https://doi.org/10.1002/9781119482260
8. White IR, Royston P, Wood AM (2011) Multiple imputation using chained equations: Issues and guidance for practice. Stat Med 30(4):377–399. https://doi.org/10.1002/sim.4067
9. Phipson B, Lee S, Majewski IJ, Alexander WS, Smyth GK (2016) Robust hyperparameter estimation protects against hypervariable genes and improves power to detect differential expression. Ann Appl Stat 10(2):946–963. https://doi.org/10.1214/16-AOAS920
10. R Core Team (2020) R: A language and environment for statistical computing. R Foundation for Statistical Computing, Vienna, Austria. https://www.R-project.org/
11. RStudio Team (2020) RStudio: Integrated development environment for R. RStudio, PBC, Boston, MA. http://www.rstudio.com/
12. Morgan M (2019) BiocManager: Access the bioconductor project package repository. R package version 1.30.10. https://CRAN.R-project.org/package=BiocManager
13. Lenth R (2020) emmeans: Estimated marginal means, aka least-squares means. R package version 1.5.2-1. https://CRAN.R-project.org/package=emmeans
14. Gianetto QG (2020) imp4p: Imputation for proteomics. R package version 1.0. https://CRAN.R-project.org/package=imp4p
15. Ritchie ME, Phipson B, Wu D, Hu Y, Law CW, Shi W, Smyth GK (2015) limma powers differential expression analyses for RNA-sequencing and microarray studies. Nucleic Acids Res 43(7):e47. https://doi.org/10.1093/nar/gkv007
16. van Buuren S, Groothuis-Oudshoorn K (2011) mice: Multivariate imputation by chained equations in r. J Stat Soft 45(3):1–67. https://doi.org/10.18637/jss.v045.i03
17. Gianetto QG, Wieczorek S, Couté Y, Burger T (2020) A peptide-level multiple imputation strategy accounting for the different natures of missing values in proteomics data. bioRxiv p 2020.05.29.122770. https://doi.org/10.1101/2020.05.29.122770
18. Smyth GK (12 Feb. 2004) Linear models and empirical Bayes methods for assessing differential expression in microarray experiments. Stat Appl Genet Mol Biol 3(1). https://doi.org/10.2202/1544-6115.1027

Chapter 8

Uncertainty-Aware Protein-Level Quantification and Differential Expression Analysis of Proteomics Data with seaMass

Alexander M Phillips, Richard D Unwin, Simon J Hubbard, and Andrew W Dowsey

Abstract

seaMass is an R package for protein-level quantification, normalization, and differential expression analysis of proteomics mass spectrometry data after peptide identification, protein grouping, and feature-level quantification. Using the concept of a blocked experimental design, seaMass can analyze all common discovery proteomics paradigms, including label-free (e.g., Waters Progenesis input), SILAC (e.g., MaxQuant input), isotope labelling (e.g., SCIEX ProteinPilot iTraq and Thermo ProteomeDiscoverer TMT input), and data-independent acquisition (e.g., OpenSWATH-PyProphet input), and is able to scale to study with hundreds of assays or more. By utilizing hierarchical Bayesian modelling, seaMass assesses the quantification reliability of each feature and peptide across assays so that only those in consensus influence the resulting protein group quantification strongly. Similarly, unexplained variation in each individual assay is captured, providing both a metric for quality control and automatic down-weighting of suspect assays. To achieve this, each protein group-level quantification outputted by seaMass is accompanied by the standard deviation of its posterior uncertainty. Moreover, seaMass integrates a flexible differential expression analysis subsystem with false discovery rate control based on the popular MCMCglmm package for Bayesian mixed-effects modelling, and also provides uncertainty-aware principal components analysis. We provide a description for using seaMass to perform an end-to-end analysis using a real dataset associated with a published clinical proteomics study.

Key words Quantitative proteomics, Protein quantification, Bayesian modelling, Differential expression analysis, False discovery rate control

1 Introduction

seaMass (https://github.com/biospi/seamass) is an R package that provides a complete protein quantification, normalization, and differential expression pipeline for discovery mass spectrometry data after prior identification and feature-level quantification. In particular, it is expected that protein grouping has been performed, so that each "protein" to be quantified represents a "protein

Thomas Burger (ed.), *Statistical Analysis of Proteomic Data: Methods and Tools*,
Methods in Molecular Biology, vol. 2426, https://doi.org/10.1007/978-1-0716-1967-4_8,

group" of accessions that cannot be unambiguously identified given the peptide identification evidence. In practice, seaMass consists of three main components: `seaMass-sigma`, which performs raw protein group-level quantification from peptide and feature-level mass spectrometry data; `seaMass-theta`, which performs protein group-level normalization across assays (label-free runs or iTraq/TMT/SILAC channels); and `seaMass-delta`, which performs differential expression analysis and false discovery rate (FDR) estimation. All three of these procedures use Bayesian hierarchical mixed-effects modelling in order to estimate the uncertainty of the estimated quantities including: peptide and protein group quantifications; normalization effects; and differential expression fold change estimates.

The mixed-effects modelling employed by `seaMass-sigma` includes random effects to account for variance at multiple levels: the variability of peptides across samples (for example, due to poor or variable digestion) and the variability of measurements across assays (for example, due to contamination or matrix). `seaMass-sigma` wraps this model within an empirical Bayes procedure that borrows strength across the population of protein groups: it uses those protein groups with a large number of peptides and measurements to estimate informative prior distributions for the distribution of the variance of peptides and features across all protein groups.

By estimating the uncertainty of each peptide, each peptide's contribution to the final protein group quantification estimate can be weighted according to their inferred variance such that highly variable peptides have a smaller contribution to the overall protein group quantification. Similarly, where peptides are observed via multiple features, each feature has its variance estimated so that more variable features contribute less to the peptide-level quantifications.

This quantification uncertainty is propagated from the feature level through the peptide and protein group levels up to the differential expression estimates. Then, seaMass wraps external methods, which leverage this additional uncertainty information to provide robust significance testing.

Also, seaMass captures assay-specific variation not explained by variation at the peptide or feature levels. In this way unreliable assays are identified and flagged during processing, and their contribution toward differential expression analysis and principal components analysis can be automatically down-weighted.

The model fitting is performed with Bayesian Markov chain Monte Carlo (MCMC) sampling using the `MCMCglmm` [1] R package. Multiple MCMC "chains" are fit for each protein group. False discovery rate (FDR) estimation is similarly provided by the `ashr` [2, 3] R package.

2 Material

2.1 Data Type

The input data generally consists of tabulated data in either comma-separated or tab-separated values from a number of different pre-processing software (*see* Subheading 2.2).

2.2 Data Format

Concretely, seaMass has functions for reading data from each of the following formats:

1. SCIEX ProteinPilot.
2. Thermo ProteomeDiscoverer.
3. Waters Progenesis QI.
4. MaxQuant [4].
5. OpenSWATH.

For ProteinPilot, seaMass requires the `PeptideSummary.txt` file output by ProteinPilot. For ProteomeDiscoverer, seaMass requires the `PSMs.txt` file output by ProteomeDiscoverer. For data output by the Progenesis QI software, seaMass requires the `pep_ion_measurements.csv` file. For data output by MaxQuant, seaMass requires both the `evidence.txt` and `proteinGroups.txt` files. For OpenSWATH, seaMass takes in either the output of PyProphet or TRIC. Import routines for other formats can be implemented on request.

2.3 Hardware Requirements

Depending on the user's needs, seaMass can be run on either a desktop computer or on a high-performance computing cluster. This tutorial focuses on running seaMass on a desktop machine. The number of samples to be analyzed determines the memory requirements of the software; at least 16GB is preferable. Multiple CPU cores can be utilized, though this will increase the memory footprint.

2.4 Software Requirements

1. Either Linux, macOS, or Windows operating systems.
2. A recent version of the R software; for version 1.0.2.0 of seaMass, version 4.0.4 or higher of R is required.

2.5 Software Installation

To install seaMass, enter the following into the R console:

```
install.packages("devtools")
devtools::install_github("biospi/seaMass",
                         ref = "v1.0.2.0", dependencies = TRUE)
```

which will install the `devtools` R package before downloading and installing seaMass v1.0.2.0 and its dependencies (*see* **Note 1**). The following R packages should be installed in this process:

```
1 ashr, data.table, bit64, doRNG, doSNOW, egg, emmeans,
extraDistr, FactoMineR, filelock, fitdistrplus, fst, ggplot2,
ggrepel, gridExtra, igraph, MCMCglmm, plotly, rmarkdown, R.
utils, utf8, uuid, zip
```

In addition, to download the example dataset used in this tutorial, the `osfr` package is required, which can be installed by running:

```
install.packages("osfr")
```

3 Methods

This section details the typical workflow of using seaMass to perform analysis of a quantitative proteomics dataset by walking through the process using data associated with a clinical study on Alzheimer's disease (AD) progression, which was first analyzed using an earlier version of seaMass in [5] (*see* **Note 2**). Tissue samples from multiple brain regions were collected from the brains of eighteen subjects: nine AD-affected patients (S1–S9) and nine age- and sex-matched controls (S10–S18). For each brain region, a pooled reference sample, R, was created by mixing equal amounts of each of the eighteen samples together. Each region was processed as a separate experiment of three iTRAQ 8-plexes. Mass spectrometry analysis was then performed using a SCIEX QSTAR Elite Q-TOF instrument. Peak extraction, peptide identification, protein grouping, and iTRAQ reporter quantification were performed using ProteinPilot v4.0. The `PeptideSummary.txt` file from ProteinPilot's output provides the quantitative feature-level data, which is input into seaMass.

Here, to illustrate the robustness of seaMass, we analyze the middle temporal gyrus brain region that was excluded from the original publication as the proteomics data for this region did not pass quality control. Notes throughout this section provide guidance for how the example data may be substituted for data from other sources.

3.1 Loading seaMass

First, load the `seaMass` package in R by entering into the R console:

```
library(seaMass)
```

3.2 Data Loading

1. The mass spectrometry data from the Alzheimer's disease study is openly available online and can be downloaded using the `osfr` R package by inputting the following into the R console:

```
osfr::osf_download(
    osfr::osf_retrieve_file("https://osf.io/vqcgz/"),
    conflicts = "skip", verbose = T, progress = T
            )
```

which will download the mass spectrometry data for the middle temporal gyrus.

2. For data processed using ProteinPilot, seaMass requires the output file to perform protein group quantification. We specify the location of the PeptideSummary file and import it into the R environment, before using seaMass's `import_ProteinPilot` function (*see* **Note 3**) to extract the feature-level data into a data frame that seaMass can use for subsequent processing:

```
file <- "PeptideSummary_MiddleTemporalGyrus.txt"
data <- import_ProteinPilot(file)
```

3.3 Fractionation

1. For data that has been fractionated, it is necessary to specify which fractions belong to which runs. First, generate a skeleton run table:

```
data.runs <- runs(data)
```

2. Next, we assign runs to each fraction:

```
data.runs$Run[1:68] <- "A"
data.runs$Run[69:152] <- "B"
data.runs$Run[153:222] <- "C"
```

In this instance, fractions 1 through 68 belong to run A, 69 through 152 to run B, and 153 through 222 to run C.

3. This fractionation information is then merged back with the imported data:

```
runs(data) <- data.runs
```

3.4 Experimental Design

1. We can now create a skeleton design matrix from our data:

```
data.design <- new_assay_design(data)
```

2. The biological sample associated with each assay can optionally be renamed. The distinction between technical and biological replicates can be made. In this instance, the pooled sample "R" is assigned to six different assays as six technical replicates of the same sample. Each of the biological samples S1–S18 are also assigned to separate assays (*see* **Note 4**):

```
data.design$Sample <- factor(c(
    "R","R","S1","S3","S7","S12","S17","S10",
    "R","R","S2","S6","S9","S13","S15","S18",
    "R","R","S4","S5","S8","S11","S14","S16"
                                              ))
```

The assays, in this case corresponding to each iTRAQ channel in each of the three runs, can also be similarly renamed through `data.design` $Assay (*see* **Note 5**).

3. The condition to which each assay belongs is assigned, the `levels` argument can be used to determine which conditions are to be compared. Here, we specify that "Ct" is the first and therefore the baseline condition. The pooled reference assays "R" should also be excluded from differential expression analysis:

```
data.design$Condition <- factor(c(
                  NA,NA,"AD","AD","AD","Ct","Ct","Ct",
                  NA,NA,"AD","AD","AD","Ct","Ct","Ct",
                  NA,NA,"AD","AD","AD","Ct","Ct","Ct"
                                  ),
                              levels = c("Ct", "AD"))
```

4. For experiments where runs are performed in batches or across multiple instruments, it may be desirable to split the assays into multiple "blocks" (*see* **Note 6**); for iTRAQ and TMT experiments with multiple runs, seaMass automatically splits each multiplex out into a separate block.
5. Additional covariates can be added to the experimental design at this stage by adding additional columns to the `data.design` table.
6. Finally, "reference weights" can be assigned to specify reference assays. Conventionally, replicated pooled sample assays are used as reference assays in each block so that protein group quantifications can be standardized in relation to them for direct comparison across blocks. As seaMass allows for multiple reference samples per block, to standardize to the average of the two pooled sample assays in each block, the reference weights are set as follows:

```
data.design$RefWeight <- c(1,1,0,0,0,0,0,0,1,1,0,0,0,
                           0,0,0,1,1,0,0,0,0,0,0)
```

For a blocked experimental design where each condition is represented in each block, seaMass also allows standardization using a suitable weighted average of the samples themselves, so that no pooled samples are necessary. For example, to standardize to the average of the AD and Ct samples (*see* **Note 7**):

```
data.design$RefWeight <- c(0,0,1,1,1,1,1,1,0,0,1,1,1,1,1,1,
                           0,0,1,1,1,1,1,1)
```

7. The complete experimental design can be viewed by typing `data.design` into the R console. The complete table for the example dataset is shown in Table 1.

3.5 Protein Group Quantification and Normalization

1. After the initial setup and addition of experimental design, protein group quantification can be performed by running `seaMass_sigma`, which takes as input the feature-level data. Optionally, the output directory can be specified using the `path` argument. The experimental design table can also be supplied; while not required at this stage, it will be used to add design metadata to the automatically generated plots:

```
fit.sigma <- seaMass_sigma(
                 data, data.design,
                 path = "MiddleTemporalGyrus",
                 control = sigma_control(nthread = 8)
                         )
```

2. After `seaMass_sigma` is finished, the derived raw protein group quantifications can be normalized within blocks and standardized across blocks by executing `seaMass_theta`:

```
fit.theta <- seaMass_theta(
                    fit.sigma,
                    norm.groups = top_groups(fit.sigma)
                       )
```

The `seaMass_theta` normalization model derives a normalization factor for each assay as to minimize protein group-level variance for the maximum number of protein groups. For computational efficiency, by default only high-quality protein groups are examined. If the user has some prior knowledge of the subset of protein groups to normalize against, this can be

Table 1
Experimental design table for the middle temporal gyrus dataset

Run	Channel	Assay	RefWeight	Sample	Condition
A	113	R1	1	R	NA
A	114	R2	1	R	NA
A	115	S1	0	S1	AD
A	116	S3	0	S3	AD
A	117	S7	0	S7	AD
A	118	S12	0	S12	Ct
A	119	S17	0	S17	Ct
A	121	S10	0	S10	Ct
B	113	R3	1	R	NA
B	114	R4	1	R	NA
B	115	S2	0	S2	AD
B	116	S6	0	S6	AD
B	117	S9	0	S9	AD
B	118	S13	0	S13	Ct
B	119	S15	0	S15	Ct
B	121	S18	0	S18	Ct
C	113	R5	1	R	NA
C	114	R6	1	R	NA
C	115	S4	0	S4	AD
C	116	S5	0	S5	AD
C	117	S8	0	S8	AD
C	118	S11	0	S11	Ct
C	119	S14	0	S14	Ct
C	121	S16	0	S16	Ct

specified by supplying a subset of `groups(fit.sigma)$}}` `Group` to the `norm.groups` option instead.

3. Configuration of seaMass processing is achieved through the `sigma_control` and `theta_control` objects. On multicore systems with sufficient amounts of RAM, multiple CPU threads can be used; the number of threads can be specified as the `nthread` option to `sigma_control` (*see* **Note 8**).

3.6 Protein Group Quantification Output

1. seaMass outputs a set of convenient CSV files of the results in the `csv` subfolder of the output folder 🗀 **`MiddleTemporalGyrus`**.

   Alternatively, results can be output within R. For instance, a table of normalized protein group quantifications can be output using:

```
proteinQuants <- group_quants(fit.theta)
```

2. In the resulting `data.frame` , seaMass outputs each protein group quantification `m` (the posterior mean) along with its uncertainty `s` (the posterior standard deviation). Subsequently, these protein group quantifications can be outputted to, for example, a CSV file:

```
write.csv(proteinQuants, file = "my_proteinQuants.csv")
```

3.7 seaMass-sigma and seaMass-theta Output Plots

As output, seaMass provides a full HTML report with a rich set of interactive plots as a zip archive in the output directory (*see* **Note 9**). Below, we instead use the R package to output specific plots. For some of the following examples, we will generate the plots for a single protein group, sp|P09211|GSTP1_HUMAN.

Each violin in the violin plots (*see* Fig. 1) spans the 90% interval of probable values for that variable (the Bayesian 90% credible interval), with the median value represented as a vertical bar. Left of the median the girth of the violin reduces in size, representing the local FDR for the variable being that value or less (posterior probability that the variable is that value or more), with the violin truncated at 5% local FDR. Conversely, right of the median also the girth reduces, representing the local FDR for the variable being that value or more and is truncated similarly. Wider violins, therefore, represent more uncertain estimations. Practically:

1. Local FDR violin plots showing the inferred normalization factors ("assay means") and unexplained assay variation ("assay standard deviations") can be generated by typing in the R console:

```
g1 <- plot_assay_means(fit.theta, output = "ggplot")
g2 <- plot_assay_stdevs(fit.sigma, output = "ggplot")
g <- gridExtra::grid.arrange(g1, g2)
ggplot2::ggsave("assay_means_stdevs.pdf", g,
                width = 7, height = 7)
```

In the above, setting the `output` option to `"plotly"` generates interactive plots, whereas setting it to `"ggplot"` generates static plots potentially more suitable for constructing publication figures. The PDF output of this code is shown in Fig. 1.

2. Principal components analysis (PCA) plots are generated for each block and for the experiment as a whole automatically. These PCA plots down-weight poorly quantified assays and

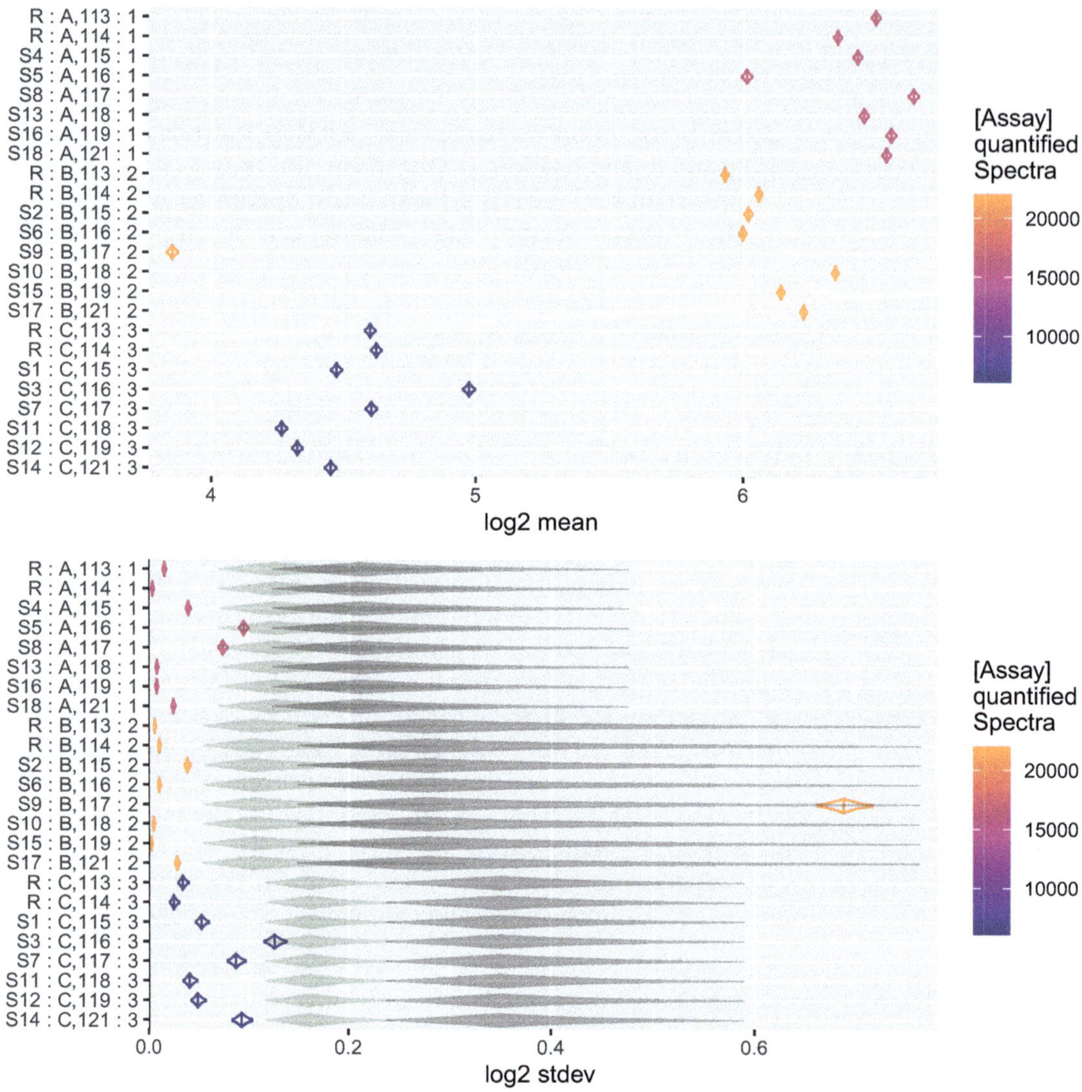

Fig. 1 Local FDR violin plots of assay means and standard deviations for the example dataset. Top: The estimated assay means are colored by the number of quantified spectra for that iTraq 8-plex. It can be seen that the assays in iTRAQ 8-plex run C (block 3) are generally underexposed relative to the assays in runs A and B, except for sample S9 in run B, which is even more underexposed. Bottom: Similarly, the estimated assay standard deviations are shown, together with the inferred distribution of "explained" peptide standard deviations (green violins) and feature standard deviations (gray violins). As a rule of thumb, unexplained assay variation should be substantively lower than the explained variation, and both should be less than the fold change of differential expression you hope to discover. Hence here it illustrates that sample S9 in run B is a significant potential quality control problem—as a result `seaMass-delta` will automatically down-weight this assay's contribution downstream

protein groups and are subsequently augmented with ellipses indicating the 95% and 68% posterior regions of uncertainty in the principal components for each assay. These can be used to determine whether any assays exhibit more variation than the others, which would be indicative of the issues in sample preparation, for example:

```
g <- plot_robust_pca(fit.theta, colour = "Condition",
                     fill = "Condition", shape = 6,
                     output = "ggplot")
ggplot2::ggsave("robust_pca.pdf", g,
                width = 7, height = 5)
```

The PDF plot generated for the example data is shown in Fig. 2.

3. Plots of the raw and normalized protein group quantifications for any particular protein group can be generated by running:

```
g <- plot_group_quants(fit.theta,
                       "sp|P09211|GSTP1_HUMAN",
                       output = "ggplot")
ggplot2::ggsave("group_quants.pdf", g,
                width = 7, height = 4)
```

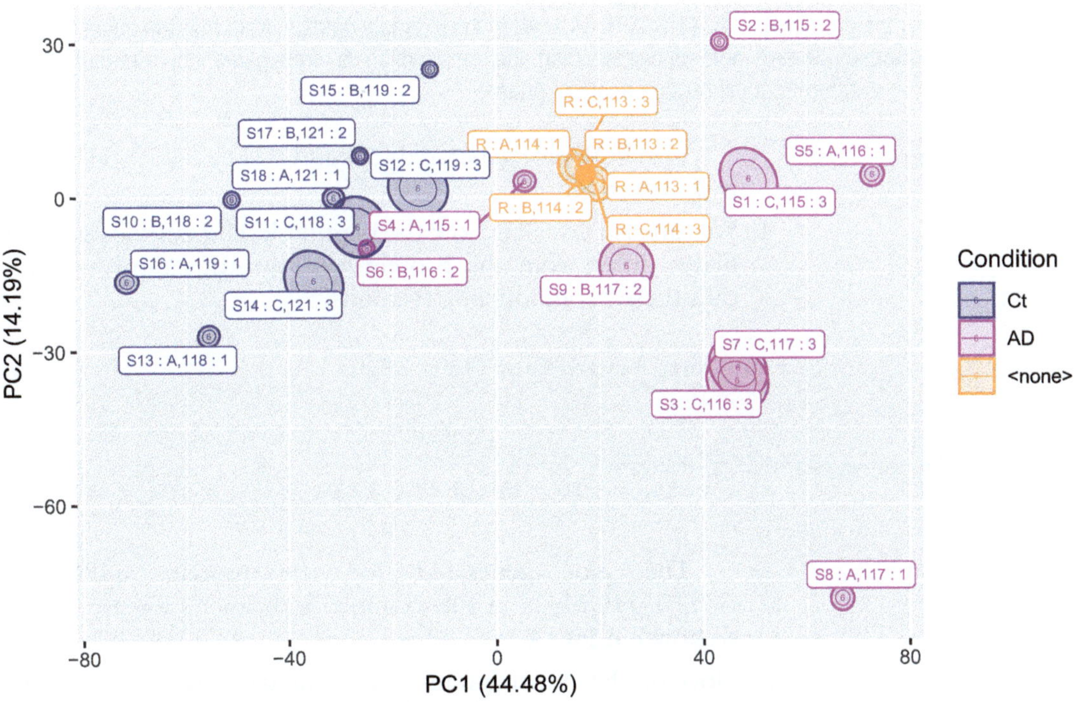

Fig. 2 Robust principal component analysis plot for the example dataset. Each assay in the experiment appears as an ellipse, colored by condition. Contours are shown for each assay indicating the uncertainty in quantifications for that assay. Here it can be seen that the potential issues with run C and sample S9 are reflected in larger uncertainty ellipses

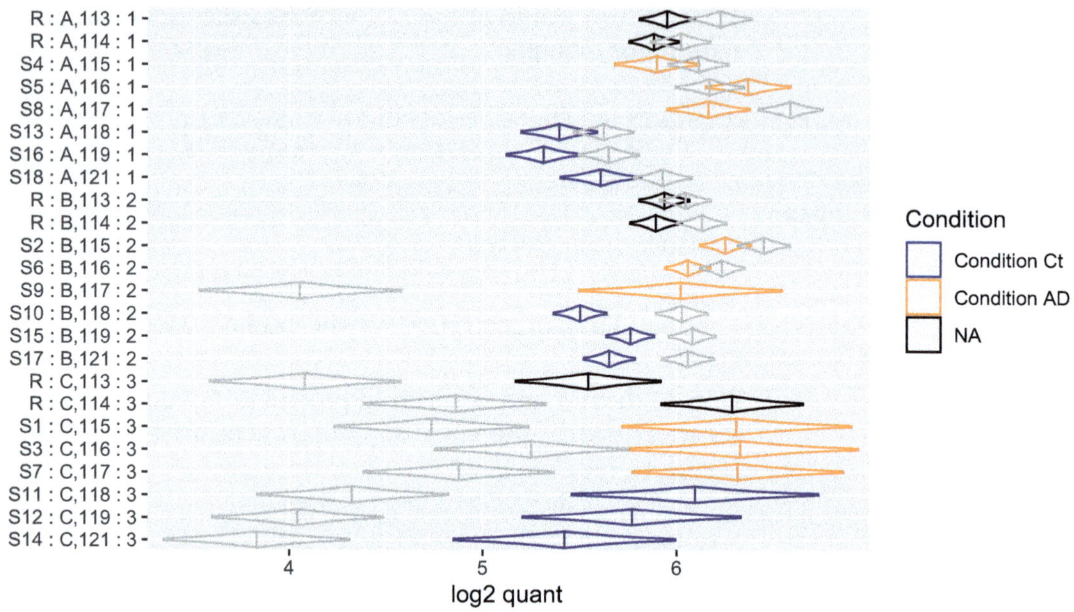

Fig. 3 Local FDR violin plots of the raw (gray) and normalized (colored by condition) protein group-level quantifications for the protein sp|P09211|GSTP1_HUMAN in the example dataset. Note that the quantifications for run C and sample S9 are more uncertain than the rest, which is propagated down-stream to the `seaMass-delta` differential expression analysis phase

The PDF plot generated for sp|P09211|GSTP1_HUMAN is shown in Fig. 3.

4. We can also visualize how the peptide-level quantification estimates differ from the protein-level quantifications; seaMass calls these "component deviations":

```
g <- plot_component_deviations(fit.sigma,
                               "sp|P09211|GSTP1_HUMAN",
                               output = "ggplot")
ggplot2::ggsave("component_deviations.pdf", g,
                width = 10, height = 12)
```

The plot generated for the protein sp|P09211|GSTP1_HUMAN in the example Alzheimer's disease dataset is shown in Fig. 4.

5. Plots of the mean intensity and standard deviation of each peptide observed for a particular protein group can be generated by typing:

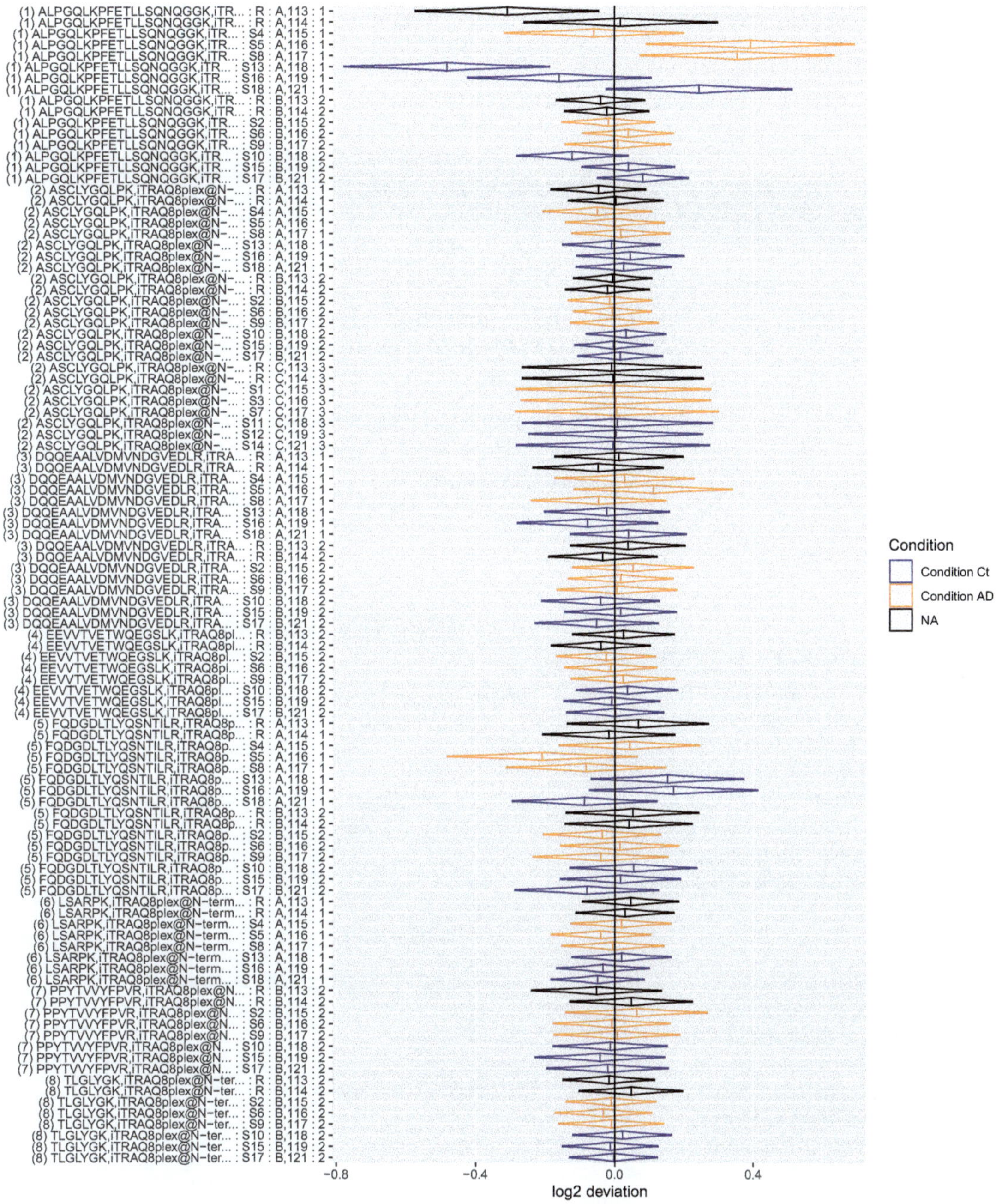

Fig. 4 Local FDR violin plots showing the peptide-level deviations from the parent protein group quantification for the protein group sp|P09211|GSTP1_HUMAN. The difference is notable particularly for the top peptide in the plot, which could be due to a systematic technical issue or be indicative of a differingly expressed proteoform

```
g1 <- plot_component_means(fit.sigma,
                           "sp|P09211|GSTP1_HUMAN",
                           output = "ggplot")
g2 <- plot_component_stdevs(fit.sigma,
                            "sp|P09211|GSTP1_HUMAN",
                            output = "ggplot")
g <- gridExtra::grid.arrange(g1, g2)
ggplot2::ggsave("component_means_stdevs.pdf", g,
                width = 7, height = 4)
```

The plot generated for the protein sp|P09211|GSTP1_HUMAN in the example Alzheimer's disease dataset is shown in Fig. 5.

6. Similar to the peptide-level plots, we can also generate violin plots of the feature-level ("measurement") mean intensities and standard deviations:

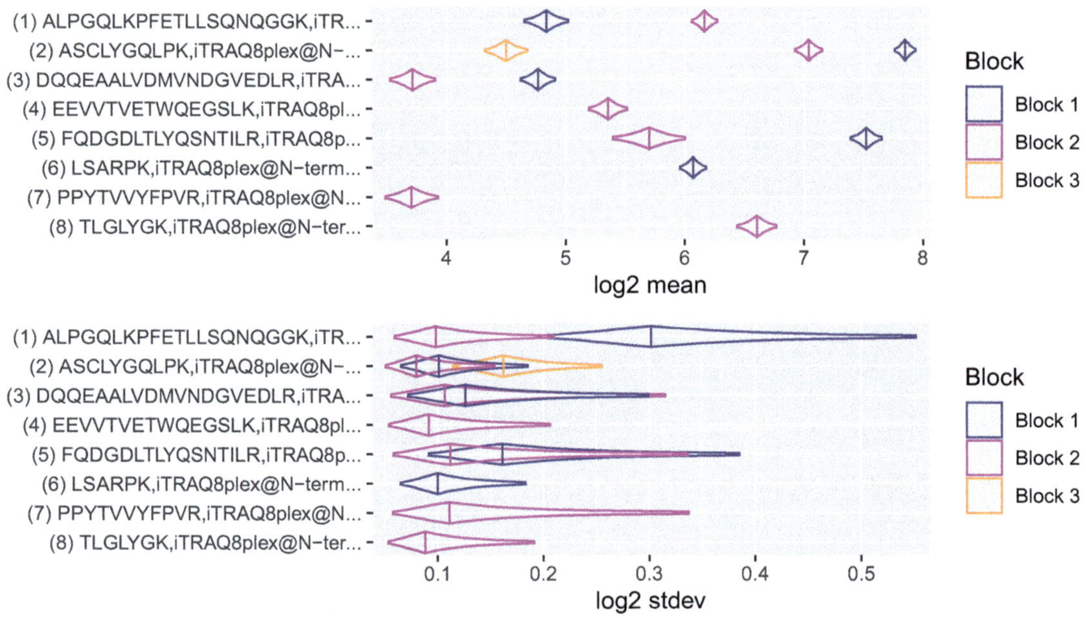

Fig. 5 Local FDR violin plots showing peptide-level means and standard deviations for the protein sp|P09211|GSTP1_HUMAN in the example dataset. Peptide-level means are a weighted average of feature-level mean intensities. Each feature is weighted by its precision; hence, more variable features contribute less to the peptide-level quantification. Peptide precisions affect the overall protein group-level quantification similarly. Here the top peptide in the plot is particularly variable in block 1 (run A)

```
g1 <- plot_measurement_means(fit.sigma,
                             "sp|P09211|GSTP1_HUMAN",
                             output = "ggplot")
g2 <- plot_measurement_stdevs(fit.sigma,
                              "sp|P09211|GSTP1_HUMAN",
                              output = "ggplot")
g <- gridExtra::grid.arrange(g1, g2)
ggplot2::ggsave("measurement_means_stdevs.pdf", g,


                width = 7, height = 9)
```

The generated PDF is shown in Fig. 6.

3.8 Differential Expression and FDR Estimation

1. The differential expression analysis and FDR estimation component of seaMass, `seaMass-delta`, can be run on the resulting `seaMass-theta` fit object:

```
fit.delta <- seaMass_delta(fit.theta)
```

2. `seaMass-delta` will proceed to fit a differential expression model to the normalized protein quantification estimates generated by `seaMass-theta`. By default, this differential expression model is equivalent to performing a Welch's t-test for each pairwise comparison of defined conditions. Different differential expression models can be configured by supplying additional arguments to `seaMass_delta` (*see* **Note 10**).
3. FDR estimation is then performed using the `ashr` R package [3]. It uses an empirical Bayes approach to perform "adaptive shrinkage" on the estimated log2 fold changes generated by `seaMass-delta`, harnessing the extra uncertainty information provided by seaMass to estimate the distribution of log2 fold changes across the dataset and, depending on the uncertainty of each log2 fold change, moderate those estimated log2 fold changes. Protein groups for which there is high uncertainty have more shrinkage applied to their log2 fold changes than those proteins whose log2 fold changes are less uncertain.

3.9 Differential Expression Output

1. Once `seaMass-delta` has completed processing, a data.frame containing estimates of log2 fold change and quantitative false discovery rate can be obtained using:

```
data.fdr <- group_quants_fdr(fit.delta)
```

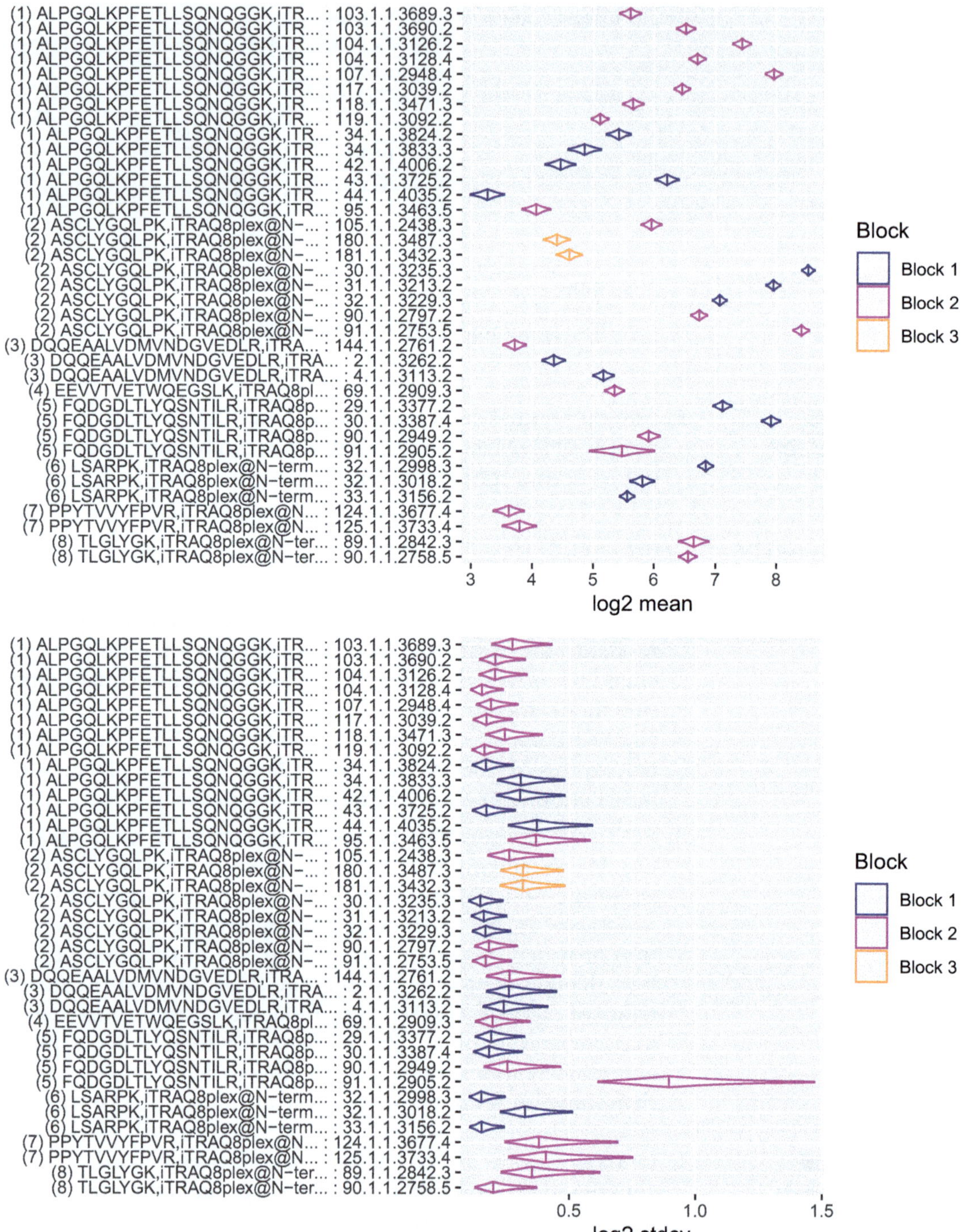

Fig. 6 Local FDR violin plots showing iTraq reporter ion-level means and standard deviations for the protein sp| P09211|GSTP1_HUMAN in the example dataset. It can be seen that reporter ion intensities range over 6 orders of magnitude and several exhibit high variance; however, `seaMass-sigma` is able to focus its quantification on the most stable

Table 2
Excerpt of the outputted results with estimated global FDR (qvalue) and posterior mean log2 fold change and posterior standard deviation of log2 fold change for each protein (Group)

Batch	Group	qvalue	PosteriorMean	PosteriorSD
Condition.AD-Ct	sp—P09211—GSTP1_HUMAN	2.120733e−09	0.5017938	0.06695361
Condition.AD-Ct	sp—Q13228—SBP1_HUMAN	1.930646e−08	0.4427758	0.07636209
Condition.AD-Ct	sp—Q9UEY8—ADDG_HUMAN	2.679536e−06	0.3380300	0.06532386
Condition.AD-Ct	sp—P10909—CLUS_HUMAN	2.246470e−05	0.4554334	0.08949953
Condition.AD-Ct	sp—Q13510—ASAH1_HUMAN	3.719622e−05	0.4149335	0.09259445
Condition.AD-Ct	sp—Q16643—DREB_HUMAN	7.131150e−05	−0.4427733	0.10096924
Condition.AD-Ct	sp—Q9NSD9—SYFB_HUMAN	1.099772e−04	−0.4478407	0.10202492
Condition.AD-Ct	sp—Q99497—PARK7_HUMAN	1.526100e−04	0.3767814	0.07312156
Condition.AD-Ct	sp—P49006—MRP_HUMAN	1.894262e−04	0.4266677	0.10103153
Condition.AD-Ct	sp—Q9NZH0—GPC5B_HUMAN	2.255137e−04	0.4376124	0.10208458

An example of the results from the Alzheimer's disease study is shown in Table 2.

2. This data frame can be saved as a CSV file, e.g., for further downstream processing:

```
write.csv(data.fdr, file = "MiddleTemporalGyrus-FDR.csv")
```

3.10 seaMass-delta Output Plots

`seaMass-delta` appends a number of interactive plots into the HTML report by default, and more can be optionally generated after the fact. These include quantitative volcano plots and FDR curves:

1. Volcano plots can be generated from the `seaMass-delta` results by inputting into the R console:

```
g <- plot_volcano(fit.delta, output = "ggplot")
ggplot2::ggsave("volcano.pdf", g,
                width = 7, height = 7)
```

A volcano plot for the "AD–Ct" comparison of the Alzheimer's disease study is given in Fig. 7.

2. A plot showing the predicted qvalue FDR against the number of discoveries at that FDR can be plotted with:

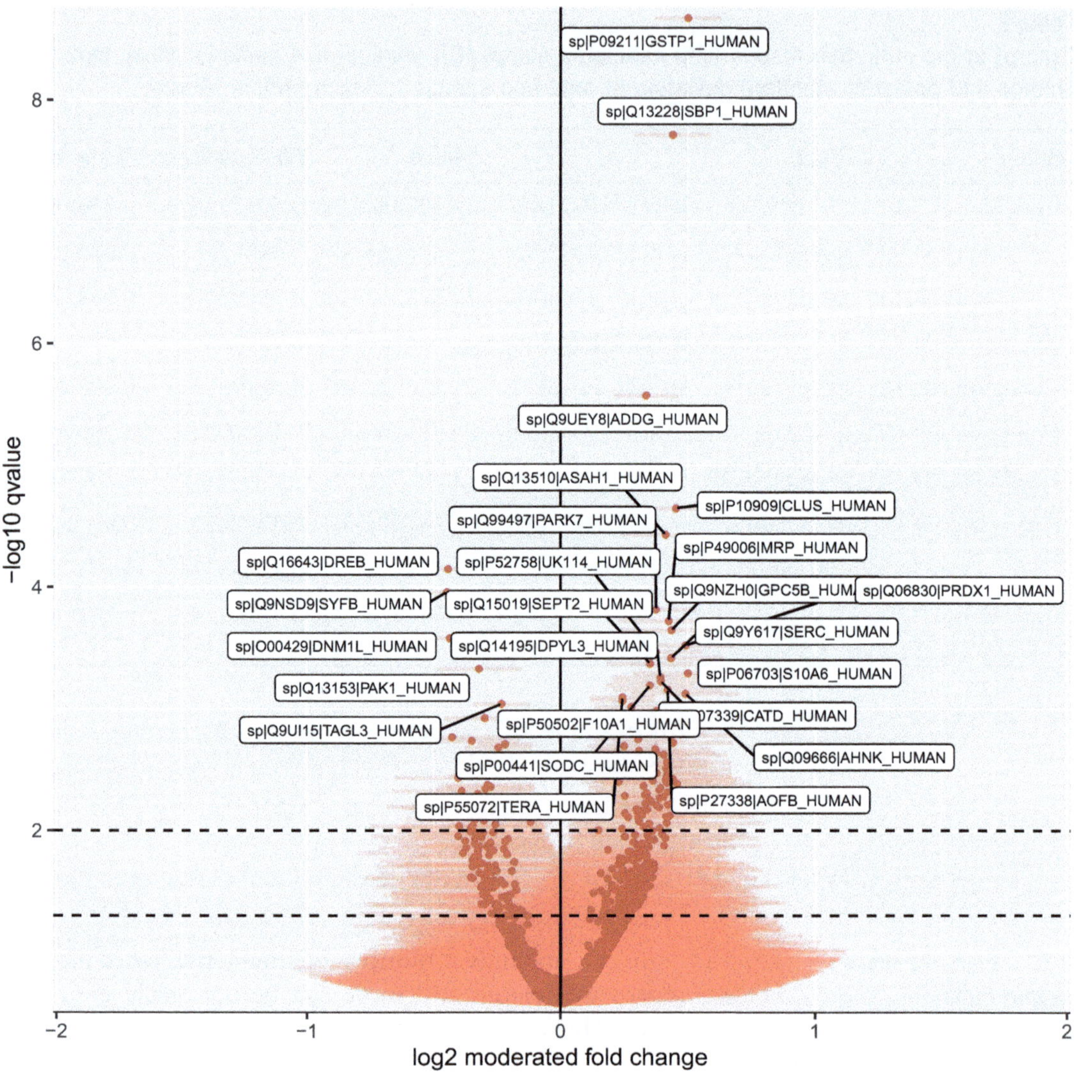

Fig. 7 Volcano plot for the AD–Ct comparison of the example dataset, with the x-axis denoting the estimated log2 moderated fold change and the y-axis denoting the FDR as −log10(qvalue). Each point has horizontal error bars denoting the 95% credible interval of the estimated fold change. The 25 protein groups with the lowest qvalues are labelled, and horizontal dashed lines are shown at FDRs of 1% and 5%. 0.05 and 0.01

```
g <- plot_fdr(fit.delta, output = "ggplot")
ggplot2::ggsave("fdr.pdf", g,
                width = 7, height = 4)
```

The resulting PDF is shown in Fig. 8.

3. Finally, local FDR violin plots of the FDR-controlled log2 fold changes can be plotted for any number of protein groups. For example, to plot the 25 most differentially expressed protein groups in the AD–Ct comparison:

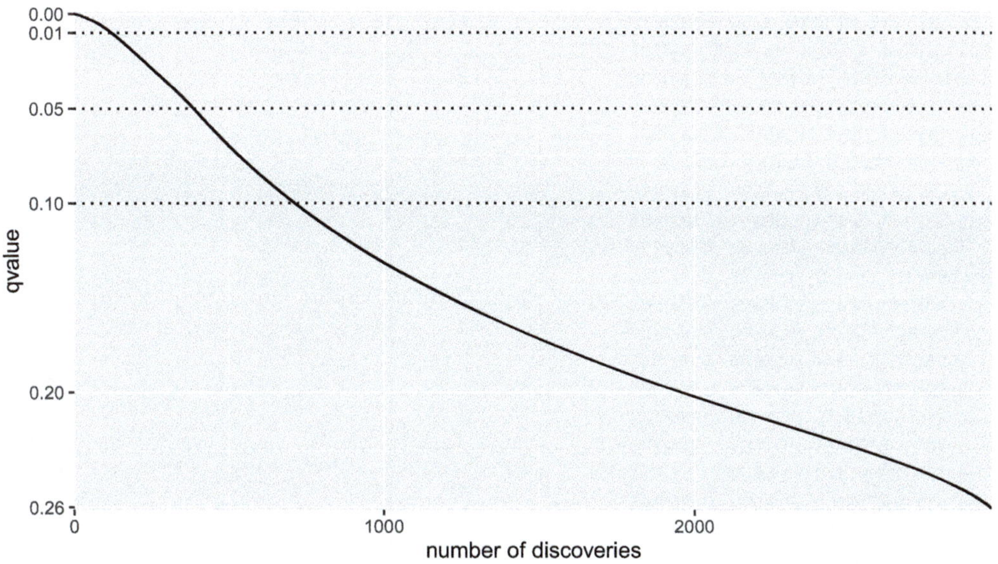

Fig. 8 qvalue FDR versus number of discoveries curve for the example dataset, showing the number of discoveries that would be declared at a given FDR cutoff. Horizontal dashed lines are shown at 1%, 5%, and 10% FDR

```
g <- plot_group_quants_fdr(
                    fit.delta,
                    group_quants_fdr(fit.delta)$Group[1:25],
                    output = "ggplot"
                          )
ggplot2::ggsave("group_quants_fdr.pdf", g,
                width = 7, height = 5)
```

The plot generated is shown in Fig. 9.

4 Notes

1. We are using v1.0.2.0 of seaMass to ensure compatibility with this tutorial. For production use, we always recommend you use the latest version of seaMass instead.
2. In [5], the version of the seaMass software used for quantitative analysis was then named "Bayesprot."
3. seaMass provides import functions for other input formats, which are similarly named, including MaxQuant (`import_MaxQuant`), Progenesis (`import_Progenesis`), ProteomeDiscoverer (`import_ProteomeDiscoverer`), OpenSWATH (`import_OpenSWATH`), and MSstats (`import_MSstats`). The details for the input files required for these import routines

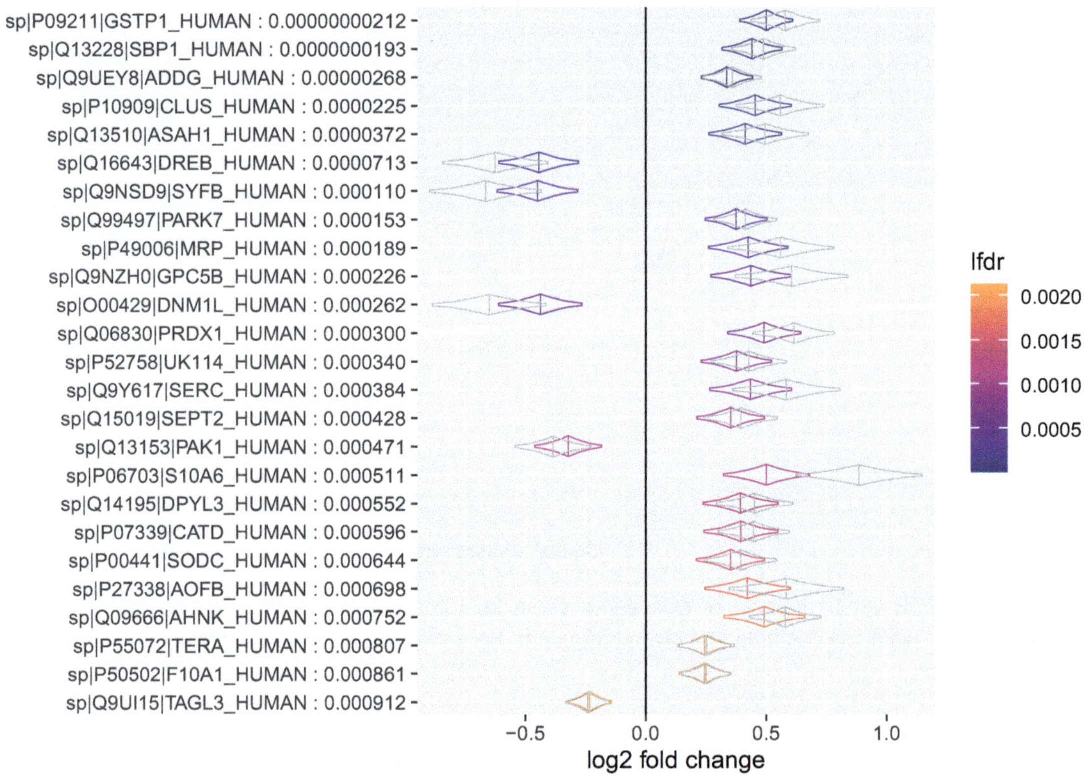

Fig. 9 Local FDR violin plots of the estimated fold change for the 25 protein groups with lowest qvalue (given in y-axis labels) in the AD–Ct comparison. The unmoderated fold changes from the individual MCMCglmm Welch's t-tests are shown in gray, while the fold changes moderated through Ash modelling of the distribution of fold changes in the study are colored by their local FDR

can be found by typing `?` and the name of the function in the R console and reading the documentation.

4. As an example, suppose that multiple technical replicates of sample "S1" were included in the experiment. In this scenario, each should be assigned as sample "S1" with different assay names.

5. It is also sometimes desirable to remove an assay from the analysis, say because only some of the iTRAQ reporter channels were filled in a particular run or a sample is identified to have been contaminated. In these scenarios, the assay can be removed by assigning its name as missing with `NA`.

6. Blocking can be specified by adding additional columns containing `TRUE` and `FALSE` values to the the experimental design table with column names of the form: `Block.1`, `Block.2`, *etc.*. Assays may appear in multiple blocks. On a HPC cluster, protein group quantification with `seaMass-sigma` is able to run in parallel across these blocks.

7. For the purposes of quality control, where, e.g., a pooled sample is available, it may be preferable to *not* use the pooled sample assays as reference assays. Then, any deviation between pooled assays in different blocks can be inferred as a measure of quality control. Conversely, when quality control has been assured, the reverse can be done; and the pooled samples can be used as the references such that protein group quantifications are calculated in relation to the reference samples.
8. Running seaMass on high-performance computing (HPC) clusters is also supported. This is achieved by specifying a scheduler in `sigma_control`. In seaMass, schedulers for SLURM-managed clusters (`schedule_slurm`), PBS-managed clusters (`schedule_pbs`), and SGE-managed clusters (`schedule_sge`) are implemented. More details are given at http://github.com/biospi/seamass.
9. Due to the large size of unzipped reports, it is preferred to mount the zip as a drive for browsing without uncompressing, as described at https://github.com/biospi/seamass.
10. By default, the differential expression model fitted is a Bayesian equivalent to a Welch's t-test, where each condition is assumed to have a separate residual variance. This model can be altered by specifying different formulae and priors for `seaMass-delta`. These formulae must comply with the syntax used by the `MCMCglmm` package [1], which is similar to the formula syntax used by the `lme4` R package [6]. For example, a Bayesian model equivalent to a student's t-test can be fit by specifying `rcov=~units` and `prior=list(R=list(V=1,nu=2e-4))`. If additional covariates were entered into the `data.design` table, these can be included in the model by overriding the `fixed` argument. For example, to include "Age" as a predictor: `fixed=~Condition+Age`. Random effects can be included by specifying a `random` formula argument. Care should be taken here to ensure that the `prior` argument is modified accordingly; details of how the prior should be specified can be found in the documentation for `MCMCglmm` [1].

Acknowledgment

The development of seaMass was supported by BBSRC grants BB/M024954/2 and BB/R021430/1, as well as MRC grant MR/N028457/1.

References

1. Hadfield JD (2010) MCMC methods for multi-response generalized linear mixed models: The MCMCglmm R package. J Stat Softw 33(2): 1–22. http://dx.doi.org/10.18637/jss.v033.i02
2. Stephens M (2017) False discovery rates: A new deal. Biostatistics 18(2):275–294. https://doi.org/10.1093/biostatistics/kxw041
3. Stephens M, Carbonetto P, Gerard D, Lu M, Sun L, Willwerscheid J, Xiao N (2019) Ashr: Methods for adaptive shrinkage, using empirical Bayes. https://cran.r-project.org/web/packages/ashr/index.html
4. Tyanova S, Temu T, Cox J (2016) The MaxQuant computational platform for mass spectrometry-based shotgun proteomics. Nature Protocols 11(12):2301–2319. https://doi.org/10.1038/nprot.2016.136
5. Xu J, Patassini S, Rustogi N, Riba-Garcia I, Hale BD, Phillips AM, Waldvogel H, Haines R, Bradbury P, Stevens A, Faull RLM, Dowsey AW, Cooper GJS, Unwin RD (2019) Regional protein expression in human Alzheimer's brain correlates with disease severity. Communications Biology 2(1):43. https://doi.org/10.1038/s42003-018-0254-9
6. Bates D, Mächler M, Bolker B, Walker S (2015) Fitting linear mixed-effects models using lme4. J Stat Softw 67(1):1–48. https://doi.org/10.18637/jss.v067.i01

Chapter 9

Statistical Analysis of Quantitative Peptidomics and Peptide-Level Proteomics Data with Prostar

Marianne Tardif, Enora Fremy, Anne-Marie Hesse, Thomas Burger, Yohann Couté, and Samuel Wieczorek

Abstract

Prostar is a software tool dedicated to the processing of quantitative data resulting from mass spectrometry-based label-free proteomics. Practically, once biological samples have been analyzed by bottom-up proteomics, the raw mass spectrometer outputs are processed by bioinformatics tools, so as to identify peptides and quantify them, notably by means of precursor ion chromatogram integration. From that point, the classical workflows aggregate these pieces of peptide-level information to infer protein-level identities and amounts. Finally, protein abundances can be statistically analyzed to find out proteins that are significantly differentially abundant between compared conditions. Prostar original workflow has been developed based on this strategy. However, recent works have demonstrated that processing peptide-level information is often more accurate when searching for differentially abundant proteins, as the aggregation step tends to hide some of the data variabilities and biases. As a result, Prostar has been extended by workflows that manage peptide-level data, and this protocol details their use. The first one, deemed "peptidomics," implies that the differential analysis is conducted at peptide level, independently of the peptide-to-protein relationship. The second workflow proposes to aggregate the peptide abundances after their preprocessing (i.e., after filtering, normalization, and imputation), so as to minimize the amount of protein-level preprocessing prior to differential analysis.

Key words Statistical software, Data processing, Differential analysis, Label-free proteomics, Relative quantification

1 Introduction

Nowadays, discovery proteomics mainly relies on liquid chromatography coupled to tandem mass spectrometry (LC-MS/MS). It is often used in a bottom-up workflow, where proteins are digested by a protease into peptides, making MS-based analysis more powerful [1]. This pipeline routinely produces huge amount of raw data, which processing with bioinformatics tools turns into long lists of identified peptides. In addition, a quantitative information for these peptides is classically computed, most often through

Thomas Burger (ed.), *Statistical Analysis of Proteomic Data: Methods and Tools*,
Methods in Molecular Biology, vol. 2426, https://doi.org/10.1007/978-1-0716-1967-4_9,

eXtracted Ion Chromatograms (XIC) [2]. This quantitative information can then be used to compare the relative abundance of peptides and proteins between samples. The relative nature of the quantification derives from the fact that a XIC amounts to the integral of an ion signal over time, which amplitude is not only correlated to peptide concentrations in the biological sample but also to numerous physicochemical phenomena, which are highly context- and peptide-dependent. As a result, peptide quantifications cannot be used to compare the abundance of different peptides within a sample, but only to compare the abundances of a same peptide within distinct yet relatively similar samples. Although there exist techniques to correct these biases and tend toward a more absolute quantification measure [3], it remains ordinary to stick to relative quantification information to answer the majority of questions addressed by discovery proteomics. Notably, a recurrent question in discovery proteomics is how to thoroughly select a set of putative biomarkers out of the long list of identified analytes, based on their variation of abundance across the samples. This question being ubiquitous in proteomics, a host of biostatistics tools have been developed over the last years, among which a significant proportion embed various R packages into well-designed graphical user interfaces based on Shiny technology [4], such as notably, Prostar [5]. However, two views can be opposed: the first one is to aggregate the peptide-level identities and quantities at protein level, and then to perform the statistical analysis at protein level. The second one is to directly work at peptide level, since peptides are the entities directly analyzed by the mass spectrometer. Although the first approach is mainstream for historical reasons, it is now well-established that, at least in theory but not only, peptide-level processing is more reliable [6]. Therefore, we have extended Prostar workflows, which used to be protein-centric [7], to propose convenient strategies to analyze proteomics data at peptide level, either in a peptidomic fashion; or by postponing the peptide-to-protein aggregation step. Although the user experience is largely unchanged throughout the various versions of Prostar, this protocol is best suited to version 1.24.X and later (*see* **Note 1**).

2 Material

2.1 Live Demo Mode

Before installing the software, any user can have a quick overview by testing its demo mode on the following URL: http://live.prostar-proteomics.org. This mode provides a direct access to the `DAPAR-data` package, where some toy datasets are available, either in tabular or MSnset formats [8]. Concretely, the demo mode is accessible in the Data manager menu: on the corresponding tab, a dropdown-menu lists the datasets that are available through

DAPARdata. After selecting a peptide-level dataset, click on Load demo dataset.

Any user can also test the website version on his own data, yet we do not recommend it since the server has limited computational capabilities that are shared between all the users connected at the same moment. Overloading is therefore possible, which would lead to data loss.

2.2 Hardware Requirements

Prostar can either be installed on a desktop machine (local installation by the user) or a server. The present protocol focuses on the former install. For the latter one, we refer to the DAPAR and Prostar user manual [9]. Depending on the data size, a recent workstation is necessary (we advise a minimum of 8GB of RAM, although there are no strict constraints).

2.3 Software Requirements

1. The operating system must either be Linux, Mas OS X, or Windows.
2. A recent version of the R software (*see* **Note 2**) must be installed in a directory where the user have the read and write permissions.
3. Optionally, an IDE (Integrated Development Environment) such as R Studio [10] may be useful to conveniently deal with the various R package installs.

2.4 Software Install

Prostar can be installed following two ways:

1. The "zero-install," which is the easiest way, as it does not need a prior install of R. However, it is so far only available for Microsoft Windows machines. "Zero-install" is a portable version of Prostar without requiring any installation. It can directly be downloaded from: http://prostar-proteomics.org/#zero-install, as a zip file referred to as `Prostar_1.24.x.zip`. Unzip it into a directory with read/write permissions (*see* **Note 3**).
2. The stand-alone Bioconductor install, which is the standard method to install Bioconductor distributed software. R must be already installed. First, install Bioconductor package manager, and then Prostar, by copy-paste of the following commands:

```
install.packages("BiocManager")
BiocManager::install(version=BiocManager::version())
BiocManager::install("Prostar")
```

This will install Prostar together with all the required dependency packages.

2.5 Data Type

The quantitative data should fit into a matrix-like representation where each line corresponds to a peptide and each column to a sample. Within the (*i*th, *j*th) cell of the matrix, one reads the abundance of peptide *i* in sample *j*.

2.6 Data Size—Number of Peptides

Although strictly speaking, there is no lower or upper bound to the number of lines, it should be recalled that the statistical tools implemented in Prostar have been chosen and tuned to fit a discovery experiment dataset with large amount of peptides, so that the result may lack of reliability on too small datasets. Conversely, very large datasets are not inherently a problem, as R algorithms are well scalable, but one should keep in mind the hardware limitations of the desktop machine on which Prostar runs to avoid overloading.

2.7 Data Size—Number of Samples

As for the number of samples (the columns of the dataset), it is necessary to have at least 2 conditions (or groups of samples) as it is not possible to perform relative comparison otherwise. Moreover, it is necessary to have at least 2 samples per condition (*see* **Note 4**), as otherwise, it is not possible to compute an intra-condition variance, which is a prerequisite to numerous processing.

2.8 Data Format

The data table should be formatted in a tabulated file where the first line of the text file contains the column names. It is recommended to avoid special characters such as "]", "@", "$", "%", etc. that are automatically removed. Similarly, spaces in column names are replaced by underscore ("_"). Dot (".") must be used as decimal separator for quantitative values. In addition to the columns containing quantitative values (*see* Subheading 2.7), the file may contain additional columns for metadata. Prostar supports any tabular file but is directly compatible with MaxQuant [11] and Proline [12] files (*see* Chapter 4 for a protocol on Proline use). Alternatively, if the data have already been processed by Prostar and saved as an MSnset file [8], it is possible to directly reload them (*see* **Note 5**).

3 Methods

3.1 Starting Prostar

1. Prostar is launched differently depending on how it was installed:
   (a) **Zero-install**: the unzipped folder contains an executable file (`Prostar.exe`) which directly launches Prostar in a webpage on the default internet navigator. At first launch, the latest version of Prostar is automatically downloaded from Bioconductor and silently installed. Once done, the user is invited to close the current page. Then, a new webpage with Prostar is automatically opened.

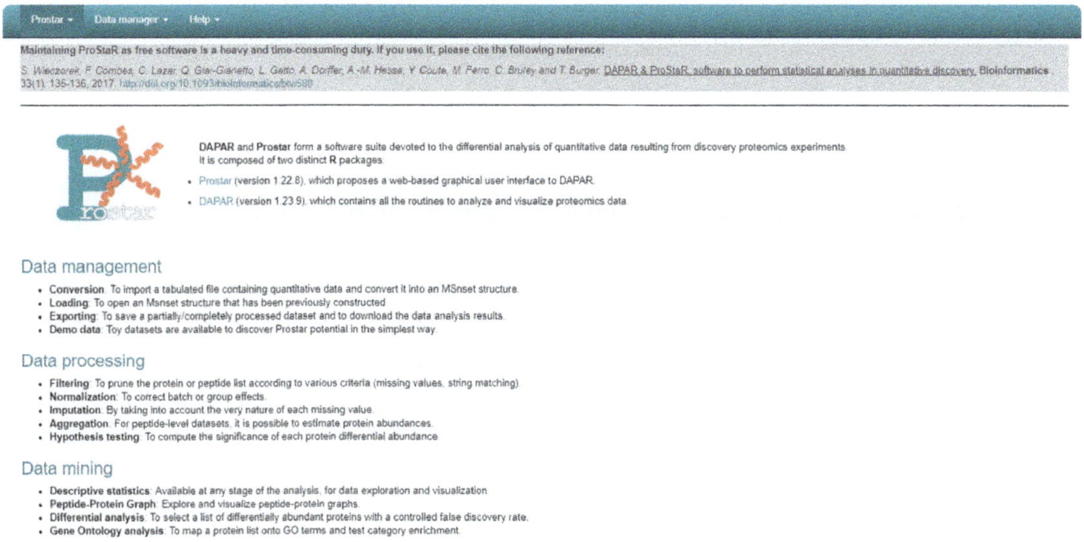

Fig. 1 Prostar welcome page

(b) **Bioconductor install**: run the following commands in a R console (Prostar will open in a new webpage):

```
library(Prostar)
Prostar()
```

2. Once Prostar is launched, it displays its welcome page (*see* Fig. 1). Since no dataset is loaded, the main menu at the top of the page contains only the following items:
   (a) `Prostar`: It contains some information about versions of Prostar and release notes.
   (b) `Data manager`: It contains the different tools to load or convert a new dataset and export a dataset analyzed by Prostar.
   (c) `Help`: It gathers FAQ and a bug report form.
3. Once a peptide-level dataset has been loaded in Prostar (*see* Subheading 3.2), the content of the menus changes:
   (a) the three Import tools (`Open MSnSet`, `Convert`, and `Demo mode`) from the menu `Data Manager` are hidden.
   (b) Two items appear in the main menu: `Data processing (peptide)` and `Data mining`.
4. If one wants to change the current dataset, reloading Prostar first is necessary to avoid cache memory issues. Restarting Prostar can be done in two ways:
   (a) Close the current webpage then restart Prostar as described previously (*see* Subheading 3.1).

(b) Restart the R session. For this, go to Data manager » Reload Prostar. Then, click on the Reload Prostar. This action will restart Prostar with a fresh R session in which import options are enabled in the Dataset manager menu.

3.2 Data Loading

If the dataset under consideration is not a peptide dataset (each line of the quantitative table does not represent a peptide, but, for instance, a protein), do not apply the present protocol. If it is a protein dataset, please refer to [7]:

1. To upload data from tabular file (*see* **Notes 6** and **7**) (i.e., stored in a file with one of the following extensions: `.txt`, `.csv`,`.tsv`, `.xls`, or `.xlsx`), go to the upper menu and click on Data manager » Convert data.
2. Go to the Select File tab.
3. Choose the software used to produce the quantitative dataset (*see* **Note 8**).
4. Click on Browse... and select the tabular file of interest.
5. Once the upload is complete, indicate it is a peptide dataset (i.e., each line of the data table should correspond to a single peptide).
6. Indicate whether the data are already log-transformed or not. If not, they will be automatically log-transformed (*see* **Note 9**).
7. If the quantification software uses "0" in places of missing values, tick the last option Replace all 0 and NaN by NA (as in Prostar, 0 is considered a value, not a missing value).
8. Move on to the Data Id tab.
9. If the dataset already contains an ID column (a column in which each cell has a unique content that can serve as an ID for the peptides), select the appropriate column name in the dropdown-menu ID definition. In any case, it is possible to use the Auto ID option, which creates an artificial index.
10. (OPTIONAL) In protein IDs (dropdown-menu), choose the column in the dataset that contains the IDs of proteins to which each peptide belongs to (*see* **Note 10**). As multiple computations of the peptide-level proteomic pipeline are based on this information, it is important to select the correct column. However, if a peptidomic analysis is anticipated (i.e., without peptide-to-protein aggregation), it is not important. To check the content of the column: when a value is selected, a small preview of the content is displayed.
11. Move on to the Exp. and feat. data tab.
12. Select the columns that contain the peptide abundances (one column for each sample of each condition). To select several

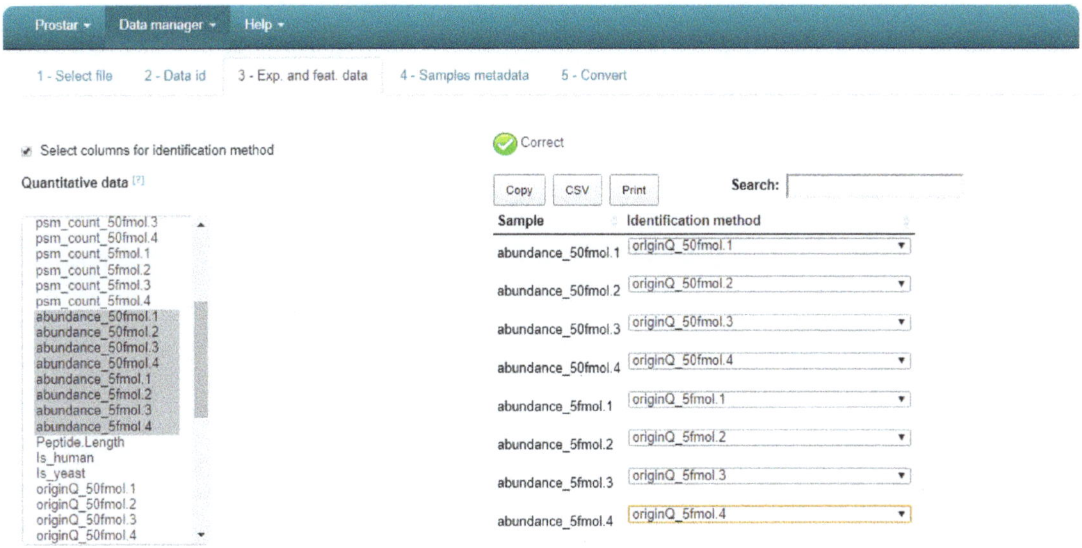

Fig. 2 Convert tool: Exp. and feat. data tab

column names in a row, click on the first one, and click-off on the last one. Alternatively, to select several names which are not continuously displayed, use the Ctrl key to maintain the selection.

13. If, for each sample, a column of the dataset provides information on the identification method, e.g., by direct MS/MS evidence, or by mapping (*see* **Note 11**), check the corresponding tick box. Then, for each sample, select the corresponding column. If none of these pieces of information is given, or, on the contrary, if all of them are specified with a different column name, a green logo appears, indicating it is possible to proceed (*see* **Note 12**) (*see* Fig. 2). Otherwise (i.e., the identification method is given only for a subset of samples, or a same identification method is referenced for two different samples), then a red mark appears, indicating some corrections are mandatory.
14. Move on to the Samples metadata tab. This tab guides the user through the definition of the experimental design.
15. Fill the empty columns with as different names as biological conditions to compare (minimum 2 conditions and 2 samples per condition, *see* Subheading 2.6) and click on Check conditions. If necessary, correct until the conditions are valid. When achieved, a green logo appears and the sample are reordered according to the conditions.
16. Choose the number of levels in the experimental design (either 1, 2, or 3), and fill the additional column(s) of the table (*see* **Note 13**).

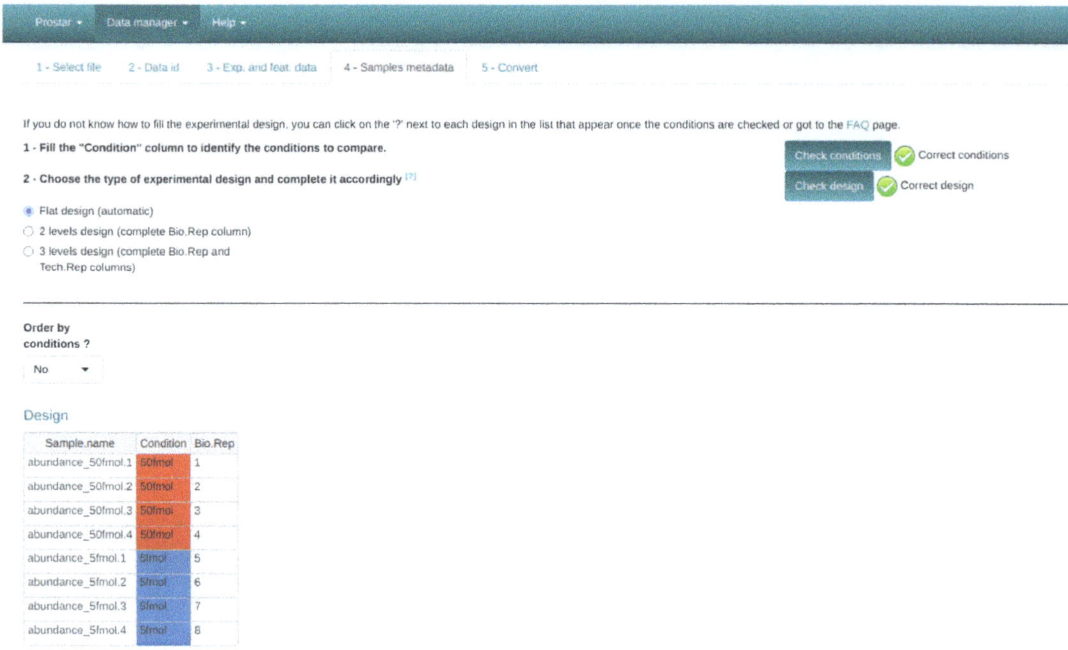

Fig. 3 Convert tool: Samples metadata tab

17. Once the design is valid, a green check logo appears (*see* Fig. 3). Then, move on to the Convert tab.
18. Provide a name to the dataset to be created and click on the Convert button. This step may take some time as Prostar computes all relationships between peptides and their proteins (to anticipate on the connected component exploration, *see* Subheading 3.5; as well as on the peptide-to-protein aggregation, *see* Subheading 3.11).
19. As a result, a new MSnset structure is created and automatically loaded (this is why Prostar returns to the welcome page). This can be checked with the name of the file appearing in the upper right-hand side of the screen, as a title to a new dropdown-menu. So far, it only contains Original – peptide, but other versions of the dataset will be added along the course of the processing.

3.3 Data Export

As importing a new dataset from a tabular file is a tedious procedure, we advise to save the dataset as an MSnset binary file right after the conversion. This makes it possible to restart the statistical analysis from scratch if a problem occurs without having to convert again the data. To do so:

1. Click on Data manager » Export.
2. Choose `MSnset` as file format and provide a name to the object (*see* **Note 14**).

Fig. 4 Descriptive statistics: metadata tab

3. Click on Download.
4. Once downloaded, store the file in the appropriate directory.
5. It is possible to reload any dataset stored as an `MSnset` structure (*see* **Note 5**).

3.4 Descriptive Statistics

By clicking on Data mining ›› Descriptive statistics, it is possible to access several tabs generating various plots (*see* **Note 15**) that provides a comprehensive and quick overview of the dataset (*see* **Note 16**):

1. On the first tab (overview), a brief summary of the quantitative data size is provided. It roughly amounts to the data summary that is displayed along with each dataset during the loading step of the demo mode.
2. On the second tab (Quantification nature), the barplots depict the distribution of the so-called cell metadata (*see* **Note 17** and Fig. 4). The user selects the label to focus on; once done, three barplots are displayed. The left-hand side barplot represents the number of peptides with the corresponding label in each sample. This number is written in the tooltip when the user hovers the mouse pointer over the bar along with the percentage of whole peptides this value corresponds to. The different colors correspond to the different conditions (or groups, or labels). The second barplot (in the middle) displays the distribution of the label of interest. The last barplot represents the same information as the previous one, yet, condition-wise.
3. The third tab (Data explorer) is dedicated to the visualization of data tables (*see* Fig. 5): it makes it possible to view the content of the MSnSet structure. It is made of three tables, which can be displayed one at a time thanks to the radio button on the left menu. The first one (Quantitative data) contains quantitative values. The missing values are represented by empty cells; a

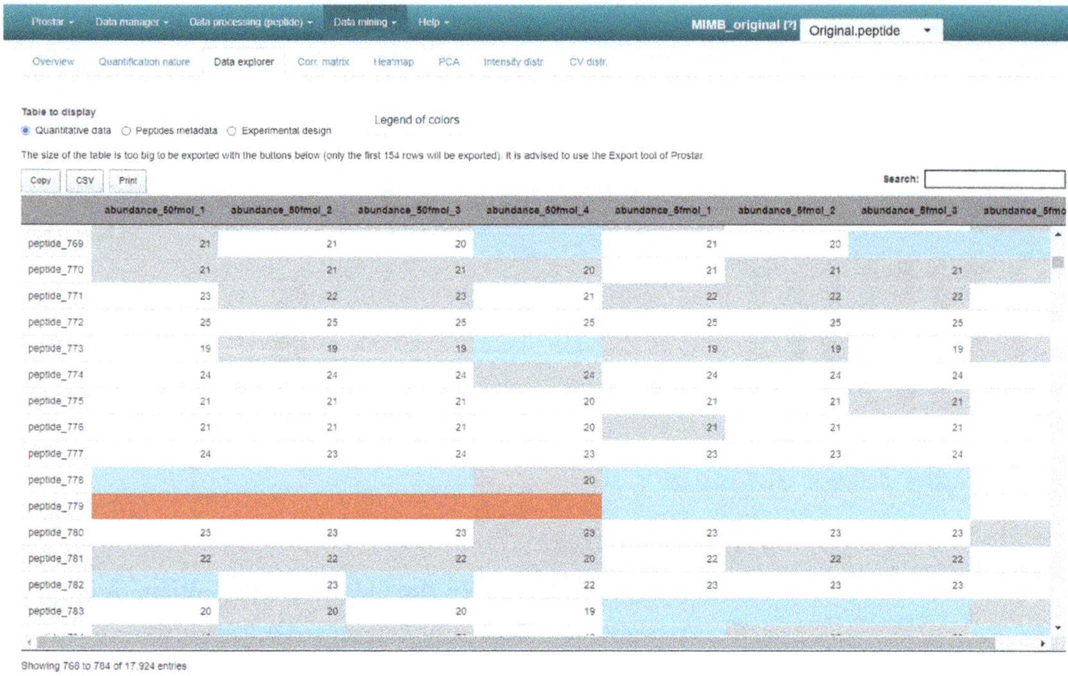

Fig. 5 Descriptive statistics: Data explorer tab

legend of the colors indicates the type of quantification metadata. The second one (Peptides metadata) contains all the column of the dataset that are not the quantitative data. The third tab (Experimental design) summarizes the design, as defined at the import step (*see* Subheading 3.2, **Step 16**).

4. In the fourth tab (Corr. matrix), it is possible to visualize to what extent the replicate samples correlates or not. The contrast of the correlation matrix can be tuned thanks to the color scale on the left-hand side menu.

5. A heatmap as well as the associated dendrogram is depicted on the fifth tab. The colors represent the intensities: Red for high intensities, green for low intensities, and white for missing values. The dendrogram shows a hierarchical classification of the samples, so as to check that they are related according to the experimental design. It is possible to tune the clustering algorithm (*see* **Note 18**) that produces the dendrogram by adjusting the distance and linkage parameters, as described in the `hclust()` function of the R package `stats` [13].

6. Tab 6 (PCA) shows different plots related to the Principal Component Analysis (*see* Fig. 6). The plots are displayed only if the dataset does not contain any missing values.

7. Tab 7 represents in various ways the same information, that is the distribution of intensity values by replicates and conditions:

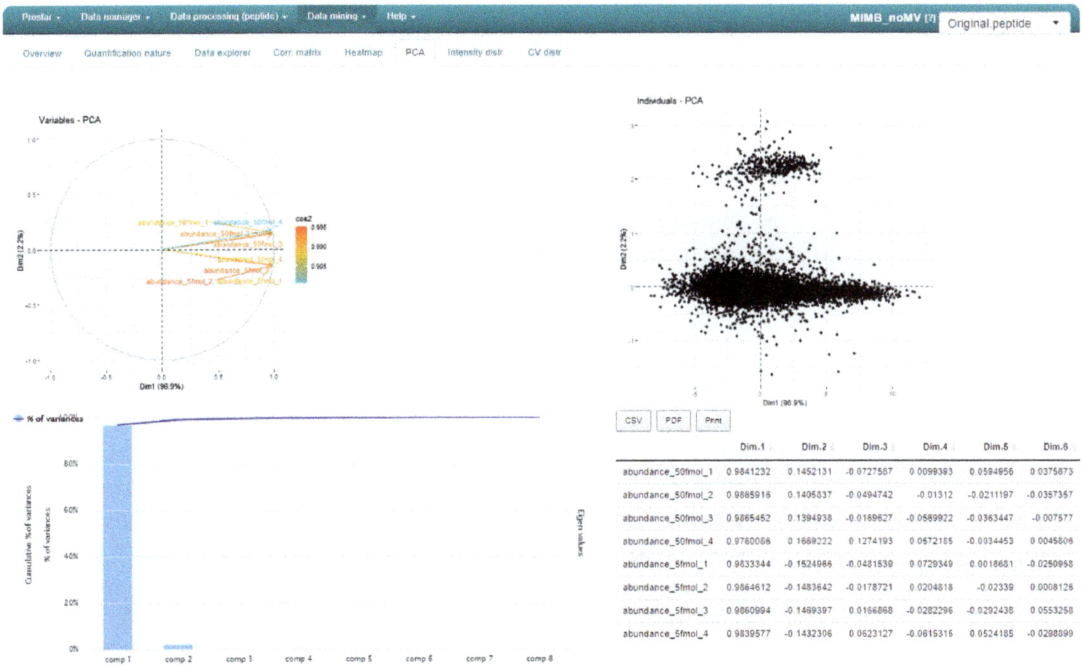

Fig. 6 Descriptive statistics: PCA tab

respectively, smoothed histograms (a.k.a. kernel density plots), boxplots, and violin-plots are used.

8. Finally, the last tab displays a density plot of the variance (within each condition) conditionally to the log-intensities (*see* **Note 19**).

3.5 Peptide-Protein Graph

Clicking on Data mining > Peptide-protein Graph gives access to a graph visualization tool displaying the relationships between the peptides and the protein(s) they belong to (*see* **Note 20**). Peptides and proteins are depicted as nodes and an edge connects a (protein, peptide) couple when the peptide belongs to the protein. The set of all possible graphs in the dataset is computed when a new dataset is loaded in Prostar (*see* Subheading 3.2, **Step 18**). This visualization tool is divided into three tabs, so as to distinguish three types of graphs:

1. One-One Connected Components (One-One CC): The table on the left hand displays all the One-One CCs, i.e., graphs containing a single protein linked to its unique (and specific) peptide (i.e., it is not shared with other proteins) (*see* **Note 21**). By selecting an item in this table, it is possible to view the quantitative values of the peptide in consideration in the table on the right hand (*see* **Note 22**). Note that due to their large number and to limit memory load, One-One CCs are not displayed.

2. One-Multi Connected Components (One-Multi CC): This interface is the same as for One-One CCs (as described above) but displays graphs containing a single protein linked to several (specific) peptides (i.e., not shared with other proteins). Note that due to their large number and to limit memory load, One-Multi CC are not displayed either.

3. Multi-Multi Connected Components (In Multi-Multi CCs, a peptide may be specific to a protein or shared between several proteins): The dropdown-menu Search for CC let the user choose if he wants to see the list of Multi-Multi CCs as a table (`"Tabular view"`) or a plot (`"Graphical view"`); each of these is displayed on the left-hand side. The table is the same as for One-One CCs and One-Multi CCs. In the graphical view, each point represents a CC where the coordinates are the number of peptides as x-axis and the number of proteins as y-axis. The selection of a particular CC is possible by clicking on the corresponding point in the plot. Once a CC has been selected, the graph is plotted on the right-hand side (*see* Fig. 7): The medium-size nodes correspond to proteins while the small nodes are for peptides (specific peptides are blue while shared peptides are green). The three tables below show information about the nodes that have been selected by clicking on them in the graph (*see* Fig. 8). The first table named `"Proteins"` lists the Ids of the proteins in the CC, while the two other tables show the quantitative values of, respectively, `"Specific proteins"` and `"Shared proteins"`. By default, all nodes are selected. Clicking on a particular node of the graph selects it as well as all of its neighbors (the other nodes—peptides or proteins—it connects to) and updates the tables below. The text-field Peptide Info allows to choose among the peptide metadata,

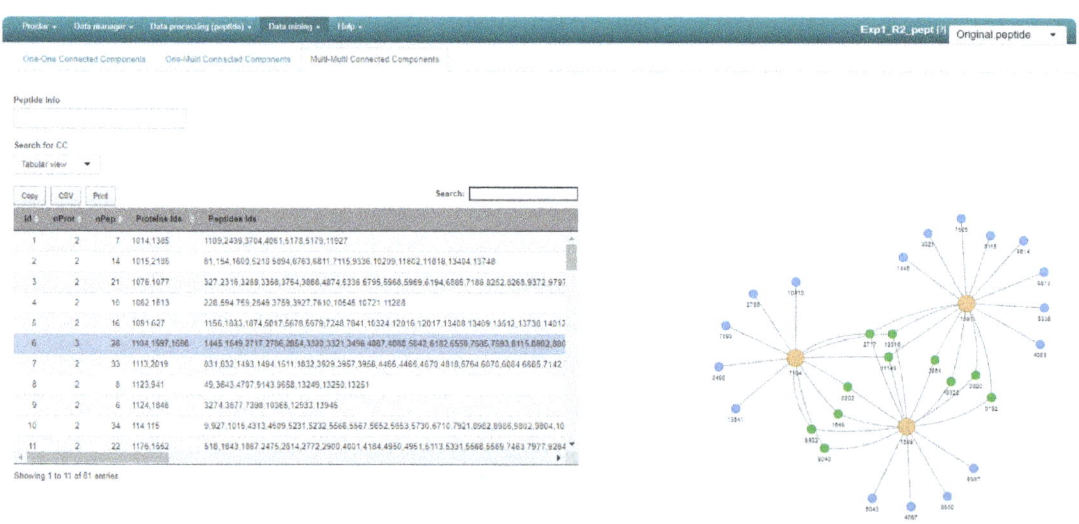

Fig. 7 Peptide-Protein Graph: Zoom on the upper part of the Multi-Multi Connected Components tab

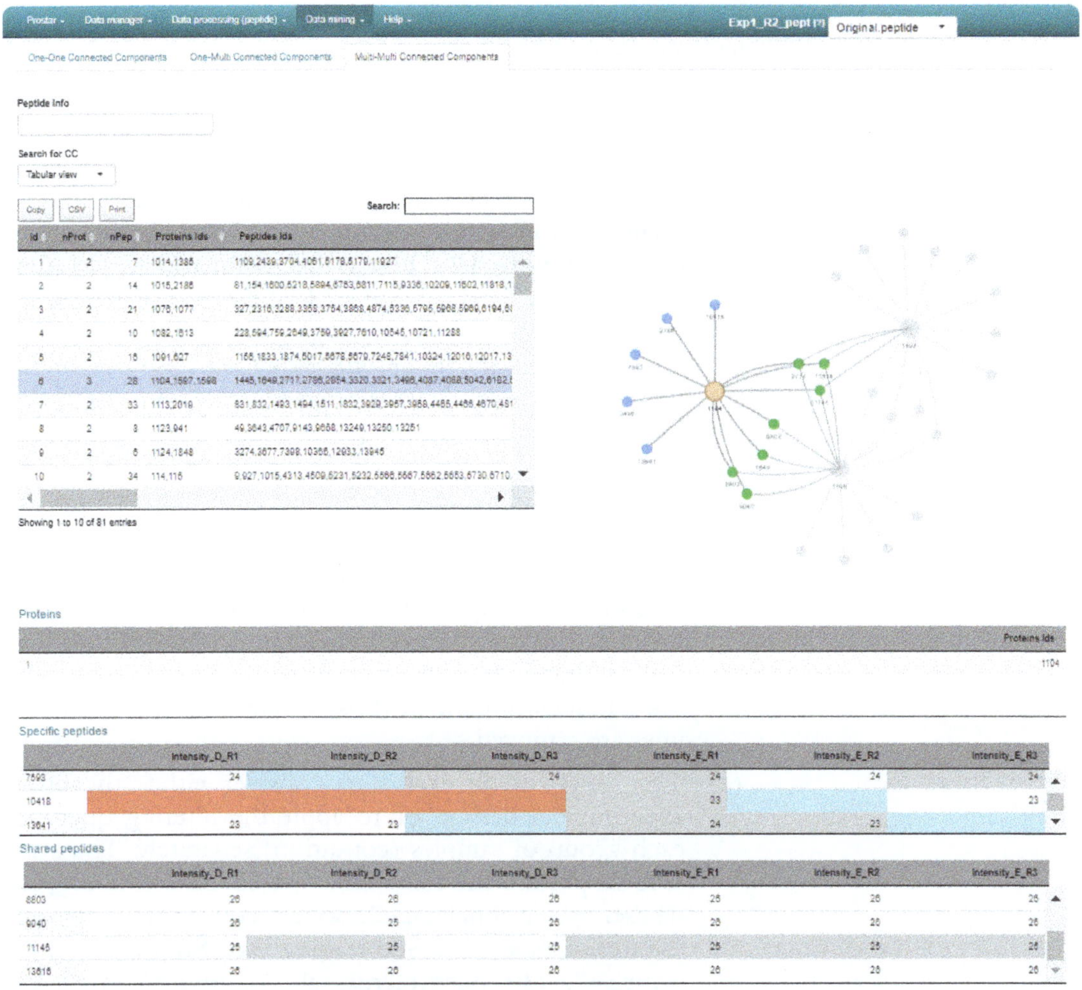

Fig. 8 Peptide-Protein Graph: Multi-Multi Connected Components tab

the one to add in the tables "`Specific proteins`" and "`Shared proteins`" beside the quantitative values.

3.6 Filtering

The first processing tool of the peptidomics (or of the peptide-level proteomics) pipeline gives the opportunity to filter out the data according to some information stored in the quantitative data or peptides metadata. Each filtering tool is based on the same behavior: the user builds a query (by choosing its parameters) that selects some lines in the dataset, to keep them or on the contrary, to remove them. Multiple filters can be run successively; in this case, the ith filtering is run on the dataset resulting from the $(i-1)$th filtering:

1. Click on Data processing » Filter data.
2. The first tab (Quanti. metadata filtering) allows to filter peptides according to the nature of the quantitative values (stored in the so-called cell metadata, *see* Subheading 3.4, **Step 2**):
   (a) Choose the data nature in the left-hand dropdown-menu (Nature of data to filter). The possible values correspond to the quantification nature metadata such as described previously (*see* **Note 17**). It may be useful to keep in mind the hierarchy of the associated labels to understand the scope of filtering. For example, if "`missing POV`" is selected, then the filter will apply only on "`missing POV`" but if "`missing`" is selected, then all missing data are concerned ("`missing POV`" as well as "`missing MEC`").
   (b) The next dropdown-menu (Type of filter operation) allows to choose whether the lines identified by the query are to be kept or deleted. If a given set of *n* lines is removed from a dataset containing *N* lines, then the resulting dataset will contain *N-n* lines. On the contrarily, if *n* lines are kept from an initial dataset containing *N* lines, then the resulting dataset will contain those *n* lines and the other *N-n* lines are removed.
   (c) The Scope menu allows to consider either the entire experimental design, or to apply the filtering query on each group of samples (condition) separately. The available scopes are the following: "`None`": No filtering, the dataset is left unchanged, so directly go to next step String-based filtering (*see* Subheading 3.6, **Step 3**). "`Whole matrix`": The lines (across all conditions) which cell metadata contain a certain number of labels fitting with the user-defined formula are flagged. "`Whole line`": lines containing only the labels referred to in the queries are concerned. This is a special case of the "`Whole matrix`" scope leading to shortcut queries (e.g., to straightforwardly remove lines with only missing values). "`Every condition`": The concerned lines are those for which each condition contains a certain number of cells with the label of interest. "`At least one condition`": The concerned lines are those for which at least one condition contains a certain number of cells with the label of interest.
   (d) Whenever a scope is selected (except for the "`Whole line`" option), a series of three dropdown-menus is displayed to refine the query by setting conditions on the number of labels: Nature of the threshold allows to choose whether to select lines on the basis of an absolute

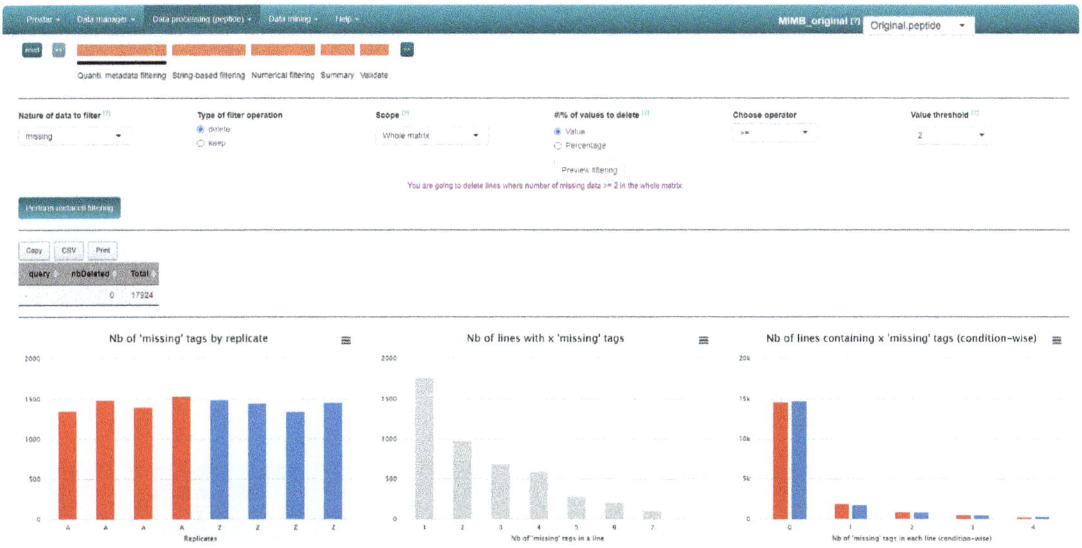

Fig. 9 Snapshot of the filtering module

counting, or on a percentage (*see* **Note 23**). Then, Operator defines the algebraic operator used to select lines associated with a threshold value (Threshold). If the threshold value is a percentage, the value must be a decimal number between 0 and 1. If it is a count value, the dropdown-menu proposes a list of several integers depending on the chosen scope: If the scope is equal to "`Whole matrix`", then the range of the threshold is from 0 to the total number of samples. If the scope is "`For each condition`" or "`At Least One Condition`", the possible values are the integers between 0 and the minimum amount of samples in all the conditions (i.e., the number of samples in the "smallest" condition) (Fig. 9).

(e) Clicking on Preview filtering button displays a popup window showing the quantitative table of a random subset of the dataset (*see* Fig. 10). This can be useful to better understand the behavior of the filters.

(f) Visualize the effect of the filtering options without changing the current dataset by clicking on Perform filtering; the barplots are the same as in the tab Descriptive Statistics › Quantification nature (*see* Subheading 3.4). If the filtering does not produce the expected effect, test another one. To do so, simply choose another method in the list and click again on Perform filtering. The plots are automatically updated. This action does not modify the dataset but offers a preview of the filtered data. Iterate this step as long as necessary. Each time a preview is run, a short summary is added in the table above the plots.

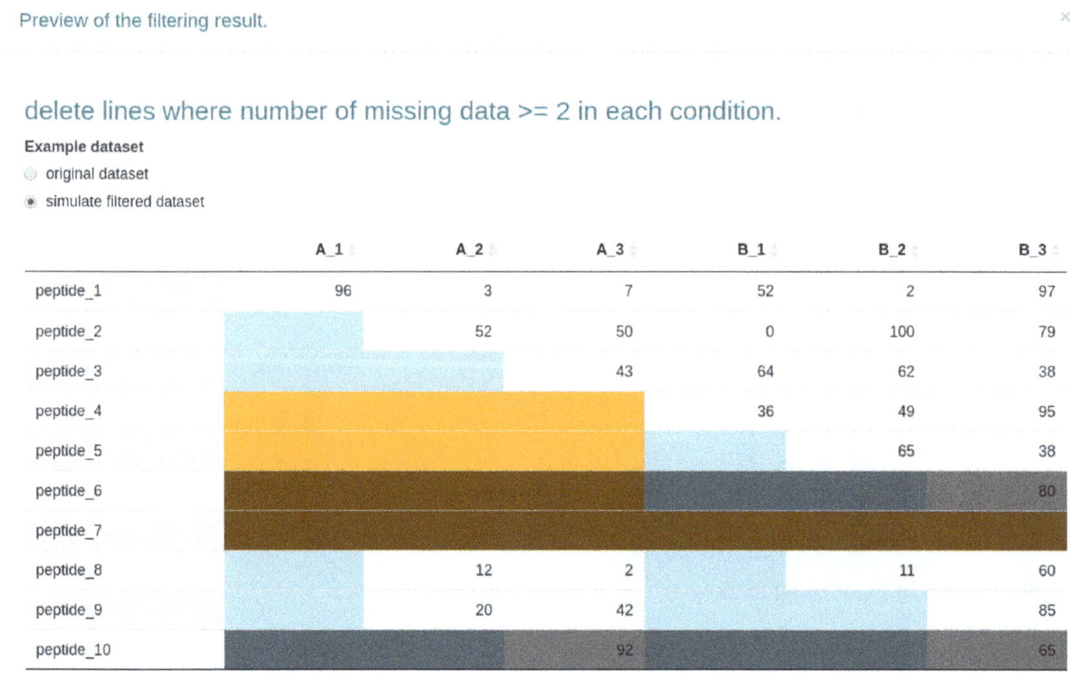

Fig. 10 Simulate filtering on a toy example: The query objective is to "delete lines where there are more than 2 missing values in each condition"

3. Move on to the second tab (String-based filtering), where it is possible to filter out peptides according to information stored as alpha-numerical strings in the peptide metadata (*see* **Note 24**):

   (a) Among the columns constituting the peptide metadata listed in the dropdown-menu, select the one containing the information (*see* **Note 25**) of interest, e.g., "`Contaminant`" or "`Reverse`" (*see* **Note 26**). Then, specify in each case the prefix chain of characters that identifies the peptides to filter (*see* **Note 27**).

   (b) Click on Perform to remove the corresponding peptides. A new line appears in the table listing all the filters that have been applied (*see* Fig. 9).

   (c) If other filters are necessary, iterate the above sub-steps.

4. Move on to the third tab (Numerical filtering), where it is possible to filter out peptides according to information stored as numerical values in the peptide metadata:

   (a) As for the filter on the first tab, fill the dropdown-menus to build a query that select the lines concerned by the filter. The first one lists the name of the columns in the metadata and tries to guess the one with numerical values.

(b) Click on Perform to remove the corresponding peptides. A new line appears in the table listing all the filters that have been applied.

(c) If other filters are necessary, iterate the above sub-steps.

5. Once all the filters have been applied, move on to the last tab (Visualize and Validate) to check the set of filtered out peptides. This visualization tools works similarly as the Data explorer (*see* Subheading 3.4, **Step 3**).

6. Finally, click on Save filtered dataset. The new dataset is accessible and referred as `"Filtered.peptide"` in the version dropdown-menu (*see* Subheading 3.7).

3.7 Navigating Through the Dataset Versions

Once the filters have been applied and the results saved, a new dataset is created, in the same way as after the data conversion (*see* Subheading 3.2, **Step 18**). It is referred to as `"Filtered--peptide"`, and its name appears right below `"Original--peptide"` in the upper right drop-down menu, beside the dataset name (as illustrated in upper right corner of Fig. 5). Unless modified, the newest created dataset is always the current dataset, i.e., the dataset on which further processing will be applied. As soon as the current dataset is modified, all the plots and tables in Prostar are automatically updated. Thus, as soon as a new dataset is created, we suggest to go back to the descriptive statistics menu (*see* Subheading 3.4) to check the influence of the latest processing on the data. It is possible to have a dynamic view of the processing steps by navigating back and forth in the dataset versions, so as to see the graphic evolutions (*see* **Note 28**).

3.8 Normalization

The next processing step proposed by Prostar is data normalization (*see* Fig. 11). Its objective is to reduce the biases introduced at any preliminary stage (such as batch effects):

1. Choose the normalization method available in the dropdown-menu (Normalization methods). For each possible normalization, a short description is provided and the interface is automatically updated to display the method parameters that must be tuned (*see* **Note 29**).

(a) `None`: No normalization is applied.

(b) `Global quantile alignment`: The Quantile of the intensity distributions of all the samples are equated.

(c) `Sum by columns`: The total (un-logged) intensity values of all the samples are equated. The rationale behind is to normalize according to the total amount of biological material within each sample.

(d) `Quantile Centering`: A given quantile of the intensity distribution is used as reference, such as the median, or a

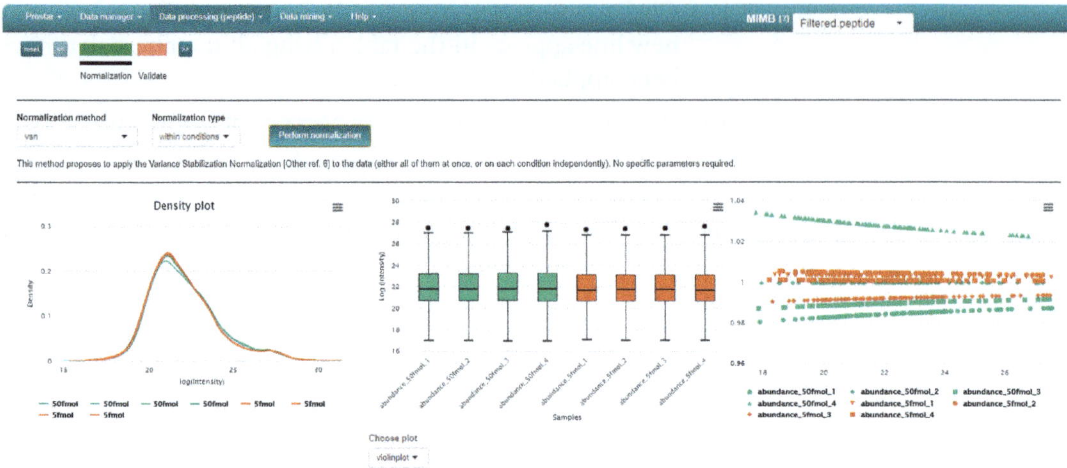

Fig. 11 Normalization tab: The intensity density plot, the violon plot (alt. box plot) and the distortion plot make it possible to visualize the influence of each normalization method

lower quantile depicting the lower limit of detection (*see* **Note 30**).

(e) `Mean Centering`: sample intensity distributions are aligned on their mean intensity values (and optionally, the variance distributions are equated to one).

(f) `LOESS`: The intensity values are normalized by means of a local regression model [14] of the difference of intensities as function of the mean intensity value (*see* [15] for implementation details).

(g) `vsn`: Variance Stabilizing Normalization, which wraps the method described in [16].

2. Possibly, indicate the normalization type (available in the dropdown-menu Normalization type). Notably, for most of the methods (in fact, all of them, but "`Global quantile alignment`"), it is necessary to indicate whether the method should apply to the entire dataset at once ("`overall`" option), or whether each condition should be normalized independently of the others ("`within conditions`" option).
3. For other parameters, which are specific to each method, the reader is referred to Prostar user manual, available through the Help menu (*see* **Note 31**).
4. Click on Perform normalization.
5. Observe the influence of the normalization method on the plots of the lower side panel. The middle graph can be switched from violin plot to box plot.

6. If the result of the normalization does not correspond to the expectations, first return to the initial situation by clicking on reset and change the normalization method (or change its tuning).
7. Once the normalization is effective, move on the Validate tab and click on Save normalization. Check that a new version appears in the dataset version dropdown-menu, referred to as "`Normalized.peptide`".

3.9 Missing Values Imputation

Prostar distinguishes two different types of missing values: POV (standing for Partially Observed Value) and MEC (standing for Missing in the Entire Condition). All the missing values for a given peptide (or protein in case of a protein dataset) in a given condition are considered POV if there is at least one observed value for this peptide in this condition. Alternatively, if all the intensity values are missing for this peptide in this condition, the missing values are considered MEC (*see* **Note 32**). At peptide level, several algorithms are proposed for the imputation of POV while the imputation of MEC is left as an option:

1. On the Peptide imputation tab (*see* Fig. 12), select the algorithm to impute missing values in the Algorithm dropdown-menu.
2. If missing values are not going to be imputed at all, select "`None`." However, this option is not recommended since it may prevent subsequent hypothesis testing (*see* Subheading 3.13) as well as strongly impact the peptide-to-protein aggregation procedure (*see* Subheading 3.11).
3. If "`imp4p`" is selected, tune the number of iterations (default is 10). We strongly recommend the imp4p algorithm as it performs a diagnosis of the nature of each POV missing value, while it is not the case with the less refined "`Basic Methods`". Indeed, missing Value can be diagnosed as MNAR (Missing Not at Random) or MCAR (Missing Completely At Random) [17] (*see* **Note 33**). The more iterations there are the more precise the imputation is. You can choose to impute the MEC values at this step (MECs are always considered MNAR missing values) by checking Impute MEC also (*see* **Note 34**).
4. If MEC imputation is decided, tune the required parameters: The first parameter is the upper bound of the MNAR imputed values, expressed as a quantile of the distribution of observed values in each sample (default is centile 2.5). The second parameter is the probability distribution defined on the range of MNAR values, either "`uniform`" or "`beta`": The latter is more conservative, as it provides more weight on values that are close to the detection limit, which reduces the risk of creating artificial differentially abundant peptides (and thus proteins).

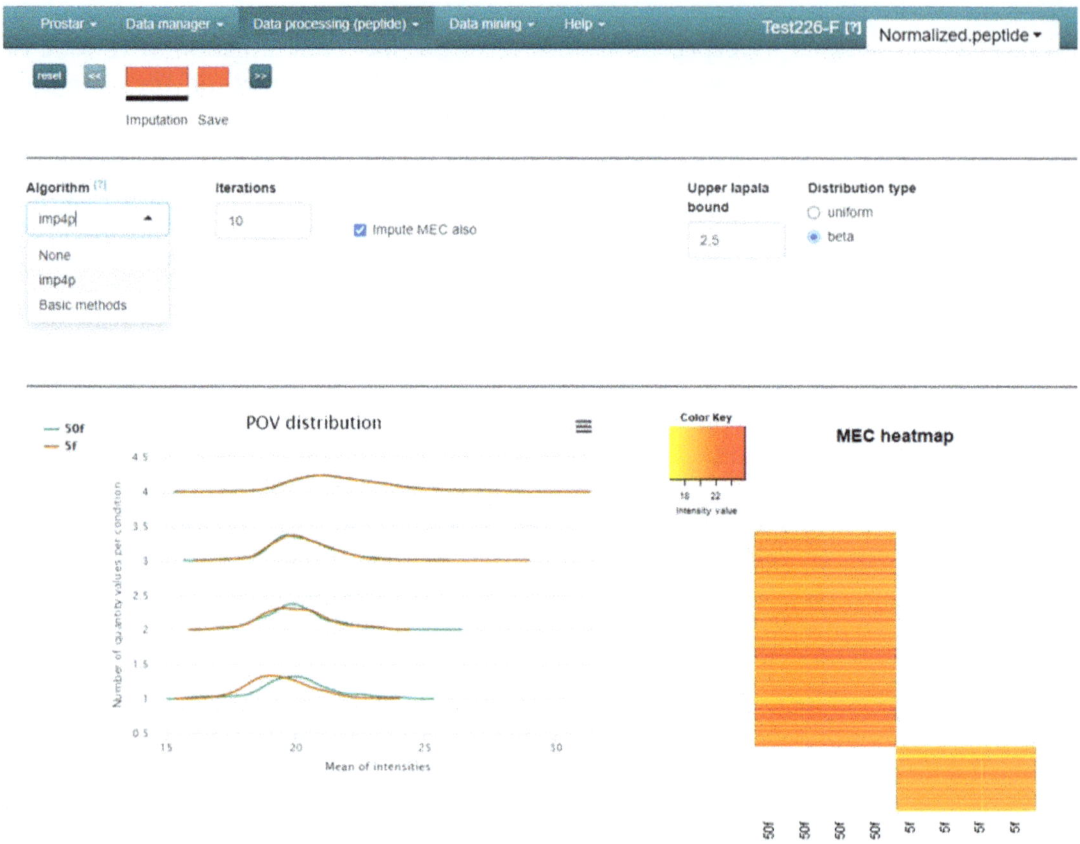

Fig. 12 Peptide imputation tab: Result of imputation of both the "Partially Observed Values" and the "Missing in the Entire Condition" missing values

5. If "`Basic Methods`" is selected, choose the appropriate method in the Methods dropdown-menu and tune its parameters. The three basic methods proposed are as follows and the user is invited to consult the manual for a detailed description. Be aware that these methods only deal with POV missing values (MECs are not imputed for now) and do not make any diagnosis of the nature of these missing values.
   (a) "`Det. quantile`": each missing value within a given sample is replaced by a deterministic value (usually a low value) (*see* **Note 35**). When using "`Det. quantile`", the list of imputation values for each sample appears above the graphics on the right panel.
   (b) "`KNN`": K-nearest neighbors estimates each missing value by the mean of the observed values of other peptides with a similar intensity pattern (called neighbors).
   (c) "`MLE`": Maximum Likelihood Estimation estimates the expected abundance value of each peptide in each condition and use this as replacement of missing values.

6. Click on Perform Imputation. If there are yet missing values in the imputed dataset (i.e., the MECs were not imputed), a message alerts the user that the aggregation of peptides into proteins may fail (*see* **Note 36**). If the user wants to avoid possible issues with further peptide-to-protein aggregation, it is necessary to clicks on Reset to reinitialize the imputation tool and impute its dataset with "`imp4p`" and the option "`Include MEC also`".
7. Observe the influence of the chosen imputation method on the plots of the lower hand side panel. If the result does not correspond to the expectations, change the imputation method (or its tuning) by clicking on reset and return to the first step of imputation.
8. Once the imputation is effective, move on the Save tab and click on Save imputation. Check that a new version appears in the dataset version dropdown-menu, referred to as "`Imputed.peptide`" (*see* **Note 37**).

3.10 Hypothesis Testing for Peptidomics Data

For peptide datasets that do not contain any missing values, or for those where these missing values have been imputed, it is possible to test whether each peptide is significantly differentially abundant between the compared conditions and next proceed to differential analysis:

1. Click on Data processing (peptide) › Hypothesis testing (*see* Fig. 13).
2. Follow the very same four steps as for hypothesis testing with protein datasets (*see* Subheading 3.13).
3. After saving, check that a new version appears in the dataset version dropdown-menu, referred to as "`HypothesisTest.peptide`".
4. Then, this new peptide dataset, containing the *p*-values and fold change (FC) cutoff for the desired contrasts, can be explored in the Differential analysis tabs available in the Data mining menu (*see* Subheading 3.14).
5. Follow the very same six steps as for differential analysis with protein datasets (*see* Subheading 3.14).
6. From the FDR tab, you may download data and save any plot/table of interest (*see* Fig. 14).

3.11 Aggregation

From a peptide dataset, it is possible to aggregate the peptide intensities to construct a new protein-level dataset. For each parent protein, aggregation can only apply to a series of non-missing values (yielding a numerical values) or to a series of missing values (yielding a missing value), but not to a combination of both (*see* **Note 38**). If missing value imputation was not performed before aggregation, a warning is displayed and the list of excluded peptides (and thus proteins) is provided to the user (*see* **Note 39**). To

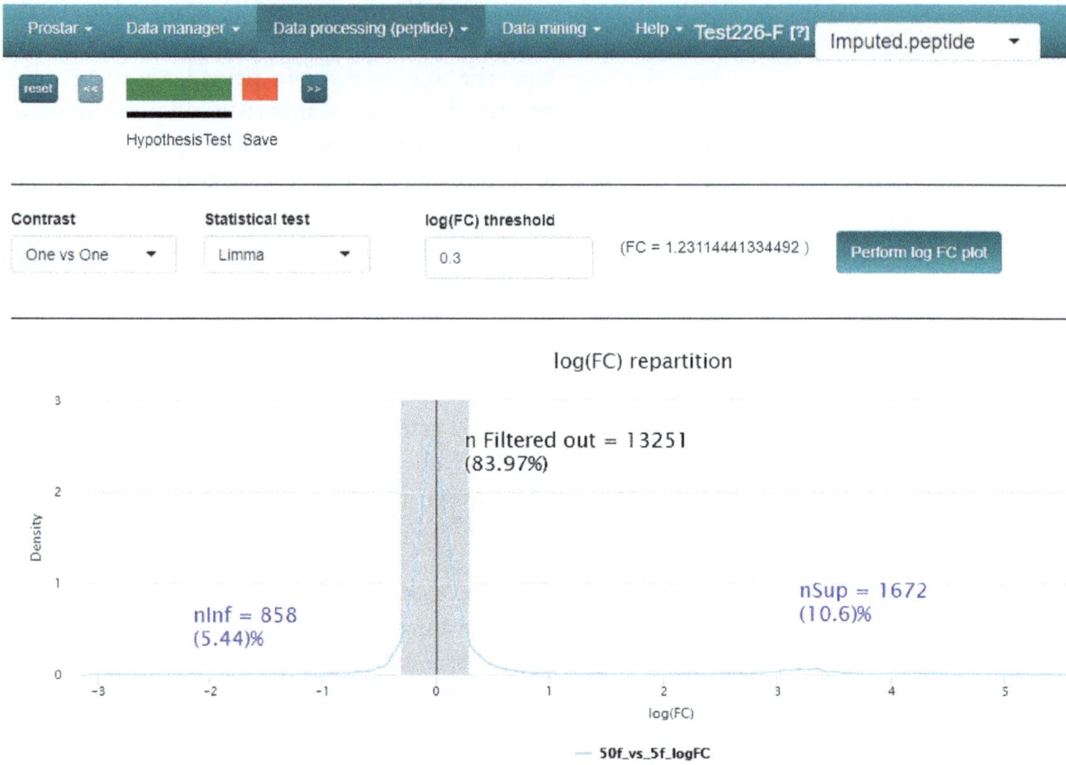

Fig. 13 Tuning of the null hypothesis significance testing (to prepare the differential analysis at peptide level)

simplify things in practice, an option is to systematically impute all the missing value upstream aggregation, however, it may lead to the imputation of peptides that the practitioner would not trust in the first place. To perform aggregation:

1. Click on the corresponding option in the Data mining › Aggregation menu. The lower panel provides two barplots depicting the protein distribution according to their number of peptides: either all of them (*see* Fig. 15, right-hand plot), or only those which are specific to a single protein (left-hand plot).
2. A key step is to decide whether and how to include shared peptides in the aggregation by clicking one of the three radio buttons under "`Include shared peptides`". Option "`No`" excludes shared peptides (*see* **Note 40**). Option "`Yes (as protein specific)`" processes shared peptides as if they were specific to each of their parent proteins (*see* **Note 41**). The option "Yes (redistribution)" computes a proportional redistribution of the intensity of shared peptides among proteins (*see* **Note 42**).

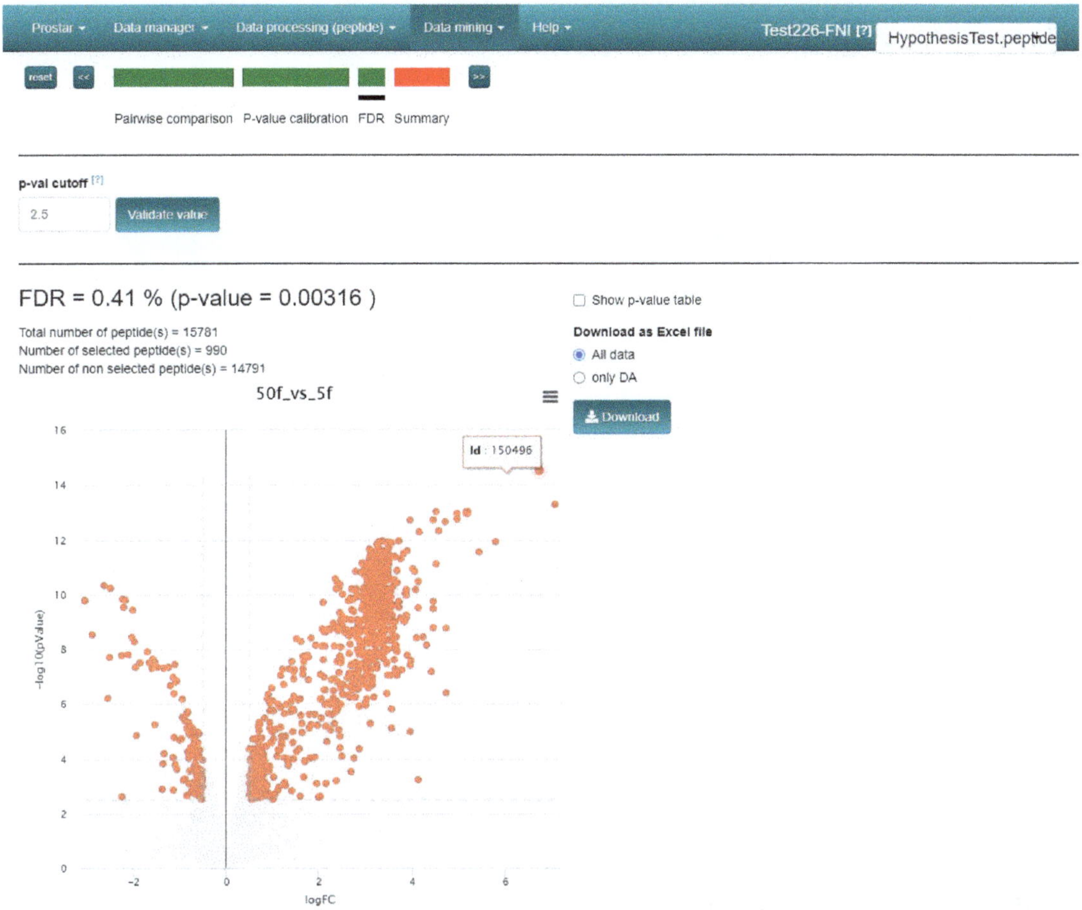

Fig. 14 Volcano plot to visualize the differentially abundant peptides according to a user-specified FDR

3. Choose whether to consider all peptides or only the first N most abundant ones for each protein by clicking the appropriate button under "`Consider`".
4. Choose the aggregation operator, by clicking either on sum or mean under "`Operator`". Note that for the redistribution option of the shared peptides, only the mean operator is allowed (*see* **Note 43**).
5. Click on Perform aggregation and wait for the message "`aggregation done`".
6. Move on to the next tab, Add metadata, and select the columns of the metadata table that must be aggregated to form protein-level metadata. This will create columns with the same headers as in the original peptide dataset. The result of metadata aggregation will be a comma-separated string concatenation of the unique values available at peptide level (duplicated values are

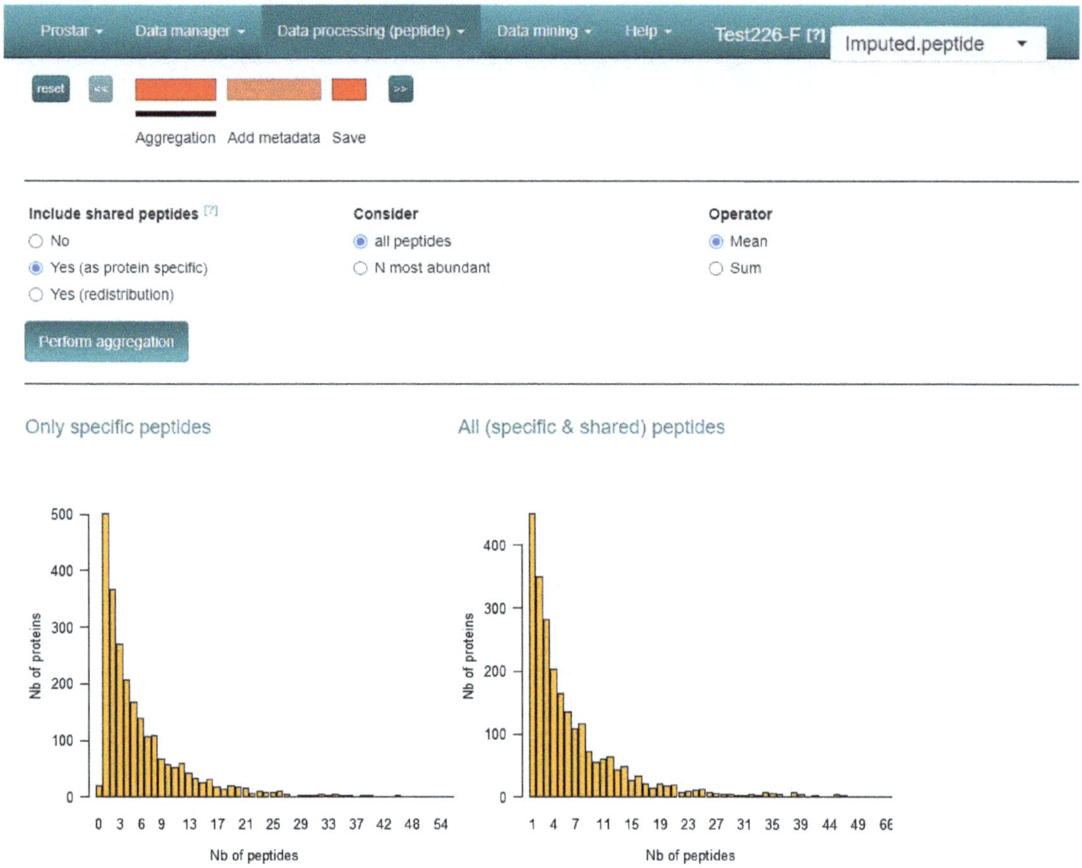

Fig. 15 Options in the Aggregation tab

not repeated). It is recommended to select at least the peptide identifier field, so that it remains possible to link protein and peptide information afterwards.

7. Move on to the Save tab and save the aggregation step. A protein dataset is created and loaded in memory for further processing. The interface automatically switches to the homepage. The new aggregated dataset is accessible and referred as "`Aggregated.protein`" in the version dropdown-menu.
8. At this stage, it may be wise to export the dataset as a MSnSet or excel file (*see* Subheading 3.3), in order to be able to carry out several different processes without having to start the aggregation step again (*see* **Notes 44** and **45**).
9. As hereafter described (*see* Subheading 3.12), the aggregated protein dataset can be explored (Descriptive Statistics menu), analyzed (Data mining menu) and processed (Data processing menu) very much the same way as a peptide dataset. However, the new dataset being a protein one, the aggregation process does not appear anymore in the Data processing menu.

3.12 Aggregated Protein Dataset Preprocessing

1. Additional filtering can be applied at the protein level if necessary. The menu displays the same interface and options as for the peptide level (*see* Subheading 3.6). In particular, if only specific peptides were accounted for during the aggregation (*see* **Note 46**), proteins with no specific peptide will thus have missing values in all conditions. It is thus necessary to filter these proteins as "`empty lines`" (*see* Subheading 3.6, **Step 2**) (*see* **Note 40**). If a protein-level filter is applied, saving is necessary (leading to a new dataset referred to as "`Filtered.protein`" in the version dropdown-menu).
2. Protein-level normalization can be applied if necessary. The menu displays the same interface and options as for the peptide level (*see* Subheading 3.8). If a protein-level normalization is applied, saving is necessary (leading to a new dataset referred to as "`Normalized.protein`" in the version dropdown-menu).
3. Depending on the options chosen during peptide imputation step (*see* Subheading 3.9), the aggregated protein dataset may still contain missing values, as previously detailed (*see* Subheading 3.11). This occurs when either no imputation was performed at peptide level; or when "`imp4p`" was applied without processing the MECs (*see* Subheading 3.9, **Step 3**); or when one of the "`Basic Methods`" was preferred. In such cases, imputation is a necessary step to proceed to hypothesis testing and differential analysis of the protein-level dataset. The interface and options for protein imputation slightly differ from those at peptide level and the user should refer to [7] for a step-by-step description (*see* **Note 37**). If a protein-level imputation is applied, saving is necessary (leading to a new dataset referred to as "`Imputed.protein`" in the version dropdown-menu).

3.13 Hypothesis Testing for Aggregated Protein Dataset

Hypothesis testing can be conducted regardless of the version of the protein dataset (either aggregated, filtered, normalized, or imputed) as long as it does not contain missing values. To do so, click on Data processing (protein) > Hypothesis testing. The steps are similar to those that can be applied to a peptide dataset as part of a peptidomics strategy (*see* Subheading 3.10):

1. Choose the test contrasts: In case of two conditions to compare, there is only one possible contrast. However, in case of N > 2 conditions, several pairwise contrasts are possible. Notably, it is possible to perform N tests of the "`1vsAll`" type, or N (N-1)/2 tests of the "`1vs1`" type.
2. Choose the type of statistical test, between limma [15] or t-test (either Welch or Student). This makes appear a density plot of the logarithmized fold-change (logFC) (as many density curves on the plot as contrasts).

3. Thanks to the logFC density plot, tune the logFC threshold (*see* **Note 47**). Refresh the plot to view the impact of the chosen threshold by clicking on Perform log FC plot. The corresponding logFC value is indicated for convenience.
4. Run the tests by clicking on Perform log FC plot, move forward to the Save tab and click Save significance test to preserve the results (i.e., all the computed *p*-values). Then, a new dataset is created, containing the *p*-values and logFC cutoff for the desired contrasts. It can be explored as any other dataset (Data mining > Differential analysis menu).
5. Check that a new version appears in the dataset version dropdown-menu, referred to as "`HypothesisTest.protein`".

3.14 Differential Analysis

The differential analysis of a "`Hypothesis.Test`" version of a protein-level dataset can be conducted in the same way as for a peptide dataset as part of a peptidomics strategy (*see* Subheading 3.10). To do so, click on Data mining > Differential Analysis:

1. Select a pairwise comparison of interest. The corresponding volcano plot is displayed. At this stage, it is possible to specify the field of the tooltip to appear on the Volcano (for example, the protein identifier for immediate recognition of the proteins of interest).
2. Possibly, swap the logFC axis with the corresponding tick box, depending on layout preferences.
3. If an imputation step for missing values has been performed at the peptide (or protein) level, some proteins may have an excellent *p*-value, while a too great proportion of their intensity values (within the two conditions of interest in this comparison) are in fact imputed/recovered values, so that they are not trustworthy. To avoid such proteins become false discoveries, it is possible to discard them (by forcing their *p*-value to 1). To do so, fill in the last parameters of the right-hand side menu, which are similar to the filtering options described earlier (*see* Subheading 3.6, **Step 2**).
4. Click on Perform p-value push and move on to the pvalue calibration tab.
5. Tune the calibration method, as indicated in [18] as well as in Prostar user manual.
6. Move on to the next tab and adjust the FDR threshold (*see* Fig. 16). Clicking on one protein in the volcano plot makes it appears in the lower table.
7. Save any plot/table of interest and move on to the next tab (Summary) to have a comprehensive overview of the differential analysis parameters.

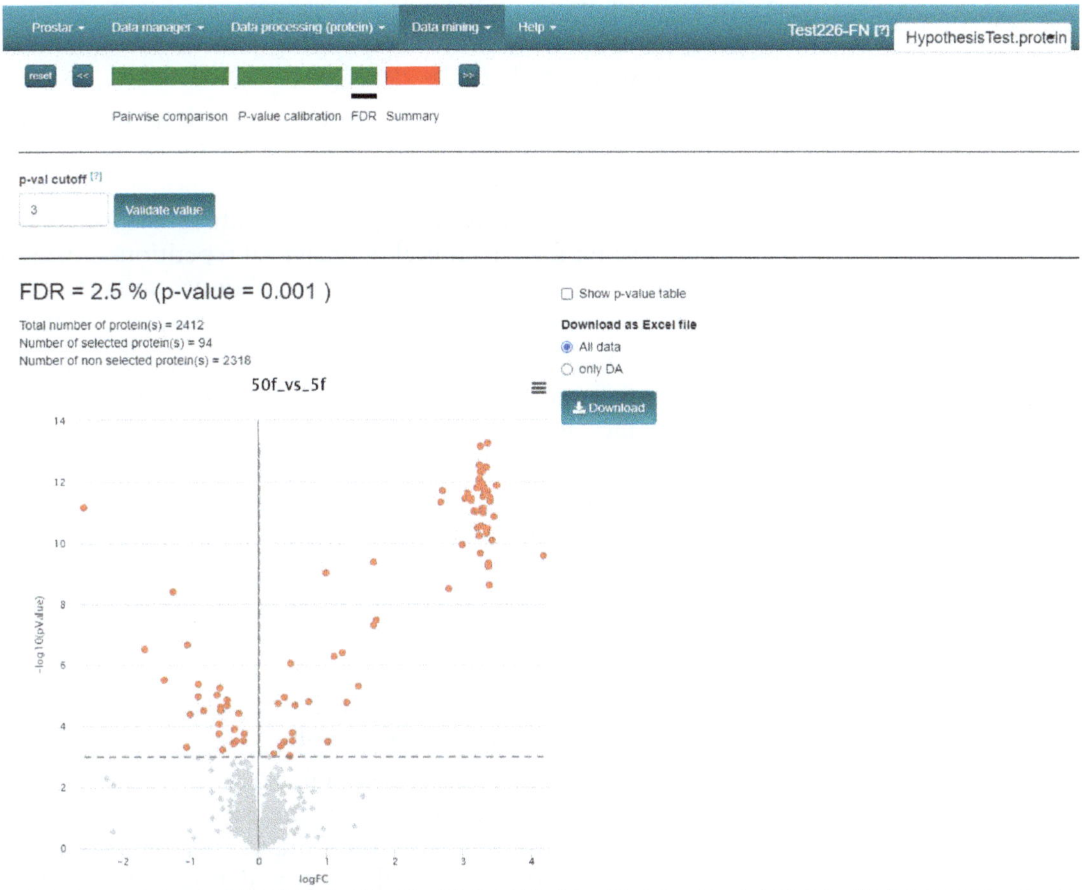

Fig. 16 Volcano plot to visualize the differentially abundant proteins according to a user-specified FDR

8. Possibly, go back to step 1 to process another pairwise comparison.

   It is possible to go on with the current protein list, and to explore the underlying functional profiles using GO analysis proposed in the Data mining menu [7].

4 Notes

1. Prostar versions which are posterior to 1.24 may slightly differ from what is described in this protocol. However, the general spirit of the graphical user interface remains unchanged allowing any user accustomed to an earlier version to easily adapt to a newer version.
2. We advise to use the latest version of R and to make regular updates, as it will guarantee the compatibility with the latest Prostar developments.

3. To ensure full compatibility and debugging, the zip file is often available later than the corresponding Bioconductor release (up to one month later).
4. With only two replicates per condition, the computations are tractable. It does not mean that statistical validity is guaranteed. Classically, 3 replicates per condition are considered a minimum in case of a controlled experiment with a small variability, mainly issuing from technical or analytical repetitions. Analysis of complex proteomes between conditions with a large biological variability requires more replicates per condition (5–10).
5. To reload a dataset that has previously been stored as an MSnset file, go to Data manager » Open MSnset file and simply browse the file system to find the desired file.
6. Before uploading a real dataset, any user can test Prostar thanks to the demo mode. This mode provides a direct access to the DAPARdata packages where some toy datasets are available, either in tabular or MSnset formats. Concretely, the demo mode is accessible in the Data manager menu: on the corresponding tab, a dropdown-menu lists the datasets that are available through DAPARdata. After selecting a peptide-level dataset, click on Load demo dataset.
7. The DAPARdata package also contains tabular versions (in txt format) of the datasets available in the demo mode. Thus, it is also possible to test the import/export/save/load functions of Prostar with these toy datasets. Concretely, one simply has to import them from the folder where the R packages are installed, in the following sub-folder: `../R/R-4.x.x/library/DAPARdata/extdata`.

   Note that each dataset is also available in the MSnset format, but these datasets should not be considered to test conversion functions from/to tabular formats.
8. It is important to specify the quantification software since this choice conditions the way with which Prostar recodes the origin of quantification. In the current version, the choice is limited to Proline and MaxQuant software. With MaxQuant, the file "peptides.txt" is appropriate. Notably, it contains a column referred to as "Identification_type", which will be instrumental (as it contains the "by Matching" and "by MS/MS" qualifiers, *see* Subheading 3.2, **Step 13**). With Proline, the export of the Display Abundances » Peptides window is appropriate (*see* Chapter 4). The user should be careful to include the "Quant. PSMs count" columns (checked by default) which will determine the origin of each peptide quantification in each sample: The numerical value amounts to the number of MS/MS identifications so that a non-zero value

indicates at least one MS/MS evidence; on the contrary, "0" means the identification has been transferred to the XIC from another run (In other words, it conveys the same information as the "Identification_type" column in MaxQuant output, and it will be used similarly, *see* Subheading 3.2, **Step 13**).

9. Prostar cannot process non-log-transformed data. Thus, do not try to cheat the software by indicating data on their original scale are log-transformed.
10. If a peptide belongs to several proteins, their respective IDs must be separated by a comma.
11. Here, "mapping" refers to the abundance value being recovered via the Match Between Run option with MaxQuant, or the Cross Assignment option with Proline.
12. However, the content of the specified columns is not checked, so that it is the user's responsibility to select the correct ones.
13. In case of difficulty, either to choose the adapted design hierarchy or to fill the table design, it is possible to click on the interrogation mark beside the sentence "Choose the type of experimental design and complete it accordingly." Except for flat design, which are automatically defined, it displays an example of the corresponding design. It is possible to rely on this example to precisely fill the design table.
14. Alternatively, it is possible to export data as excel spreadsheets or as a zip containing text files. This has no interest in case of a preliminary export; however, it may be useful to share a dataset once the statistical analysis is completed.
15. The user can download the plots showed in Prostar by right-clicking on the plot. A contextual menu appears and let the user choose either Save image as as or Copy image. In the latter case, the user has to paste the image in appropriate software.
16. It is essential to regularly go back to these tabs, so as to check that each processing step has produced the expected results.
17. Cell metadata are metadata associated with each cell of the quantitation table, and encoding the nature of the quantification value for each peptide and each sample. For a peptide, there are three main types of quantification, each contains several subtypes: *(i)* Quantitative data ("`quanti`"): The peptide under consideration is either directly `"identified"` (e.g., `"By MS/MS"` tag in MaxQuant) or `"recovered"` (indirectly found, e.g., `"By matching"` tag in MaxQuant). If the very status (`"identified"` or `"recovered"`) is unknown, it is simply labelled `"quanti"`. *(ii)* Missing value ("`missing`"): The peptide may be a Partially Observed Value (`"missing POV"`) or Missing in Entire Condition (`"missing MEC"`).

(iii) Imputed value ("`imputed`"): Imputed value are of two types, "`imputed POV`" or "`imputed MEC`".

18. Computing the heatmap and the dendrogram may be very computationally demanding, depending on the dataset.
19. As is, the full-scale plot is often difficult to read, as the high variances are concentrated on the lower intensity values. For this reason, the plot is pre-zoomed and focused on the lower intensity values. It is possible to reset the zoom by clicking on Reset zoom and to interactively zoom in on any part of the plot by clicking and dragging on the plot.
20. This functionality is available only if for a peptide-level proteomics analysis. In fact, for a peptidomic analysis, the parent protein column was not necessarily defined (*see* Subheading 3.2, **Step 10**), making peptide-protein relationship analysis useless.
21. The first column, named "`Proteins Ids`" corresponds to the value selected in the dropdown-menu called protein IDs (*see* Subheading 3.2, **Step 10**), while the column named "`Peptides Ids`" corresponds to the value selected in the dropdown-menu ID definition in the loading process of a peptide dataset (*see* Subheading 3.2, **Step 9**).
22. The color legend is the same as in the Data Explorer tab (*see* Subheading 3.4, **Step 9**).
23. The latter case may be more adapted in case of unbalanced designs, i.e., datasets in which conditions have not the same number of samples.
24. Peptide metadata are additional columns of the dataset which contains metadata attached to each line, i.e., each peptide, as opposed to cell metadata.
25. To work correctly, the selected column must contain information encoded as a string of characters. For each peptide, the beginning of the corresponding string is compared to a given prefix. If the prefix matches, the peptide is filtered out. Otherwise, it is retained. Note that the filter only operates a prefix search (at the beginning of the string), not a general tag match search (anywhere in the string). Similarly, filters based on regular expressions are not implemented.
26. In datasets outputted from MaxQuant, metadata indicates under a binary form which peptides are reversed sequences (resulting from a target-decoy approach) and which are potential contaminants. Both of them are indicated by a "+" in the corresponding column (the other peptides having an NA instead). It is thus possible to filter both reversed and contaminants out by indicating "+" as the prefix filter. However, if

adequately encoded, filtering on other type of information is possible.

27. If one has no idea of the prefixes, it is possible to switch to the Descriptive Statistics > Data Explorer (*see* Subheading 3.4, **Step 3**), so as to visualize the corresponding metadata.
28. It is not possible to keep in parallel several datasets or multiple versions of a dataset at a particular level (for instance, a dataset filtered using various rules). Thus, if one goes on with the next processing steps on an older dataset, or if one goes back to a previous step and restart it, the new results will overwrite the previously saved ones at this same step, without updating other downstream processing, leading to possible inconsistencies.
29. This list corresponds to the normalization methods available in Prostar version 1.24; future versions will possibly propose slightly different methods.
30. This normalization method should not be confused with Global quantile alignment.
31. It should be noted that the choice of a normalization method and its tuning is highly data dependent, so that a single protocol cannot be proposed. The data analyst should gather expertise on the normalization methods, so as to be able to choose soundly. Thus, we advise to refer to Prostar user manual, as well as to the literature describing the normalization methods (to be found in the Help menu).
32. As a result, all the missing values are either POV or MEC. Moreover, for a given peptide across several conditions, the missing values can be both POV and MEC, even though within a same condition they are all of the same type.
33. The routine used for imputation with imp4p depends on the final MNAR or MCAR diagnosis.
34. Considering that imputation of MECs is always risky, the checkbox lets the alternative to impute MECs at the protein level (i.e., after the aggregation process), despite the risk that the missing value distribution makes the aggregation impossible (*see* Subheading 3.11).
35. If "`Det. quantile`" is used, we advise to use a small quantile of the intensity distribution to define the imputation value, for instance, 1–5% depending on the stringency you want to apply on peptides quantified only in one condition of a pairwise comparison. In case of a dataset with too few peptides, the lower quantile may amount to instable values. In such a case, we advise to use a larger quantile value (for instance, 10% or 20%) but to use a smaller multiplying factor so as to keep the imputation value reasonably small with respect to the detection limit.

36. During the aggregation process, it may be necessary to aggregate quantified peptides and missing valued peptides. As in such cases it is impossible to correctly estimate the protein abundance, the aggregation is cancelled.
37. When exporting the data in the Microsoft Excel format, imputed values are displayed in a colored cell so that their origin POV and MEC can easily be distinguished.
38. The reason why combining missing and non-missing values is not authorized is given in [17]: Broadly speaking, it would amount to implicitly impute the missing value(s) with a value that does not change the aggregated value.
39. The rules established in the current 1.24 version to manage the aggregation of missing values may be refined in future versions.
40. If shared peptides are not included in the aggregation step, one faces the risk of losing some proteins (those which do not have a single specific peptide). This is visible on the "`Only specific peptides`" barplot where a few proteins may appear to have 0 peptides.
41. Thus, the intensity of each shared peptide is fully accounted for all the proteins it belongs to. This straightforward solution is obviously not the most rigorous aggregation model.
42. To achieve this proportional redistribution, the intensities of shared peptides are iteratively split and shared proportionally to the parent protein abundances. The iteration goes on until the redistribution process does not change the protein abundances any more.
43. The reason why the sum operator is not allowed is the following: an iterative sum would endlessly inflate the protein abundance and convergence would never be reached.
44. By consulting the Excel export, you will notice several columns that provide information on the number and nature of the aggregated peptides, either at the whole experiment level, or at each single replicate level. Moreover, the missing values are colored according to their type (POV and MEC, if any) using the same color code as in Prostar.
45. Each protein abundance value is associated with a cell metadata directly inherited from those of the peptides. In general, it contains the same label as all the children peptides, namely "`identified`", "`recovered`", "`imputed POV`" or "`imputed MEC`." However, if all the children peptides do not have the same labels, more generic labels are used: a mix of "`identified`" and "`recovered`" peptides leads to a protein-level quantitation value, referred to as "`quanti`". In the case of peptides tagged with a mix of "`imputed POV` and "`imputed MEC`, the protein is simply referred to as "`imputed`". Finally, in

the case of a mix of quantified peptides (i.e., "`identified`" or "`recovered`"), and of imputed peptides (i.e., "`imputed POV`" or "`imputed MEC`" or "`imputed`"), the protein is labelled as "`combined`".

46. Depending on the chosen option: "`Include shared peptides`" set to "`No`" (*see* Subheading 3.11, **Step 2**).

47. It is important to tune the logFC threshold to a small enough value, to avoid discarding too many proteins [19, 20]: it is essential to keep enough remaining proteins for the next coming FDR computation step (*see* Subheading 3.14), as (1) FDR estimation is more reliable with many proteins, (2) FDR, which relates to a percentage, does not make sense on too few proteins.

Acknowledgements

This work was supported by grants from Agence Nationale de la Recherche under: ProFI project (Proteomics French Infrastructure, ANR-10-INBS-08), GRAL project, a program from the Chemistry Biology Health (CBH) Graduate School of University Grenoble Alpes (ANR-17-EURE-0003), DATA@UGA and SYMER projects (ANR-15-IDEX-02) and MIAI @ Grenoble Alpes (ANR-19-P3IA-0003).

References

1. Zhang Y, Fonslow BR, Shan B, Baek MC, Yates III JR (2013) Protein analysis by shotgun/bottom-up proteomics. Chem Rev 113(4): 2343–2394. https://doi.org/10.1021/cr3003533
2. Ong SE, Foster LJ, Mann M (2003) Mass spectrometric-based approaches in quantitative proteomics. Methods 29(2):124–130. https://doi.org/10.1016/s1046-2023(02)00303-1
3. Schwanhäusser B, Busse D, Li N, Dittmar G, Schuchhardt J, Wolf J, Chen W, Selbach M (2011) Global quantification of mammalian gene expression control. Nature 473(7347): 337–342. https://doi.org/10.1038/nature10098
4. Beeley C (2013) Web application development with R using Shiny. Packt Publishing Ltd. https://github.com/PacktPublishing/Web-Application-Development-with-R-Using-Shiny-third-edition
5. Wieczorek S, Combes F, Lazar C, Giai Gianetto Q, Gatto L, Dorffer A, Hesse AM, Coute Y, Ferro M, Bruley C, Burger T (2017) Dapar & prostar: software to perform statistical analyses in quantitative discovery proteomics. Bioinformatics 33(1):135–136. https://doi.org/10.1093/bioinformatics/btw580
6. Goeminne LJ, Argentini A, Martens L, Clement L (2015) Summarization vs peptide-based models in label-free quantitative proteomics: performance, pitfalls, and data analysis guidelines. J Proteome Res 14(6):2457–2465. https://doi.org/10.1021/pr501223t
7. Wieczorek S, Combes F, Borges H, Burger T (2019) Protein-level statistical analysis of quantitative label-free proteomics data with prostar. In: Proteomics for biomarker discovery. Springer, New York, pp 225–246. https://doi.org/10.1007/978-1-4939-9164-8_15
8. Gatto L, Lilley KS (2012) MSnbase-an R/Bioconductor package for isobaric tagged mass spectrometry data visualization, processing and quantitation. Bioinformatics 28(2): 288–289. https://doi:10.1093/bioinformatics/btr645
9. Wieczorek S, Combes F, Burger T (2018) DAPAR and ProStaR user manual. In: Bioconductor. https://www.bioconductor.org/

packages/release/bioc/vignettes/Prostar/inst/doc/Prostar_UserManual.pdf?attredirects=0
10. RStudio Team (2015) RStudio: Integrated Development for R. RStudio, Inc., Boston, MA. http://www.rstudio.com/
11. Cox J, Mann M (2008) Maxquant enables high peptide identification rates, individualized p.p. b.-range mass accuracies and proteome-wide protein quantification. Nat Biotechnol 26: 1367–1372. https://doi.org/10.1038/nbt.1511
12. Bouyssié D, Hesse AM, Mouton-Barbosa E, Rompais M, Macron C, Carapito C, Gonzalez de Peredo A, Couté Y, Dupierris V, Burel A, et al. (2020) Proline: an efficient and user-friendly software suite for large-scale proteomics. Bioinformatics 36(10):3148–3155. https://doi.org/10.1093/bioinformatics/btaa118
13. R-Core-Team (2020) stats package. URL https://www.rdocumentation.org/packages/stats/versions/3.6.2/topics/hclust, r package version 3.6.2
14. Cleveland WS (1979) Robust locally weighted regression and smoothing scatterplots. J Am Stat Assoc 74(368):829–836, https://doi.org/10.1080/01621459.1979.10481038
15. Smyth GK (2005) Limma: linear models for microarray data. In: Bioinformatics and computational biology solutions using R and bioconductor. Springer, New York, pp 397–420. https://doi.org/10.1007/0-387-29362-0_23
16. Huber W, Von Heydebreck A, Sültmann H, Poustka A, Vingron M (2002) Variance stabilization applied to microarray data calibration and to the quantification of differential expression. Bioinformatics 18(suppl_1):S96–S104. https://doi.org/10.1093/bioinformatics/18.suppl_1.S96
17. Lazar C, Gatto L, Ferro M, Bruley C, Burger T (2016) Accounting for the multiple natures of missing values in label-free quantitative proteomics data sets to compare imputation strategies. J Proteome Res 15(4):1116–1125. https://doi.org/10.1021/acs.jproteome.5b00981
18. Giai Gianetto Q, Combes F, Ramus C, Bruley C, Couté Y, Burger T (2016) Calibration plot for proteomics: A graphical tool to visually check the assumptions underlying FDR control in quantitative experiments. Proteomics 16(1):29–32. https://doi.org/10.1002/pmic.201500189
19. Giai Gianetto Q, Couté Y, Bruley C, Burger T (2016) Uses and misuses of the fudge factor in quantitative discovery proteomics. Proteomics 16(14):1955–1960. https://doi.org/10.1002/pmic.201600132
20. Wieczorek S, Gianetto QG, Burger T (2019) Five simple yet essential steps to correctly estimate the rate of false differentially abundant proteins in mass spectrometry analyses. J Proteomics 207:103441. https://doi.org/10.1016/j.jprot.2019.103441

Chapter 10

msmsEDA & msmsTests: Label-Free Differential Expression by Spectral Counts

Josep Gregori, Àlex Sánchez, and Josep Villanueva

Abstract

`msmsTests` is an `R/Bioconductor` package providing functions for statistical tests in label-free LC-MS/MS data by spectral counts. These functions aim at discovering differentially expressed proteins between two biological conditions. Three tests are available: Poisson GLM regression, quasi-likelihood GLM regression, and the negative binomial of the `edgeR` package. The three models admit blocking factors to control for nuisance variables. To assure a good level of reproducibility a post-test filter is available, where (1) a minimum effect size considered biologically relevant, and (2) a minimum expression of the most abundant condition, may be set. A companion package, `msmsEDA`, proposes functions to explore datasets based on msms spectral counts. The provided graphics help in identifying outliers, the presence of eventual batch factors, and check the effects of different normalizing strategies. This protocol illustrates the use of both packages on two examples: A purely spike-in experiment of 48 human proteins in a standard yeast cell lysate; and a cancer cell-line secretome dataset requiring a biological normalization.

Key words Biomarker discovery, Secretomes, Spectral counts, Label free, Batch effects, Normalization, Reproducibility, Bioconductor, msmsEDA, msmsTests

1 Introduction

This chapter focuses on the use of two companion packages, namely `msmsTests` [1] and msmsEDA [2], for proteomics biomarker discovery; that is the detection of protein differential expression between two biomedical conditions, using spectral counts (SpC, [3]) as a quantification measure, in label-free LC-MS/MS experiments. Inference with spectral counts (SpC) is done with linear models able to account for covariates in a distribution specific context [4]. Label-free experiments offer great flexibility but require balanced experimental designs, and the use of tools for the detection and correction of potential batch effects, as well as effects due to noncontrolled factors [5–7].

In biomarker discovery reproducibility is of the biggest importance. Besides statistical significance, biological relevance is

Thomas Burger (ed.), *Statistical Analysis of Proteomic Data: Methods and Tools*,
Methods in Molecular Biology, vol. 2426, https://doi.org/10.1007/978-1-0716-1967-4_10,

required. This means that differentially expressed proteins should not be declared solely on a *p*-value basis. They should also show a minimal signal and a minimum effect size [8, 9].

Data files and the scripts in this chapter are available in the following GitHub repository https://github.com/JosepGregori/SPH.msms.

1.1 GLMs for Inference with Counts

The inference methods used in `msmsTests` are based on generalized linear models (GLM), which is a natural choice when dealing with counts [4]. These GLMs may be based on two distributions, the Poisson one with a single parameter—as mean and variance are equal—or the Negative Binomial with two parameters—and more flexible. A third possibility is to rely on quasi-likelihood inference, using a Poisson-like regression which estimates the variance from the data [4, 10].

When the only source of variation comes from the sampling process, as when running technical replicates, the Poisson distribution works very well. Nevertheless, when doing biological experiments, to the typical variation in sampling technical replicates we have to add the biological variability expected of individuals belonging to the same biological condition. In these circumstances, the Poisson model may underestimate the variance and the inference may report false positives in differential expression (this phenomenon is known as overdispersion [4]). The immediate alternative is the negative binomial (NB) distribution, which allows for overdispersion, or the quasi-likelihood which makes abstraction of the true distribution, and estimates the variance from the data. The *p*-value for differential expression is obtained from the log-likelihood ratio comparing the full model against the null model. Multitest *p*-value adjustment with FDR control is done by the Benjamini-Hochberg method [11].

1.2 Batch Effects in Label-Free Proteomics

In biomarker discovery, we seek for differential expression between two biological conditions, generally disease and control. To ensure unbiased comparison, labeled procedures have been employed both in transcriptomics and in proteomics. Labeled procedures allow the early mix of the samples to be compared. Since they are pooled, they are processed together and any noncontrolled factor affects equally both conditions. The labels allow the posterior identification of features belonging to each condition. Complex preparation steps, requirements for increased sample concentration and incomplete labeling are the main issues. Labeling usually requires fractionation since all samples are analyzed together, causing a drop in sensitivity. The key of this approach is that all samples to be compared are processed together. In this way the experiments are fully dedicated to a unique comparison and the acquired data, in most cases, may not be reused with different purposes.

Label-free proteomic analysis provides a more flexible alternative. This means that each sample is processed and analyzed separately. The counterpart is higher risk of bias due to noncontrolled factors affecting differently one condition with respect to another [12–15]. This requires carefully planned and designed experiments, to minimize confounding and bias as much as possible. Nevertheless, given the sample sizes attainable, not all factors potentially contributing to the results may be controlled. And this brings almost unavoidably to the expression of batch effects in the label-free LC-MS/MS dataset, mainly when the period of collection and analysis of different samples spans over a long period of time [7].

The batch effect is defined to be systematic and unintentional, in contrast with the experimental noise which is random in nature. It refers exclusively to systematic technical differences when samples are processed and measured in different batches or in different times. Because of its systematic nature, the worst manifestation of batch effects is bias. These effects may be visualized by multidimensional techniques like Principal Components Analysis (PCA), Singular Value Decomposition (SVD), Hierarchical Clustering (HC), or Heatmaps (HM) [5, 7, 16]. Ideally the samples should cluster by treatment level, with independence of the time in which they were treated and measured. Once identified, a number of methods are proposed to minimize these effects [5, 6, 16].

1.3 Normalization Strategies

Omics differential expression is generally based on some basic assumptions, not always explicitly given. When these assumptions are roughly fulfilled, the comparisons between two biological states may be considered as nearly unbiased, provided that a good experimental design is used [17]. These assumptions are usually taken for granted and receive no criticism in most, if not all, studies. In transcriptomics and proteomics, where equal amounts of substance gathered from two biological conditions are measured, it is considered that the cells produce globally almost equal quantity of total substance. Under this assumption comparing equal amounts of substance corresponds to comparing the substance produced by equal number of cells: this is a cell-to-cell comparison, where the cell is the biological unit of interest. Depending on the experiment, this assumption translates in different normalization methods:

1. **Sample size normalization:** The most extended method of normalization assumes that for a given amount of total protein, the sum of SpC should be the same. In general when GLM models are used [18–20] the normalization is implicitly done with the help of an offset term in the model [4].

$$E[y] = \mu \tag{1}$$

$$\log\left(\frac{\mu}{size}\right) = \alpha + \beta x \tag{2}$$

$$\log(\mu) = \log(size) + \alpha + \beta x \quad (3)$$

where μ is the expected expression of a given protein, *size* is the normalizing condition, and α and β are the model parameters, with x equal to 0 for the control condition, or equal to one for the disease condition. The term $\log(size)$ is the *offset*, where *size* is the total number of SpC. Different covariates or blocking factors may be added to this simple model. This is independent of the underlying distribution for the expression level y of this protein. The underlying assumption is that the measured substance has been produced by equal number of cells in either state. This assumption could be accepted when analyzing plasma or interstitial fluid. But it is more questionable when analyzing cell-line secretomes.

2. **Secretome normalization:** When studying secretomes the above assumption deserves a careful consideration, as the number of involved proteins is drastically reduced. On the other hand no structural proteins or proteins related to the metabolism are expected, which could be used with normalization purposes. One biological state may be globally stimulated or depressed, in secretome terms, with respect to the other, and hence the basic assumption may fail, giving rise to biased comparisons [21]. The biological state of secretion may be estimated by counting the cells, and measuring the total protein secreted by them. Introducing the ratio number of cells to total secreted protein as an additional offset in the model produces a correct normalization [21].

1.4 Reproducibility

In biomarker discovery, reproducibility is of the maximum concern. As evidenced by the MAQC-1 study [8], the main reason for the lack of reproducibility of lists of differentially expressed genes found with oligonucleotide microarray (MA) experiments, was its sole relying on p-values [22]. The recommendation was to rank the lists of differentially expressed genes by fold change instead of by p-value, giving relevance to the effect size beyond the significance. This may be extended by saying that significant features with low signal will be scarcely reproducible because in different samples they may elude detection, both because of the sampling process and because of technical reasons related to the noise level. Significant features with a low effect size will be scarcely reproducible because they will obtain a significant p-value only when the observed variance is low enough. Significant features with low signal and low effect size will show very bad reproducibility. The combination of good signal and effect size is the best guarantee for reproducibility.

This has also been evidenced with SpC label-free data [9], where we have to add the effect caused by the coarse granularity at low expression levels. The next value to 1SpC is 2SpC, with this smallest increase the signal doubled.

2 Material

We developed two R/Bioconductor [23, 24] packages which offer solutions to the discussed challenges in SpC label-free projects. Although optimized to work with secretomes, they are equally useful in the analysis of any SpC label-free project. msmsEDA provides functions and tools to assess the quality of a LC-MS/MS project, and to detect and visualize outliers or batch effects. msmsTests includes functions to test the differential expression of proteins in SpC matrices by GLM, using either the Poisson or the negative binomial (NB) distribution, or the quasi-likelihood (QL) extension [4]. These functions take the formulas of the two models to be compared as parameters, and allow for blocking factors to correct observed batch effects. The NB method in msmsTests uses the implementation in the package edgeR [25] which is an empirical Bayes approach to share information across proteins and allows the estimation of the dispersion with fewer replicates. Flexible normalization factors are incorporated into the GLM model by means of offsets. Another function flags the proteins as significant and likely reproducible (DEP), based on multiple test adjusted *p*-values, signal strength and size-effect. Both packages provide utility functions to help in the interpretation of the results.

2.1 Requirements

1. The two packages depend of MSnbase [26].
2. Other packages imported are MASS [27], gplots [28], RColorBrewer [29], edgeR [30], and qvalue [31].
3. R [23] version should be ≥3.0.1.
4. There are no special requirements concerning system or computer capabilities.

2.2 Packages Installation

The libraries may be installed in the usual Bioconductor way:

```
###  Install packages from Bioconductor
install.packages("BiocManager")
BiocManager::install("msmsEDA")
BiocManager::install("msmsTests")
```

2.3 Data Format

Both packages are based on the S4 class MSnSet defined in the MSnbase package [26]. This class has different slots to support all experiment data. The most important ones are, the expression matrix assayData and the metadata phenoData, as most computations can be carried out with the data stored in these two slots only.

An MSnSet object may be created from a matrix of spectral counts, and a data.frame with meta data. For instance, (*see*

Subheading 2.4), the file `UPS1_Yeast500_counts.txt` contains a matrix of spectral counts, with 34 samples in the columns, and 691 accessions in the rows, and the file `UPS1-samples.txt` has a data.frame with two columns, `Sample` with 18 sample IDs, and `Treat` with the corresponding treatment levels (*see* **Note 1**):

```
###  Read the file with spectral counts
cm <- read.table('UPS1_Yeast500_counts.txt',header=TRUE,
                 sep='\t',stringsAsFactors=FALSE)
###  Inspect the object just read
str(cm)
```

```
'data.frame':   691 obs. of  34 variables:
 $ Y500U100_001: int  151 154 64 161 157 96 23 53 52 76 ...
 $ Y500U100_002: int  195 244 89 155 161 109 28 58 53 53 ...
 $ Y500U100_003: int  188 237 128 158 173 113 27 64 54 62 ...
 $ Y500U100_004: int  184 232 109 172 175 115 27 63 44 74 ...
 $ Y500U150_001: int  263 165 161 132 120 101 98 62 49 53 ...
 $ Y500U150_002: int  225 213 162 140 110 151 100 92 64 52 ...
 $ Y500U150_003: int  156 145 138 89 82 75 116 59 45 43 ...
 $ Y500U150_004: int  148 154 159 103 97 83 95 60 44 45 ...
 $ Y500U200_001: int  221 190 116 164 177 119 48 66 73 63 ...
 $ Y500U200_002: int  201 187 119 165 164 121 44 70 62 65 ...
 $ Y500U200_003: int  187 215 139 161 176 141 34 78 67 67 ...
 $ Y500U200_004: int  194 189 114 184 174 124 36 64 75 68 ...
 $ Y500U200_005: int  186 210 139 160 123 157 58 88 92 57 ...
 $ Y500U200_006: int  120 130 119 97 86 88 61 54 49 42 ...
 $ Y500U200_007: int  158 155 160 98 100 113 83 62 59 49 ...
 $ Y500U200_008: int  131 144 149 98 109 100 67 62 55 48 ...
 $ Y500U200_009: int  330 220 319 268 174 121 237 78 89 111 ...
 $ Y500U200_010: int  325 215 271 272 167 145 219 91 97 125 ...
 $ Y500U200_011: int  272 194 141 164 179 153 75 66 62 83 ...
 $ Y500U200_012: int  172 172 93 127 169 118 31 65 47 59 ...
 $ Y500U400_001: int  243 203 184 152 126 135 113 93 73 72 ...
 $ Y500U400_002: int  149 147 146 100 96 95 93 63 60 53 ...
 $ Y500U400_003: int  153 150 165 98 110 109 107 61 60 50 ...
 $ Y500U400_004: int  104 138 140 98 96 78 50 54 53 42 ...
 $ Y500U600_001: int  275 177 218 248 140 120 110 80 72 74 ...
 $ Y500U600_002: int  259 174 214 241 145 125 111 85 71 70 ...
 $ Y500U600_003: int  302 197 189 186 131 120 143 70 77 94 ...
 $ Y500U600_004: int  294 193 179 192 132 111 140 73 73 86 ...
 $ Y500U600_005: int  229 178 227 147 148 108 153 81 63 64 ...
 $ Y500U600_006: int  238 183 221 152 145 115 143 82 68 70 ...
 $ Y500U750_001: int  210 189 118 155 125 134 97 85 81 66 ...
 $ Y500U750_002: int  143 139 119 99 99 77 85 53 52 49 ...
 $ Y500U750_003: int  120 135 127 106 101 84 84 59 61 46 ...
 $ Y500U750_004: int  115 141 131 93 86 84 75 50 52 47 ...
```

```
###  Read the file with sample IDs and treatment levels
samples <- read.table('UPS1-samples.txt',header=TRUE,
                      sep='\t',stringsAsFactors=FALSE)
###  Inspect the object just read
str(samples)
```

```
'data.frame':   18 obs. of  2 variables:
 $ Sample: chr  "Y500U200_001" "Y500U200_002" "Y500U200_003" ...
 $ Treat : chr  "Y5U200" "Y5U200" "Y5U200" "Y5U200" ...
```

The constructor `MSnSet(exprs, fData, pData)` can be used to create `MSnSet` object instances. Argument `exprs` is a data matrix and `fData` and `pData` must be of class `data.frame` or `"AnnotatedDataFrame"` and all must meet the dimensions and name validity constrains (*see* the `eSet` class in Bioconductor's package `Biobase` [24], from which `MSnSet` is derived). For instance, with the above data a corresponding `MSnSet` object may be created as follows:

```
### The levels of treatment for each sample
pdata <- samples[,'Treat',drop=FALSE]
rownames(pdata) <- samples$Sample

### Construct an MSnSet from the parts
msms <- MSnSet(data.matrix(cm[,samples$Sample]), pData=pdata)
### Inspect the object
msms
```

```
MSnSet (storageMode: lockedEnvironment)
assayData: 691 features, 18 samples
  element names: exprs
protocolData: none
phenoData
  sampleNames: Y500U200_001 Y500U200_002 ... Y500U600_006 (18 total)
  varLabels: Treat
  varMetadata: labelDescription
featureData: none
experimentData: use 'experimentData(object)'
Annotation:
- - - Processing information - - -
 MSnbase version: 2.10.1
```

The data items in a `MSnSet` object may be recovered by the usual `eSet` methods:

```
### Get the sample names
sampleNames(msms)
```

```
 [1] "Y500U200_001" "Y500U200_002" "Y500U200_003" "Y500U200_004"
 [5] "Y500U200_005" "Y500U200_006" "Y500U200_007" "Y500U200_008"
 [9] "Y500U200_009" "Y500U200_010" "Y500U200_011" "Y500U200_012"
[13] "Y500U600_001" "Y500U600_002" "Y500U600_003" "Y500U600_004"
[17] "Y500U600_005" "Y500U600_006"
```

```
### Get the factor name
varLabels(msms)
```

```
[1] "Treat"
```

```
### Get the levels in the treatment factor
pData(msms)$Treat
```

```
 [1] "Y5U200" "Y5U200" "Y5U200" "Y5U200" "Y5U200" "Y5U200" "Y5U200"
 [8] "Y5U200" "Y5U200" "Y5U200" "Y5U200" "Y5U200" "Y5U600" "Y5U600"
[15] "Y5U600" "Y5U600" "Y5U600" "Y5U600"
```

```
### Get the expression matrix
str(exprs(msms))
```

```
int [1:691, 1:18] 221 190 116 164 177 119 48 66 73 63 ...
- attr(*, "dimnames")=List of 2
 ..$ : chr [1:691] "YKL060C" "YDR155C" "YOL086C" "YJR104C" ...
 ..$ : chr [1:18] "Y500U200_001" "Y500U200_002" "Y500U200_003"
```

2.4 Example Data

The package `msmsEDA` contains a dataset already formatted as an MSnSet, `msms.dataset`. It is a spiking dataset that consists in a set of replicates of 500 ng of standard yeast lysate, spiked with an equimolar mix of 48 human proteins (UPS1, Sigma-Aldrich) (*see* **Note 2**) at 200 and 600 fmols, which corresponds to 0.81 and 2.16% of the total mixture [7]. Three replicates of each condition were analyzed on 03/02, and four additional replicates were analyzed a few weeks later, on 25/02:

```
### Load library and dataset
library(msmsEDA)
data(msms.dataset)
```

The package msmsTests contains another spiking dataset of the same type as the previous one, with 100, 200, 400, and 600 fmols of UPS1 in 500ng of yeast cell lysate:

```
###  Load library and dataset
library(msmsTests)
data(msms.spk)
```

The last dataset is given in the more general form of two text files, the first with spectral counts, A549-EMT_SamplesReport.txt and the second with corresponding metadata A549-EMT_-samples.txt. It consists in three technical replicates of two biological replicates of the treatment of A549 cells with TGFβ [21]. This is a cell-line biological experiment requiring of a cell-to-cell normalization, and with batches as biological replicates.

3 Methods

3.1 Experimental Design and EDA

In label-free proteomics, as in most experimental situations, a good experimental design is of utmost importance. A single run with all required samples is not always possible, and samples have sometimes to be processed in different batches. The splitting in batches should be done in MS runs as balanced as possible. Each run should include samples of both conditions to be compared. The experimental design should include a balanced preprocessing of samples as well (*see* **Note 3**). Exploratory data analysis (EDA) is the means to guarantee that the design was correct and no biases are observed. By analogy, it could be seen as the forensic means to diagnose the cause of death of an experiment:

1. To illustrate the use of the EDA tools in package msmsEDA we will use a spiking dataset contained in the package:

```
###  Load library and dataset
library(msmsEDA)
data(msms.dataset)

### Inspect object
msms.dataset
```

```
1 MSnSet (storageMode: lockedEnvironment)
2 assayData: 697 features, 14 samples
3   element names: exprs
4 protocolData: none
5 phenoData
6   sampleNames: U2.2502.1 U2.2502.2 ... U6.0302.3 (14 total)
7   varLabels: treat batch
8   varMetadata: labelDescription
9 featureData: none
10 experimentData: use 'experimentData(object)'
```

```
  pubMedIds: http://www.ncbi.nlm.nih.gov/pubmed/22588121
Annotation:
- - - Processing information - - -
 MSnbase version: 1.8.0
```

The 14 samples are distributed in two conditions, `U200` and `U600`, and in two batches `0302` and `2502`:

```
### Cross tabulate treatments and batches in dataset
table(pData(msms.dataset)$treat,pData(msms.dataset)$batch)
```

```
        0302 2502
  U200     3    4
  U600     3    4
```

2. Given an MSnSet, possibly subsetted from a bigger dataset, as it is the case, it is always a good idea to preprocess it by removing rows with all zeroes, and replacing NAs by zeroes (usually NAs in spectral counts matrices correspond to proteins not identified). This is the task of function `pp.msms.data()` (*see* **Note 4**).
3. A first inspection of raw counts may be done by function `count.stats()` which computes the Tukey's five number summary, as `R` function `fivenum()` - min, low hinge, median, high hinge and maximum—in the distribution of counts for each sample, along with the corresponding number of proteins identified and the total number of counts (*see* Table 1):

```
### Preprocess dataset
msms <- pp.msms.data(msms.dataset)
### Stats in distribution of counts
count.stats(msms)
```

4. The distributions of counts may be visualized by the help of functions `spc.boxplots()` and `spc.densityplots()` (*see* Fig. 1):

```
###  Densityplots of counts by samples
spc.densityplots(exprs(msms),
                 fact=pData(msms)$treat,
                 main='U200 vs U600')
```

5. Ideally the total number of spectral counts by sample should be the same, and this may be visualized by function `spc.barplots()`. In Fig. 2 the ordinate of 1 corresponds to the median number of total counts by sample, so the figure shows the fraction by which each sample is off the median,

Table 1
Results of function count.stats()

	Proteins	Counts	Min	lwh	Med	Hgh	Max
U2.2502.1	590	5398	0	2	3	8.0	183
U2.2502.2	593	5501	0	2	3	7.0	205
U2.2502.3	586	5477	0	1	3	8.0	202
U2.2502.4	586	5251	0	1	3	7.0	203
U6.2502.1	582	5692	0	1	3	8.5	194
U6.2502.2	577	5686	0	1	3	8.0	208
U6.2502.3	578	5552	0	1	3	8.0	215
U6.2502.4	560	5601	0	1	3	8.0	217
U2.0302.1	512	5629	0	1	2	7.0	409
U2.0302.2	499	5840	0	0	2	7.0	384
U2.0302.3	513	5726	0	1	2	7.0	364
U6.0302.1	491	5975	0	0	2	7.5	395
U6.0302.2	474	5739	0	0	1	7.5	358
U6.0302.3	474	5891	0	0	2	8.0	355

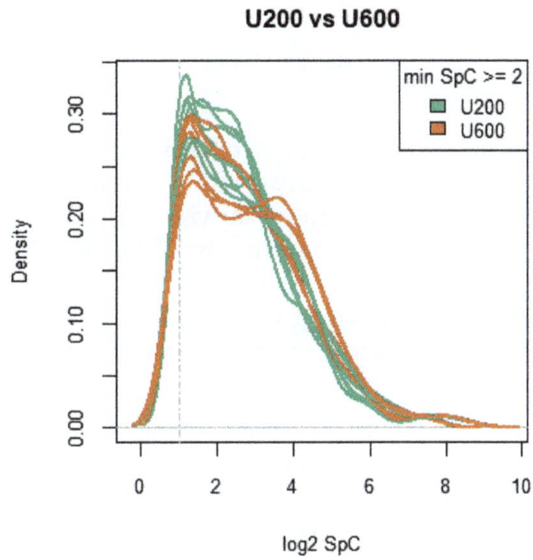

Fig. 1 Distribution of spectral counts by sample, visualized as densityplots

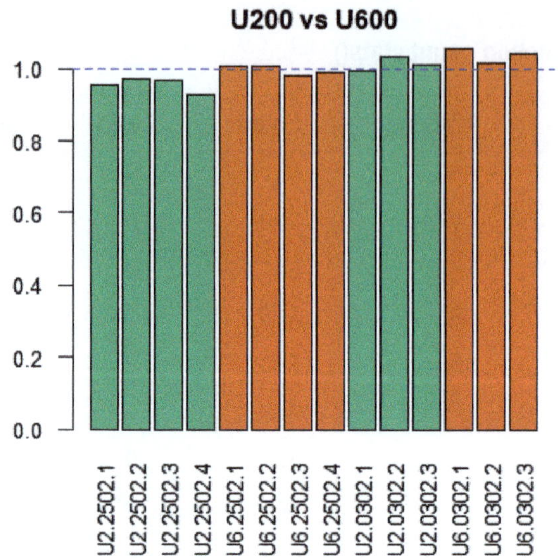

Fig. 2 Counts by sample, normalized to the median

and consequently the impact of the normalization by size on each sample:

```
###  Barplot of total counts by sample, normalized to the
   median
spc.barplots(exprs(msms),
             fact=pData(msms)$treat,
             main='U200 vs U600')
```

6. The similarities between the samples in the dataset may be visualized by means of a hierarchical clustering based on the pairwise distance between samples, using function `counts.hc()` (*see* Fig. 3):

```
### Plot dendrogram of hierarchical clustering of samples
counts.hc(msms, facs=pData(msms)[,'batch',drop=FALSE])
```

We see from Fig. 3 how the samples cluster first by batch and afterward by condition. This is typical of the presence of batch effects. In this example, as the experimental design was balanced, and we still see the samples of each condition clustering together within each batch, the effect is an increase of variance and hence a decrease in sensitivity. In these cases, the inclusion of a blocking factor term in the model will help in correcting the effect. If the samples were processed in two batches, with each batch having all the samples of a single

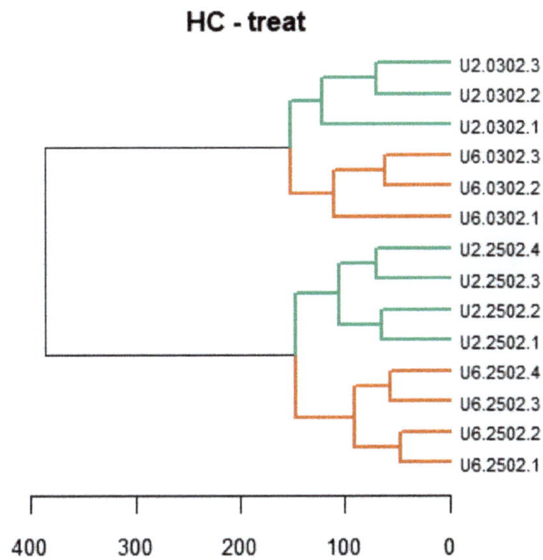

Fig. 3 Hierarchical clustering of samples from the spectral counts matrix. Dendrogram branches are colored by treatment level

condition, the clustering of samples in conditions could be perfect, but the result highly biased and not reproducible. Before proceeding to any MSMS experiment make sure that the design is balanced and possibly randomized.

7. The reader is invited to visualize the batch effects in this dataset with a PCA by means of function `counts.pca()` or with a heatmap with function `counts.heatmap()`.
8. Although the correction will be made later, by using a blocking factor in the GLM model, we may visualize the impact of the batch correction by mean centering the two batches. This is done by `batch.neutralize()` which accepts a matrix of spectral counts and a blocking factor as arguments, and returns a transformed matrix of spectral counts (*see* Fig. 4):

```
### Correct batch effects
spcm <- batch.neutralize(exprs(msms), pData(msms)$batch,
                         half=TRUE, sqrt.trans=TRUE)
msms.bc <- msms
exprs(msms.bc) <- spcm
### Visualize the effect of the batch correction
counts.hc(msms.bc, facs=pData(msms)[,'treat',drop=FALSE])
```

Figure 4 shows the samples clustering by treatment, and with the distances between the two clusters much reduced with respect to raw counts in Fig. 3.

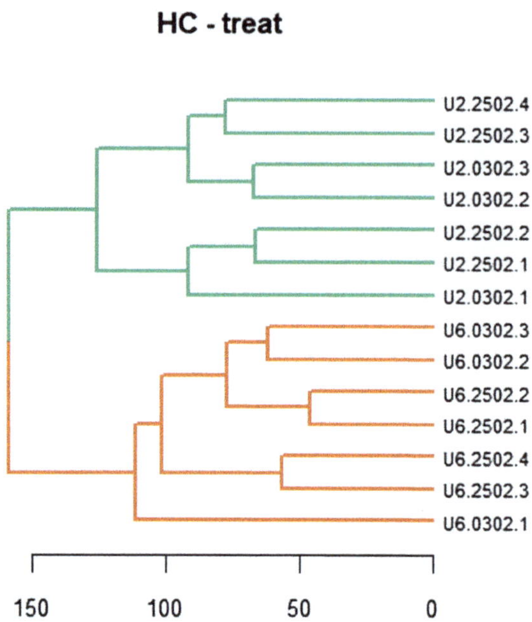

Fig. 4 Hierarchical clustering of samples from the batch corrected spectral counts matrix. Dendrogram branches are colored by treatment level

9. The impact of the batch correction may be seen also from the distribution of differences between the spectral counts matrix before and after correction:

```
###  Statistics of the distribution of differences
round(summary(as.vector(exprs(msms) - spcm)),3)
```

```
1      Min.  1st Qu.   Median     Mean  3rd Qu.     Max.
2 -115.552   -0.476   -0.007    0.210    0.881  137.911
```

10. A first comparative glance may be obtained by the use of function `spc.scatterplot()` as in Fig. 5, where the red crosses, outside the borders, correspond to proteins with log-fold change expressions higher than 1 or lower than -1, and with a signal for the most abundant condition not below 2SpC, showing informative proteins with putative differential expression (*see* **Note 5**):

```
###  Comparative Scatterplot
spc.scatterplot(exprs(msms),treat=pData(msms)$treat,
                trans="sqrt")
```

In summary the functions in `msmsEDA` are useful to inspect the dataset before any inference, and take acknowledge of possible flaws

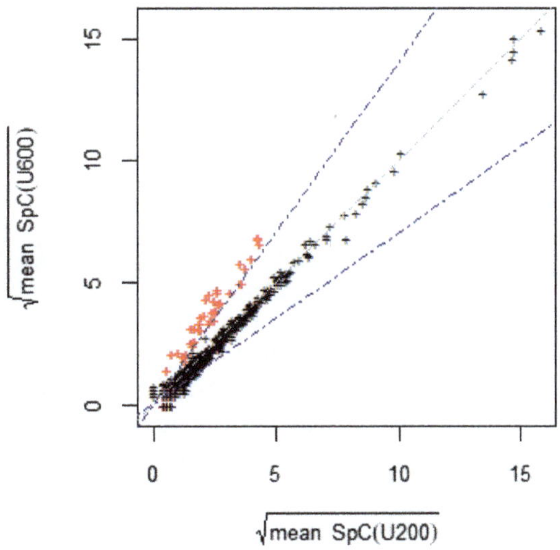

Fig. 5 Scatterplot of square root transformed mean expressions. Each cross corresponds to a protein in the dataset. The crosses outside the dashed borders, in red, are putative differentially expressed proteins with an absolute log-fold change higher than 1

in the experimental process. It is always good to start knowing what you have at hands before getting any *p*-value. Batch effects should be corrected, and outliers removed, to start with a clean and confident dataset. Similarly, it is wise to limit biases as much as possible.

3.2 Inference

The models in `msmsTests` are Generalized Linear Models (*see* **Note 6**), where the model is of the form;

$$\log \mu_i = \sum_j \beta_j x_{ij} \tag{4}$$

where the x_{ij} are the design matrix elements and β_j the model parameters. The probability distribution of the response is either the Poisson distribution (*see* **Note 7**) or the Negative Binomial (*see* **Note 8**).

An extension of the GLM, making abstraction of the true distribution, which allows for overdispersion is the quasi-likelihood (*see* **Note 9**). Deciding which of the three available models to use depends of the number of samples in each condition, and of the variance to mean relationship. Each distribution has a different law:

$$\text{Poisson}: \quad Var(\Upsilon) = \mu \tag{5}$$

$$\text{NegativeBinomial}: \quad Var(\Upsilon) = \mu + \phi\mu^2 \tag{6}$$

$$\text{Quasi} - \text{likelihood}: \quad Var(\Upsilon) = \psi\mu \tag{7}$$

where ϕ and ψ are parameters to estimate.

The Poisson distribution depends of a single parameter and considers a variance equal to the mean. The Negative Binomial distribution depends of two parameters and allows greater flexibility in the variance. The quasi-likelihood depends also of two parameters and has flexibility in the variance. The variance to mean ratio is known as dispersion factor:

$$\psi = Var(\Upsilon)/\mu \tag{8}$$

1. To help us in deciding the optimal distribution for our dataset function `disp.estimates()` estimates the dispersion factor for each protein in the dataset and shows a density plot of the dispersion values, and a scatterplot mean versus variance in log scale. The function returns a set of quantiles of dispersion values for each factor in `pData(msms)`:

```
###  Dispersion estimates
dsp <- disp.estimates(msms.bc)
dsp
dsp <- disp.estimates(msms.bc,
            facs=pData(msms.bc)[,'treat',drop=FALSE],
            do.plot=TRUE, etit='Dispersion', wait=TRUE)
round(dsp,3)
```

```
1        0.25   0.5  0.75   0.9  0.95  0.99     1
2 treat 0.217 0.382 0.579 0.808 1.037 1.494 4.567
3 batch 0.249 0.445 0.666 1.158 1.907 5.313 8.140
```

With technical replicates, that is without biological variability, as is the case with this spiking experiment, we observe most of the dispersion values below 1, a sign of underdispersion, and aligned along the diagonal in a mean versus residual variance plot, as in Fig. 6. In these cases both models are useful, with Poisson the most indicated when the number of replicates is very limited.

2. The beauty of a spiking experiment is that we know the ground truth, and hence we may evaluate the results of any test:

```
###  Accession names
acc.nms <- rownames(exprs(msms))
###  Flag human proteins
fl.human <- grepl('_HUMAN$',acc.nms)
###  Truth table
table(fl.human)
```

```
1 fl
2 FALSE  TRUE
3   629    46
```

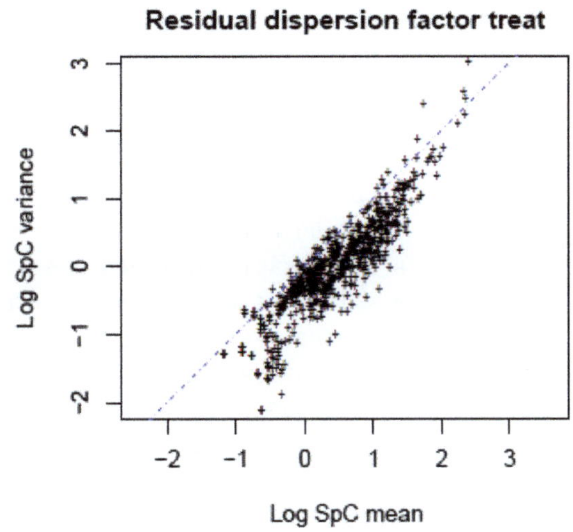

Fig. 6 Scatterplot comparing mean versus residual variance. The blue dashed line corresponds to variance equal to mean, as expected in Poisson distributions

3. Let us now apply a Poisson test to detect the spiked human proteins:

```
###  Load library
library(msmsTests)
###  Null and Alternative models
null.f <- "y~1+batch"
alt.f <- "y~1+batch+treat"
###  Size normalizing offsets
div <- apply(exprs(msms),2,sum)
###  Poisson GLM - the following instruction may trigger warn
pois.res <- msms.glm.pois(msms,alt.f,null.f,div=div)
str(pois.res)
```

```
'data.frame':   675 obs. of  3 variables:
 $ LogFC   : num  -0.304 -0.455 0.115 -0.6 -1.327 ...
 $ D       : num  0.269 5.584 10.271 2.594 5.759 ...
 $ p.value: num  0.60387 0.01812 0.00135 0.10726 0.01641 ...
```

The return value is a data.frame with three columns, the log-fold change in `LogFC`, the residual deviance of the test in `D`, and the p-values in column `p.values`, and with a row for each row in the expression set. Sometimes, one gets warning messages from this GLM regression (*see* **Note 10**):

```
###  Top 20 features, sorted by p-value
o <- order(pois.res$p.value)
head(pois.res[o,],20)
```

```
                               LogFC        D      p.value
sp|P02787|TRFE_HUMAN   0.93568486 84.43629 3.967972e-20
sp|P04040|CATA_HUMAN  -0.06125810 82.20029 1.229719e-19
sp|P08758|ANXA5_HUMAN -0.10071494 66.82651 2.964834e-16
sp|P02144|MYG_HUMAN   -0.13091423 66.80697 2.994377e-16
sp|P02788|TRFL_HUMAN   1.31170189 66.72445 3.122389e-16
sp|P00167|CYB5_HUMAN   0.72838482 62.89724 2.177767e-15
sp|P15559|NQO1_HUMAN  -0.22341691 59.88300 1.006673e-14
sp|P02768|ALBU_HUMAN   0.91359958 53.80383 2.215404e-13
sp|P06396|GELS_HUMAN   0.25324175 52.54276 4.209811e-13
sp|P01008|ANT3_HUMAN   0.87632750 44.70765 2.287615e-11
sp|P00915|CAH1_HUMAN   0.29836364 44.27115 2.858981e-11
sp|P55957|BID_HUMAN   -0.54632246 37.07149 1.138763e-09
sp|P41159|LEP_HUMAN   -0.39575867 34.27004 4.797103e-09
sp|P01375|TNFA_HUMAN   0.38589325 33.77986 6.171453e-09
sp|P63279|UBC9_HUMAN   0.76675464 29.65264 5.168232e-08
sp|P06732|KCRM_HUMAN   0.00941951 29.55172 5.444419e-08
sp|P99999|CYC_HUMAN    0.36452534 29.09425 6.894134e-08
sp|P08263|GSTA1_HUMAN  1.06454350 28.39686 9.882545e-08
sp|P10636|TAU_HUMAN    0.04235621 27.88645 1.286475e-07
sp|Q06830|PRDX1_HUMAN  0.14253601 27.59798 1.493347e-07
```

4. We may obtain a confusion matrix (*see* **Note 11**) for these results, comparing the truth versus the significant, which by now we take all proteins with a *p*-value ≤ 0.05:

```
fl.signif.pois <- pois.res$p.value <= 0.05
###  Confusion matrix
table(fl.human,fl.signif.pois)
```

```
         fl.signif.pois
fl.human FALSE TRUE
   FALSE   611   18
   TRUE      4   42
```

As we did a test for each of the rows in the expression matrix, that is 675 tests, a significance of level of 0.05 means that on average we expect to get 5% false positives (FP) by chance, that is 33–34.

5. The *p*-values may be adjusted to a given false discovery rate (FDR) with the Benjamini-Hochberg method [11] thus adjusting for multiple tests. This may be done with function `p.adjust()` (*see* **Note 12**):

```
### Adjusting p-values for multitest
adjp <- p.adjust(pois.res$p.value,method='BH')
fl.adj.signif.pois <- adjp <= 0.05
###  Confusion matrix
table(fl.human,fl.adj.signif.pois)
```

```
         fl.adj.signif.pois
fl.human FALSE TRUE
   FALSE   627    2
   TRUE      9   37
```

After the adjustment one loses 5 true positives (TP), but at the same time, one removes 16 FP. In other words, the adjustment led to lower sensitivity in favor of specificity, which is required for high reproducibility.

6. Let us see now the results of the quasi-likelihood method:

```
###  QLL GLM
ql.res <- msms.glm.qlll(msms,alt.f,null.f,div=div)
str(ql.res)
```

```
'data.frame':   675 obs. of  3 variables:
 $ LogFC   : num  -0.304 -0.455 0.115 -0.6 -1.327 ...
 $ D       : num  0.269 5.584 10.271 2.594 5.759 ...
 $ p.value: num  0.67461 0.10237 0.00524 0.0751 0.25088 ...
```

```
fl.signif.ql <- ql.res$p.value <= 0.05
###  Confusion matrix
table(fl.human,fl.signif.ql
```

```
         fl.signif.ql
fl.human FALSE TRUE
   FALSE   531   98
   TRUE      0   46
```

```
### Adjusting p-values for multitest
adjp <- p.adjust(ql.res$p.value,method='BH')
fl.adj.signif.ql <- adjp <= 0.05
###  Confusion matrix
table(fl.human,fl.adj.signif.ql)
```

```
         fl.adj.signif.ql
fl.human FALSE TRUE
   FALSE   611   18
   TRUE      5   41
```

Before adjusting the p-values this test has detected all TP but at the expenses of a high number of FP. After adjusting the FDR, the FP have been reduced, but there are still 18 of them, on the other hand we lost 5 TP. The quasi-likelihood is not robust to very low variances [20] and tends to produce a higher number of FP. The reduced variance in this case may be caused by the batch effects correction on technical replicates.

7. Finally let us see the results of the negative binomial in `edgeR`:

```
nb.res <- msms.edgeR(msms,alt.f,null.f,div=div,fnm='treat')
str(nb.res)
```

```
'data.frame':   675 obs. of  3 variables:
 $ LogFC   : num  0.0269 -0.1264 -0.1878 -0.085 -0.1185 ...
 $ LR      : num  0.269 5.584 10.271 2.594 5.759 ...
 $ p.value: num  0.60387 0.01812 0.00135 0.10726 0.01641 ...
```

```
###  Confusion matrix
fl.signif.nb <- nb.res$p.value <= 0.05
table(fl.human,fl.signif.nb)
```

```
         fl.signif.nb
fl.human FALSE TRUE
   FALSE   611   18
   TRUE      4   42
```

```
### Adjusting p-values for multitest
adjp <- p.adjust(nb.res$p.value,method='BH')
fl.adj.signif.nb <- adjp <= 0.05
table(fl.human,fl.adj.signif.nb)
```

```
         fl.adj.signif.nb
fl.human FALSE TRUE
   FALSE   627    2
   TRUE      9   37
```

This is a very fast function (faster than the other two), which, in this example, produces the same results as the Poisson tests.

8. Assembling the confusion matrices of the three tests in a single matrix we may compare easily the results (*see* **Note 13**):

```
### Comparative table
ct <- rbind( as.vector(table(fl.human,fl.adj.signif.pois)),
             as.vector(table(fl.human,fl.adj.signif.ql)),
             as.vector(table(fl.human,fl.adj.signif.nb)))
rownames(ct) <- c('Poisson','QLL','NB')
colnames(ct) <- c('TN','FN','FP','TP')
ct
```

```
1           TN FN FP TP
2 Poisson 627  9  2 37
3 QLL     611  5 18 41
4 NB      627  9  2 37
```

The conclusion is that with technical replicates both the Poisson or the NB are good options, and that the QL tends to produce a higher number of FP. With a single replicate use Poisson, with over two replicates `edgeR` could be more suited (*see* **Note 14**).

3.3 Reproducibility

Although the UPS1 mix is equimolar in its components, the signal we obtain for each of them is different, spanning two orders of magnitude. This range of signals will help us in studying the balance between sensitivity and specificity. Our concern now is having some experimental evidence about the best way to obtain reproducible results, with very limited false positives but the highest sensitivity:

1. In this example we will use a dataset from the `msmsTests` package with samples of yeast lysate spiked with UPS1, similar to that from the `msmsEDA` package. In this case, the experiments were not batch balanced in all conditions, allowing for extra variability beyond the technical replicates (*see* Fig. 7):

```
###  Library and data
library(msmsTests)
data(msms.spk)
treat <- pData(msms.spk)$treat
table(treat)
```

```
1 treat
2 U100 U200 U400 U600
3    4    6    3    6
```

2. First we will compare samples spiked with 400 fmols of UPS1 vs samples spiked with 200 fmols, giving a real log-fold change of 1. The key function here is `test.results()` which based on the expression set and the test results returns a table sorted by FDR adjusted *p*-value with the observed average SpC by condition, the log-fold change from the averages, the log-fold change estimated from the model, the likelihood ratio between

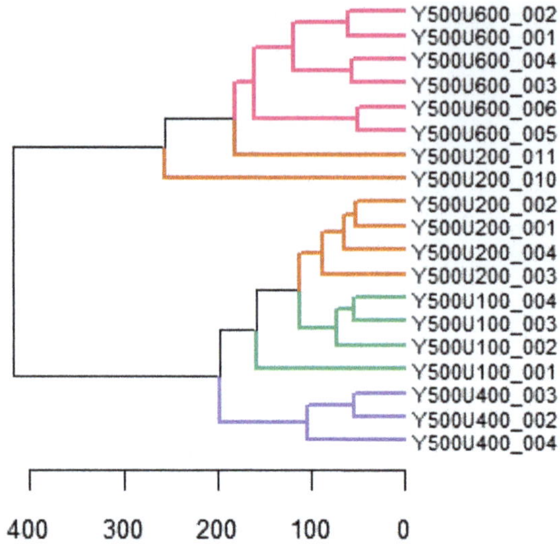

Fig. 7 Hierarchical clustering of samples in dataset *msms.spk*. Dendrogram branches are colored by treatment level

alternative and null model, the *p*-value, the FDR adjusted *p*-value, and a Boolean indicating whether the feature is considered differentially expressed in reproducible terms (DEP). This Boolean results of the logical AND of three conditions, adjusted *p*-value below a threshold, absolute log-fold change above a given threshold, and signal of the most abundant condition above a given threshold in SpC. In the example below the thresholds are 0.05, 1, and 2. The comments in the code will guide you through the process:

```
###  Subset to U200 and U400
fl <- (treat=='U200' | treat=='U400')
msms <- msms.spk[,fl]
pData(msms)$treat <- droplevels(pData(msms)$treat)
msms <- pp.msms.data(msms)
treat <- pData(msms)$treat

###  Ground truth
acc.nms <- rownames(exprs(msms))
fl.human <- grepl('_HUMAN$',acc.nms)

###  Null and Alternative models
null.f <- "y~1"
alt.f <- "y~1+treat"

###  Size normalizing offsets
div <- apply(exprs(msms),2,sum)

###  Differential expression tests
nb.res <- msms.edgeR(msms,alt.f,null.f,div=div,fnm='treat')
head(nb.res)
```

```
1                 LogFC         LR      p.value
2 YKL060C -0.6671077 15.8449425 6.875003e-05
3 YDR155C -0.3374664 17.8138902 2.435984e-05
4 YOL086C  0.1400088  0.4173946 5.182399e-01
5 YJR104C -0.7825334 30.8585964 2.775300e-08
6 YGR192C -0.6672964 42.5783675 6.790625e-11
7 YLR150W -0.3961689 13.6527176 2.199228e-04
```

```
###  Flags of significance based on p-values
###     raw p-values
fl.05 <- nb.res$p.value <= 0.05
fl.01 <- nb.res$p.value <= 0.01
###     FDR adjusted p-values
adjp <- p.adjust(nb.res$p.value,method='BH')
fl.adj.05 <- adjp <= 0.05
fl.adj.01 <- adjp <= 0.01

###  Significance, signal and effect size thresholds
alpha.cut <- 0.05    ###  significance
SpC.cut <- 2         ###  signal threshold
lFC.cut <- 1         ###  effect size threshold

###  Post-test filter
nb.tbl <- test.results(nb.res,msms,treat, 'U400','U200',div,
                   alpha=alpha.cut, minSpC=SpC.cut,
                   minLFC=lFC.cut,method='BH')
head(nb.tbl$tres,10)
```

```
1                 U400  U200  lFC.Av   LogFC    LR   p.value      adjp   DEP
2 ALBU_HUMAN      29.0   9.3  1.7710  1.7370 49.90 1.615e-12 1.069e-09  TRUE
3 TRFL_HUMAN      37.7  16.2  1.3300  1.3280 44.05 3.200e-11 1.059e-08  TRUE
4 YGR192C        100.7 172.8 -0.6742 -0.6673 42.58 6.791e-11 1.284e-08 FALSE
5 TRFE_HUMAN      43.3  20.5  1.1880  1.1890 42.32 7.760e-11 1.284e-08  TRUE
6 YJR104C         98.7 185.0 -0.7784 -0.7825 30.86 2.775e-08 3.674e-06 FALSE
7 YBL087C (+1)    29.0  60.0 -0.9322 -0.9315 30.16 3.987e-08 4.399e-06 FALSE
8 ANT3_HUMAN      17.0   5.8  1.6420  1.6360 27.42 1.636e-07 1.547e-05  TRUE
9 SYH_HUMAN        5.0   0.3  4.0020  3.5850 23.57 1.204e-06 9.960e-05  TRUE
10 TAU_HUMAN       8.0   2.0  2.0980  2.0460 18.28 1.903e-05 1.400e-03  TRUE
11 YDR155C       145.0 198.3 -0.3403 -0.3375 17.81 2.436e-05 1.613e-03 FALSE
```

```
fl.rep.05 <- nb.tbl$tres[rownames(nb.res),'DEP']

### Comparative table

ct <- rbind( as.vector(table(fl.human,fl.05)),
             as.vector(table(fl.human,fl.01)),
             as.vector(table(fl.human,fl.adj.05)),
             as.vector(table(fl.human,fl.adj.01)),
             as.vector(table(fl.human,fl.rep.05)))
rownames(ct) <- c('alpha 0.05','alpha 0.01',
                  'adj.p 0.05','adj.p 0.01',
                  'repr. 0.05')
colnames(ct) <- c('TN','FN','FP','TP')
ct
```

1		TN	FN	FP	TP
2	alpha 0.05	585	13	42	22
3	alpha 0.01	602	17	25	18
4	adj.p 0.05	619	20	8	15
5	adj.p 0.01	620	24	7	11
6	repr. 0.05	626	23	1	12

From the comparative table we see that the unadjusted *p*-values lead to a high number of FP, bigger than the number of TP, even with $\alpha < 0.01$. The FDR adjusted *p*-values reduce drastically the number of FP at the cost of sensitivity (TP). In addition, the post-test filter by thresholds allows a bigger correction of FP with a lower cost in sensitivity. The result is already of interest, considering the true log-fold change of 1 (*see* **Note 15**).

3. Let us repeat the exercise, now with U600 vs U200, with a bigger effect size, giving a real log-fold change of 1.585 (FC of 3):

```
###  U200 vs U600
treat <- pData(msms.spk)$treat
fl <- (treat=='U200' | treat=='U600')
msms <- msms.spk[,fl]
pData(msms)$treat <- droplevels(pData(msms)$treat)
msms <- pp.msms.data(msms)
treat <- pData(msms)$treat

###  Ground truth
acc.nms <- rownames(exprs(msms))
fl.human <- grepl('_HUMAN$',acc.nms)

###  Null and Alternative models
null.f <- "y~1"
alt.f <- "y~1+treat"

###  Size normalizing offsets
div <- apply(exprs(msms),2,sum)

nb.res <- msms.edgeR(msms,alt.f,null.f,div=div,fnm='treat')
head(nb.res)
```

```
1                LogFC           LR      p.value
2 YKL060C  0.12791153  1.314098277 2.516540e-01
3 YDR155C -0.18010309  8.796037198 3.018856e-03
4 YOL086C  0.42637517  6.944057521 8.409813e-03
5 YJR104C  0.01216915  0.006711829 9.347058e-01
6 YGR192C -0.37236189 27.215119071 1.820296e-07
7 YLR150W -0.26918331 12.937495818 3.220656e-04
```

```
fl.05 <- nb.res$p.value <= 0.05
fl.01 <- nb.res$p.value <= 0.01
adjp <- p.adjust(nb.res$p.value,method='BH')
fl.adj.05 <- adjp <= 0.05
fl.adj.01 <- adjp <= 0.01

###  Significance, signal and effect size thresholds
alpha.cut <- 0.05
SpC.cut <- 2
lFC.cut <- 1

###  Post-test filter
nb.tbl <- test.results(nb.res,msms,treat, 'U600','U200',div,
                 alpha=alpha.cut, minSpC=SpC.cut,
                 minLFC=lFC.cut,method='BH')
head(nb.tbl$tres,10)
```

```
1              U600  U200  lFC.Av   LogFC     LR  p.value      adjp   DEP
2 TRFL_HUMAN   60.3  16.2  1.8240  1.8230 150.00 1.765e-34 1.172e-31  TRUE
3 ALBU_HUMAN   38.2   9.3  1.9760  1.9490 104.30 1.722e-24 5.719e-22  TRUE
4 TRFE_HUMAN   53.3  20.5  1.3060  1.3050  81.29 1.945e-19 4.304e-17  TRUE
5 CAH2_HUMAN   19.0   3.7  2.2810  2.2670  63.76 1.409e-15 2.339e-13  TRUE
6 CATA_HUMAN   25.8   8.5  1.5290  1.5210  50.05 1.497e-12 1.988e-10  TRUE
7 CAH1_HUMAN    7.7   0.3  4.4170  4.0280  47.49 5.519e-12 6.108e-10  TRUE
8 MYG_HUMAN    15.3   3.3  2.1340  2.0920  45.65 1.411e-11 1.338e-09  TRUE
9 KCRM_HUMAN   21.7   7.7  1.4050  1.4150  36.30 1.694e-09 1.406e-07  TRUE
10 ANT3_HUMAN  16.2   5.8  1.3850  1.3830  27.31 1.737e-07 1.209e-05  TRUE
11 YGR192C    140.2 172.8 -0.3785 -0.3724  27.22 1.820e-07 1.209e-05 FALSE
```

```
fl.rep.05 <- nb.tbl$tres[rownames(nb.res),'DEP']

### Comparative table
ct <- rbind( as.vector(table(fl.human,fl.05)),
             as.vector(table(fl.human,fl.01)),
             as.vector(table(fl.human,fl.adj.05)),
             as.vector(table(fl.human,fl.adj.01)),
             as.vector(table(fl.human,fl.rep.05)))
rownames(ct) <- c('alpha 0.05','alpha 0.01',
                  'adj.p 0.05','adj.p 0.01',
                  'repr. 0.05')
colnames(ct) <- c('TN','FN','FP','TP')
ct
```

```
1              TN FN FP TP
2 alpha 0.05 574  4 51 35
3 alpha 0.01 613 11 12 28
4 adj.p 0.05 619 15  6 24
5 adj.p 0.01 622 20  3 19
6 repr. 0.05 624 16  1 23
```

We see the same pattern as with the previous example. The post-test filter allows us to go beyond *p*-values and better balance sensitivity with specificity.

Biomarker discovery is just the initial step in the more complex process of finding a model able to predict biomedical outcomes (*see* **Note 16**).

3.4 Visualizing and Checking the Results

It is always good practice to inspect the results before taking for granted any list of DEPs. The `msmsTest` package offers different resources for doing it:

1. Let us first get some results:

```
library(msmsEDA)
library(msmsTests)
### Load dataset
data(msms.dataset)
###  Preprocess dataset
msms <- pp.msms.data(msms.dataset)
###  Null and alternative models
null.f <- "y~1+batch"
alt.f <- "y~1+batch+treat"
###  Sample sizes for normalization by offsets
div <- apply(exprs(msms),2,sum)
###  Tests
nb.res <- msms.edgeR(msms,alt.f,null.f,div=div)
###  Post-test filters to improve reproducibility
nb.tbl <- test.results(nb.res, msms,pData(msms)$treat, 'U600'
                         div, alpha=0.05, minSpC=2, minLFC=1, m
                             )
```

2. Function `pval.by.fc()` returns a table with the cumulative number of features at different *p*-value thresholds by bins of log-fold change. This is a simple table that allows a quick visualization of the distribution of size-effects and significance level:

```
###  Cummulated distribution of p-values by log-fold change
   bins
pval.by.fc(nb.tbl$tres$adjp,nb.tbl$tres$LogFC)
```

```
                      p.vals
LogFC                 <=0.001 <=0.005 <=0.01 <=0.05 <=0.1 <=0.2 <=1
  (-Inf,-10]                0       0      0      0     0     0   0
  (-10,-2]                  0       0      0      0     0     1   5
  (-2,-1]                   0       0      0      0     0     0  39
  (-1,-0.848]               0       0      0      0     0     0  14
  (-0.848,-0.585]           0       0      0      0     0     0  43
  (-0.585,0]                1       1      1      2     2     2 395
  (0,0.585]                 0       0      0      0     0     0 120
  (0.585,0.848]             0       0      0      0     0     0   8
  (0.848,1]                 2       3      3      3     3     3   5
  (1,2]                    24      28     28     31    33    34  41
  (2,10]                    1       2      2      3     3     3   5
  (10, Inf]                 0       0      0      0     0     0   0
  Tot                      28      34     34     39    41    43 675
```

3. The graphic counterpart may be a volcano plot, using function `res.volcanoplot()`, (*see* Fig. 8):

```
###   Volcanoplot
res.volcanoplot(nb.tbl$tres)
```

4. Another graphic means to inspect the results is the heatmap. The next code snippet uses function `counts.heatmap()` in package `msmsEDA` to show graphically the matrix of spectral counts with only the differentially expressed features (*see* Fig. 9):

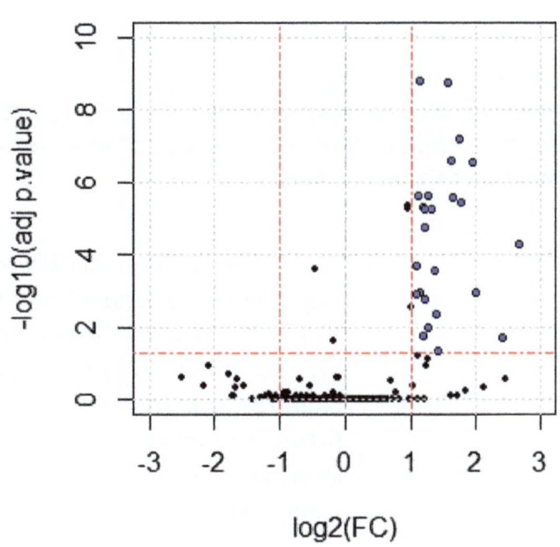

Fig. 8 Volcanoplot of tests results: The navy dots, located above -$\log_{10}$(0.05) and outside of the (− 1,+ 1) log-fold region, are considered putative DEPs

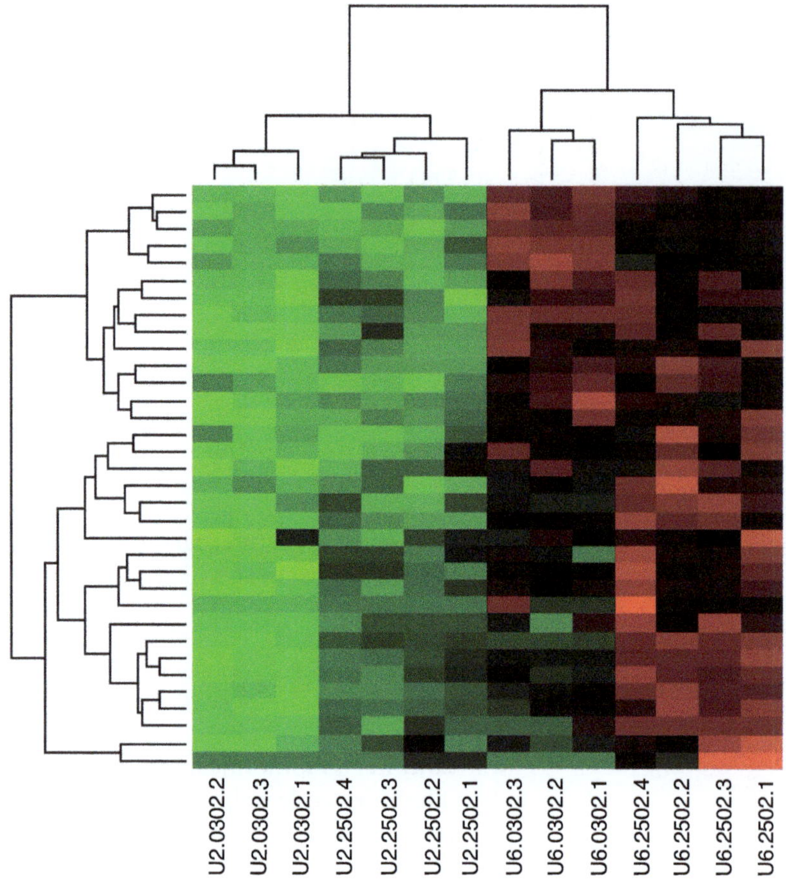

Fig. 9 Heatmap of the spectral counts matrix with only DEPs

```
### Heatmap with DEPs only
spcm <- exprs(msms)[rownames(nb.tbl$tres)[nb.tbl$tres$DEP],]
msn <- MSnSet(spcm, pData=pData(msms))
counts.heatmap(msn)
```

5. When the number of differentially expressed features is not too big, the inspection of the table (returned by function `test.results()` with the post-test filter results) is very informative. We see the mean SpC observed in each condition, the log-fold change estimated from the means, the log-fold change from fitting the model, the log-likelihood ratio of the two models, alternative vs null, the raw *p*-value and the FDR adjusted *p*-value, and the flags with the filters result (*see* Table 2).

Table 2
Top 50 features sorted by FDR adjusted *p*-value. Part of the results of the post-test filters done by function `test.results()`

	U600	U200	lFC.Av	LogFC	LR	*p*.value	adjp	DEP
TRFE_HUMAN	47.0	18.3	1.300	1.300	84.40	3.97e–20	2.68e–17	TRUE
CATA_HUMAN	46.4	18.3	1.290	1.290	82.20	1.23e–19	4.15e–17	TRUE
ANXA5_HUMAN	34.0	12.6	1.380	1.380	66.80	2.96e–16	4.22e–14	TRUE
MYG_HUMAN	21.0	5.3	1.920	1.920	66.80	2.99e–16	4.22e–14	TRUE
TRFL_HUMAN	44.1	18.9	1.170	1.170	66.70	3.12e–16	4.22e–14	TRUE
CYB5_HUMAN	19.0	4.6	2.000	1.980	62.90	2.18e–15	2.45e–13	TRUE
NQO1_HUMAN	22.6	6.7	1.700	1.680	59.90	1.01e–14	9.71e–13	TRUE
ALBU_HUMAN	36.6	15.9	1.150	1.150	53.80	2.22e–13	1.87e–11	TRUE
GELS_HUMAN	21.6	6.9	1.610	1.590	52.50	4.21e–13	3.16e–11	TRUE
ANT3_HUMAN	31.4	13.9	1.120	1.120	44.70	2.29e–11	1.54e–09	TRUE
CAH1_HUMAN	18.6	6.0	1.570	1.560	44.30	2.86e–11	1.75e–09	TRUE
BID_HUMAN	13.1	3.7	1.760	1.740	37.10	1.14e–09	6.41e–08	TRUE
LEP_HUMAN	13.7	4.3	1.620	1.600	34.30	4.80e–09	2.49e–07	TRUE
TNFA_HUMAN	10.1	2.4	2.000	1.960	33.80	6.17e–09	2.98e–07	TRUE
UBC9_HUMAN	17.4	7.0	1.250	1.250	29.60	5.17e–08	2.30e–06	TRUE
KCRM_HUMAN	21.6	9.7	1.090	1.090	29.60	5.44e–08	2.30e–06	TRUE
CYC_HUMAN	11.3	3.4	1.660	1.640	29.10	6.89e–08	2.74e–06	TRUE
GSTA1_HUMAN	9.9	2.7	1.800	1.760	28.40	9.88e–08	3.71e–06	TRUE
TAU_HUMAN	25.6	12.7	0.955	0.953	27.90	1.29e–07	4.57e–06	FALSE
PRDX1_HUMAN	17.9	7.6	1.180	1.180	27.60	1.49e–07	5.04e–06	TRUE
RASH_HUMAN	14.9	5.7	1.320	1.310	27.30	1.71e–07	5.40e–06	TRUE
SYUG_HUMAN	17.1	7.1	1.210	1.200	27.30	1.76e–07	5.40e–06	TRUE
SYH_HUMAN	25.1	12.6	0.950	0.946	27.10	1.97e–07	5.77e–06	FALSE
HBB_HUMAN	15.4	6.4	1.220	1.200	24.70	6.75e–07	1.90e–05	TRUE
HBA_HUMAN	4.4	0.6	2.880	2.660	22.70	1.91e–06	5.16e–05	TRUE
UBE2C_HUMAN	14.6	6.6	1.090	1.090	19.90	8.29e–06	2.15e–04	TRUE
YBL087C (+1)	46.1	61.6	−0.470	−0.465	19.50	1.00e–05	2.51e–04	FALSE
PPIA_HUMAN	9.7	3.6	1.380	1.360	19.10	1.22e–05	2.93e–04	TRUE
CAH2_HUMAN	11.4	5.0	1.130	1.120	16.60	4.72e–05	1.10e–03	TRUE
CRP_HUMAN	4.6	1.0	2.120	2.010	16.40	5.00e–05	1.12e–03	TRUE

(continued)

Table 2
(continued)

	U600	U200	IFC.Av	LogFC	LR	*p*.value	adjp	DEP
NQO2_HUMAN	12.0	5.4	1.100	1.080	16.20	5.61e–05	1.22e–03	TRUE
FABPH_HUMAN	9.4	3.9	1.230	1.210	15.50	8.06e–05	1.70e–03	TRUE
GSTP1_HUMAN	12.4	6.0	0.987	0.985	14.50	1.41e–04	2.88e–03	FALSE
IFNG_HUMAN	6.7	2.4	1.410	1.370	13.60	2.24e–04	4.45e–03	TRUE
RETBP_HUMAN	6.9	2.7	1.270	1.250	12.00	5.47e–04	1.06e–02	TRUE
THIO_HUMAN	6.9	2.9	1.210	1.180	10.90	9.70e–04	1.82e–02	TRUE
IL8_HUMAN	2.1	0.3	2.860	2.420	10.70	1.06e–03	1.93e–02	TRUE
YDR155C	165.0	181.0	−0.190	−0.188	10.30	1.35e–03	2.40e–02	FALSE
B2MG_HUMAN	4.1	1.4	1.470	1.410	9.00	2.70e–03	4.68e–02	TRUE
LALBA_HUMAN	5.9	2.6	1.120	1.100	8.32	3.91e–03	6.60e–02	FALSE
PDGFB_HUMAN	4.4	1.7	1.310	1.250	7.97	4.76e–03	7.83e–02	FALSE
YDR163W	0.3	1.6	−2.490	−2.090	7.17	7.41e–03	1.19e–01	FALSE
CO5_HUMAN	4.0	1.6	1.280	1.230	7.04	7.99e–03	1.25e–01	FALSE
YBR162C	0.4	1.7	−2.050	−1.770	6.06	1.38e–02	2.12e–01	FALSE
YOL086C	237.0	250.0	−0.120	−0.118	5.76	1.64e–02	2.45e–01	FALSE
YPR185W	0.0	0.6	−Inf	−2.510	5.71	1.69e–02	2.45e–01	FALSE
YJL014W	0.0	0.6	−Inf	−2.500	5.69	1.70e–02	2.45e–01	FALSE
YGL030W	0.0	0.6	−Inf	−2.490	5.64	1.76e–02	2.47e–01	FALSE
YKL060C	202.0	213.0	−0.131	−0.126	5.58	1.81e–02	2.50e–01	FALSE
YDR192C	0.6	0.0	Inf	2.450	5.38	2.03e–02	2.74e–01	FALSE

3.5 Normalizing Secretomes

When comparing two biological conditions with secretomes in cell-line cultures, we aim to compare the proteins secreted by one single cell in each of two given biological states of a cell line, even when one state is globally stimulated or depressed with respect to the other (*see* **Note 17**). So with cell-line secretomes we may compare two biological conditions unbiasedly by counting the cells and measuring the total secreted protein, and by normalizing accordingly. This dataset consists of three technical replicates of two biological replicates in the treatment of A459 lung cancer cells with TGFβ to induce the epithelial-to-mesenchymal transition [21]:

1. First, let us load the files and create an instance of an MSnSet with the experiment data:

```
library(RColorBrewer)
library(stringr)
library(msmsEDA)
library(msmsTests)

### Read metadata
samples <- read.table(file="A549-EMT_RB3i4_samples.txt",sep="\
    t",
                  dec=',',header=TRUE,stringsAsFactors=
                     FALSE)
cnms <- colnames(samples)
samples$Sample <- as.character(samples$Sample)
samples
```

```
1          Sample Treat BR cell.no sc.prot
2 1    EMT.RB3.01   EMT B3    5.87    50.5
3 2    EMT.RB3.02   EMT B3    5.87    50.5
4 3    EMT.RB3.03   EMT B3    5.87    50.5
5 4    EMT.RB4.01   EMT B4    4.05    40.5
6 5    EMT.RB4.02   EMT B4    4.05    40.5
7 6    EMT.RB4.03   EMT B4    4.05    40.5
8 7   Ctrl.RB3.01  Ctrl B3    8.37    37.6
9 8   Ctrl.RB3.02  Ctrl B3    8.37    37.6
10 9   Ctrl.RB3.03  Ctrl B3    8.37    37.6
11 10 Ctrl.RB4.01  Ctrl B4    6.00    27.6
12 11 Ctrl.RB4.02  Ctrl B4    6.00    27.6
13 12 Ctrl.RB4.03  Ctrl B4    6.00    27.6
```

```
pdata <- samples[,c('Treat','BR')]
rownames(pdata) <- samples$Sample

###  Read expression counts
data.flnm <- "A549-EMT_RB3i4_240513_R.txt"
msms <- read.table(file=data.flnm,header=TRUE,sep="\t",
                   quote="",stringsAsFactors=FALSE)

###  Subset to the selected samples
msms.counts <- data.matrix(msms[ ,samples$Sample])
rownames(msms.counts) <- msms$Accession

### Accession to gene names dictionary
patt <- "GN=[A-Z0-9_]+"
gn.dic <- str_extract(msms$Protein,patt)
gn.dic <- substring(gn.dic,4)
names(gn.dic) <- msms$Accession

### Construct an MSnSet from the parts
msms <- MSnSet(msms.counts,pData=pdata)
validObject(msms)
```

```
1 [1] TRUE
```

We will compare the results of normalizing only by library size to the results of normalizing when the secretion rate of each biological condition is taken into account (*see* **Note 18**).

2. The next code snippet computes the scaled normalizing factors by library size (*see* Fig. 10):

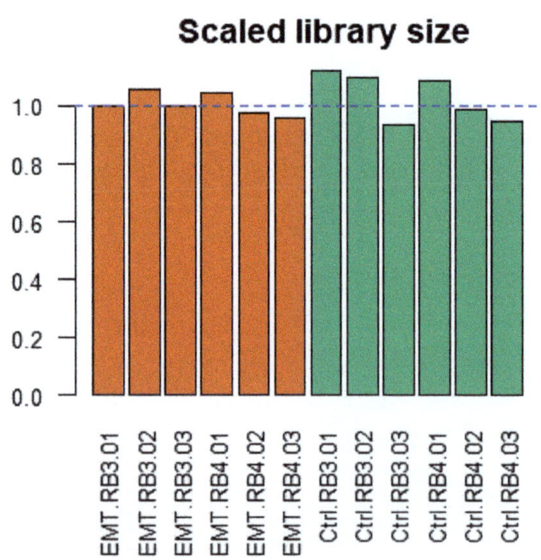

Fig. 10 Barplot with scaled library size offsets

```
###  Pre-process
msms <- pp.msms.data(msms)

###  Size offsets
div.sz <- apply(exprs(msms),2,sum)
###  Scale size offsets to the median
div.sz <- div.sz/median(div.sz)
###  Plot scaled size offsets
spc.barplots(exprs(msms),fact=pData(msms)$Treat,
             cex.axis=0.8,cex.names=0.8)
title(main="Scaled library size",line=3)
```

3. The biological replicates are taken as blocks. So, the null model takes into account the variability due to the replicates, while the alternative model considers both factors, replicates and treatment:

```
###  Tests, normalizing just by library size
null.f <- "y~1+BR"
alt.f <- "y~1+BR+Treat"
nb.res <- msms.edgeR(msms,alt.f,null.f,div=div.sz)
###  Post-test filters to improve reproducibility
nb.tbl <- test.results(nb.res, msms,pData(msms)$Treat, 'EMT',
                       div.sz, alpha=0.05, minSpC=2, minLFC=1
                           BH')
head(nb.tbl$tres,10)
```

```
1                EMT  Ctrl lFC.Av   LogFC     LR   p.value      adjp  DEP
2 MUC5A_HUMAN    0.0 199.8   -Inf -10.630 1638.0 0.000e+00 0.000e+00 TRUE
3 LAMC2_HUMAN  255.5  21.8  3.582   3.572 1420.0 9.574e-311 5.653e-308 TRUE
4 CO5_HUMAN      0.0 129.8   -Inf -10.010 1063.0 4.306e-233 1.695e-230 TRUE
5 BGH3_HUMAN   508.8 190.5  1.442   1.449  940.3 1.682e-206 4.966e-204 TRUE
6 MMP2_HUMAN   132.7   5.3  4.671   4.635  893.5 2.551e-196 6.026e-194 TRUE
7 CO4A2_HUMAN  202.0  44.5  2.212   2.210  674.2 1.230e-148 2.420e-146 TRUE
8 NPTX1_HUMAN  183.3  36.5  2.350   2.355  660.4 1.229e-145 2.074e-143 TRUE
9 CFAH_HUMAN     0.7  84.2 -6.944  -6.705  647.8 6.796e-143 1.003e-140 TRUE
10 TICN1_HUMAN 152.7  26.0  2.583   2.579  613.0 2.433e-135 3.192e-133 TRUE
11 LAMA5_HUMAN  34.7 172.0 -2.288  -2.277  580.4 3.022e-128 3.569e-126 TRUE
```

```
### Record the list of DEPs
DEPs.sz <- rownames(nb.tbl$tres)[nb.tbl$tres$DEP]
length(DEPs.sz)
```

```
1 [1] 295
```

```
lfc.sz <- nb.tbl$tres$LogFC
names(lfc.sz) <- rownames(nb.tbl$tres)
```

4. Now let us use the cell counts and total secreted proteins (*see* **Note 19**) to compute the secretion rate (*see* **Note 20**), and the offsets for biological normalization (*see* Table 3 and Fig. 11):

Table 3
Samples description. Samples ID, treatment levels, block levels, cell count, and secreted protein. Counts and quantity in arbitrary units. Normalization factors are previously scaled to the median

Sample	Treatment	Replica	Cells	Secreted
EMT.RB3.01	EMT	B3	5.87	50.5
EMT.RB3.02	EMT	B3	5.87	50.5
EMT.RB3.03	EMT	B3	5.87	50.5
EMT.RB4.01	EMT	B4	4.05	40.5
EMT.RB4.02	EMT	B4	4.05	40.5
EMT.RB4.03	EMT	B4	4.05	40.5
Ctrl.RB3.01	Ctrl	B3	8.37	37.6
Ctrl.RB3.02	Ctrl	B3	8.37	37.6
Ctrl.RB3.03	Ctrl	B3	8.37	37.6
Ctrl.RB4.01	Ctrl	B4	6.00	27.6
Ctrl.RB4.02	Ctrl	B4	6.00	27.6
Ctrl.RB4.03	Ctrl	B4	6.00	27.6

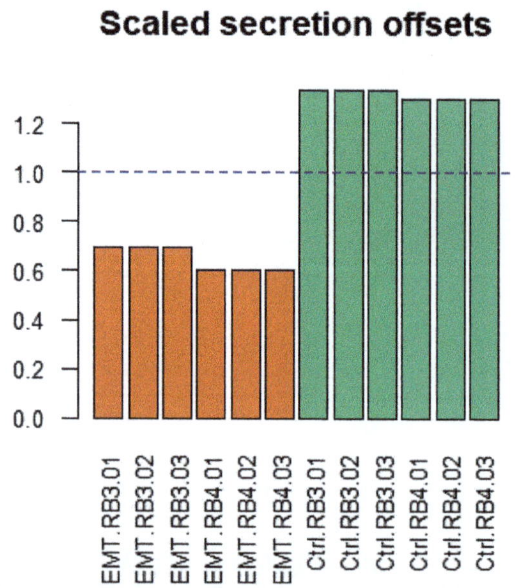

Fig. 11 Barplot with scaled secretion offsets

```
###  Cell offsets
div.scr <- samples$cell.no/samples$sc.prot
##  Scale offset to the median
div.scr <- div.scr/median(div.scr)
###  Plot scaled offsets
pal <- brewer.pal(8,'Dark2')
barplot(div.scr, col=pal[factor(samples$Treat)], las=2,
        names.arg=samples$Sample,cex.axis=0.8,cex.names=0.8)
title(main="Scaled secretion offsets")
abline(h=1,lty=2,col='navy')
```

5. Then, let us perform the tests under the biological normalization:

```
###  Tests, normalizing by size and secretion rate
null.f <- "y~1+BR"
alt.f <- "y~1+BR+Treat"
div.gbl <- div.sz * div.scr
nb.res <- msms.edgeR(msms,alt.f,null.f,div=div.gbl)
###  Post-test filters to improve reproducibility
nb.tbl <- test.results(nb.res, msms,pData(msms)$Treat,
                       'EMT','Ctrl',div.gbl, alpha=0.05,
                       minSpC=2, minLFC=1, method='BH')
head(nb.tbl$tres,10)
```

```
1                EMT  Ctrl lFC.Av LogFC   LR      p.value          adjp  DEP
2 BGH3_HUMAN   508.8 190.5  2.472 2.476 2853 0.000000e+00  0.000000e+00 TRUE
3 LAMC2_HUMAN  255.5  21.8  4.626 4.602 2643 0.000000e+00  0.000000e+00 TRUE
4 CO4A2_HUMAN  202.0  44.5  3.245 3.241 1546 0.000000e+00  0.000000e+00 TRUE
5 MMP2_HUMAN   132.7   5.3  5.703 5.657 1550 0.000000e+00  0.000000e+00 TRUE
6 NPTX1_HUMAN  183.3  36.5  3.390 3.391 1470 1.274689e-321 3.010984e-319 TRUE
7 TICN1_HUMAN  152.7  26.0  3.618 3.611 1300 1.114000e-284 2.193000e-282 TRUE
8 FINC_HUMAN   387.5 277.5  1.545 1.544 1124 1.792000e-246 3.023000e-244 TRUE
9 TIMP2_HUMAN  218.2  92.2  2.299 2.304 1119 2.324000e-245 3.431000e-243 TRUE
10 TSP1_HUMAN  192.7  73.2  2.460 2.459 1075 9.575000e-236 1.256000e-233 TRUE
11 FLNA_HUMAN  336.2 231.7  1.600 1.591 1018 2.096000e-223 2.475000e-221 TRUE
```

```
### Record the list of DEPs
DEPs.gbl <- rownames(nb.tbl$tres)[nb.tbl$tres$DEP]
length(DEPs.gbl)
```

```
1 [1] 477
```

```
lfc.gbl <- nb.tbl$tres$LogFC
names(lfc.gbl) <- rownames(nb.tbl$tres)
```

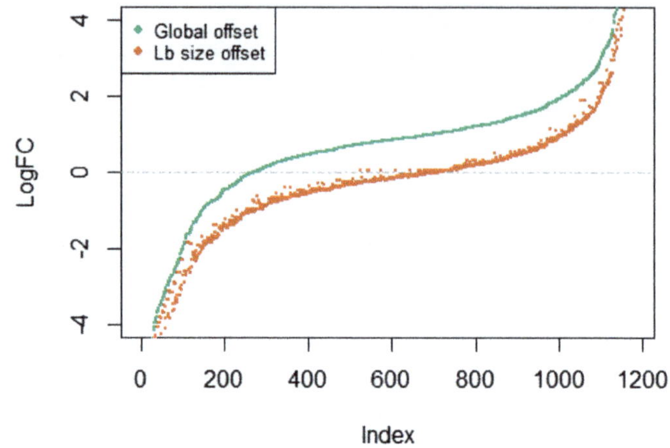

Fig. 12 Log-fold change bias when ignoring secretion rate

6. Figure 12 shows the sorted log-fold changes for all proteins in the dataset under each normalization. The library size normalized line follows the features order that result from sorting the biological normalization results. Features on any abscissa value are the same. This plot shows the bias incurred in log-fold change, for each protein, when the secretion rate is not taken into account:

```
### Show log-fold change bias between methods
o <- order(lfc.gbl)
lfc.gbl <- lfc.gbl[o]
lfc.sz <- lfc.sz[names(lfc.gbl)]

plot(lfc.gbl,pch=19,type='n',ylim=c(-4,4),ylab='LogFC')
abline(h=0,lty=4,col='gray')
points(lfc.gbl,pch=19,cex=0.3,col=pal[1])
points(lfc.sz,pch=19,cex=0.3,col=pal[2])
legend('topleft',pch=19,col=pal,cex=0.8,
       legend=c('Global offset','Lb size offset'))
```

3.6 Inspecting the Results

It is worth inspecting the results of this dataset, of a higher complexity than we have seen with the spiking datasets:

1. The biological signal is strong enough to overcome the normalization bias in 201 features that would be declared DEPs the same. However, 276 features would be considered not differentially expressed. On the other hand, 94 features would be falsely declared as DEPs:

```
###  Intersection of both sets of DEPs
length(intersect(DEPs.gbl,DEPs.sz))
```

```
1 [1] 201
```

```
### DEPs with the global normalization that were not
###    DEPs with the library size normalization
length(setdiff(DEPs.gbl,DEPs.sz))
```

```
1 [1] 276
```

```
### DEPs with the size normalization that were not
###    DEPs with the global normalization
length(setdiff(DEPs.sz,DEPs.gbl))
```

```
1 [1] 94
```

2. Distribution of log-fold changes by FDR adjusted *p*-values, table and volcanoplot (*see* Fig. 13):

```
### p-value by LFC distribution
pval.by.fc(nb.tbl$tres$adjp,nb.tbl$tres$LogFC)
```

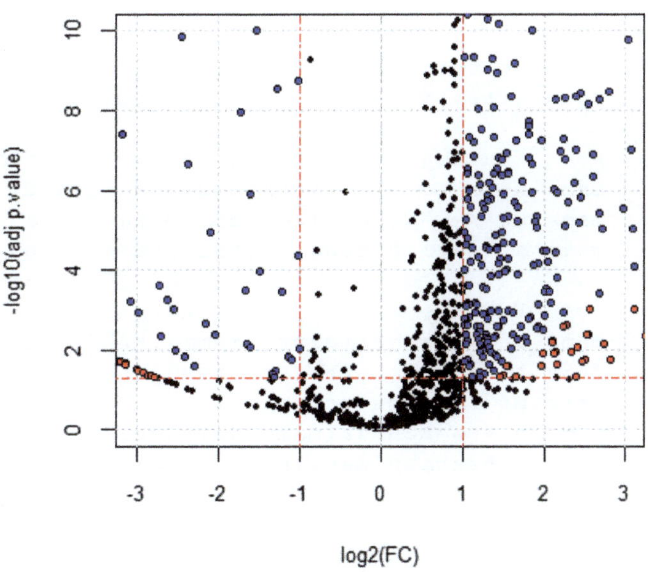

Fig. 13 A459 secretome volcanoplot: The navy dots are DEPs passing the post-test filters, while the red dots are significant features that do not show enough signal or effect size

```
1                  p.vals
2 LogFC            <=0.001 <=0.005 <=0.01 <=0.05 <=0.1 <=0.2  <=1
3   (-Inf,-10]           1       1      1      1     1     1    1
4   (-10,-2]            49      59     67     88    96   100  100
5   (-2,-1]             11      11     14     19    26    33   49
6   (-1,-0.848]          1       1      1      2     3     6   11
7   (-0.848,-0.585]      2       3      3     10    13    16   33
8   (-0.585,0]           2       2      3      6     8     9   79
9   (0,0.585]           14      17     21     39    51    69  190
10    (0.585,0.848]     52      63     71     95   112   123  143
11    (0.848,1]         54      58     63     75    81    94   99
12    (1,2]            200     235    246    272   293   304  306
13    (2,10]           137     146    150    162   166   166  166
14    (10, Inf]          0       0      0      0     0     0    0
15    Tot              523     596    640    769   850   921 1177
```

```
### Volcanoplot
res.volcanoplot(nb.tbl$tres)
```

3. To see the impact of the biological normalization, and of the post-test filters we may plot the heatmaps of the spectral counts matrix, before normalizing, after applying the global divisors to the same matrix, and selecting only the normalized rows corresponding to DEPs according to the post-test filters (*see* Fig. 14):

```
### Heatmap, no normalization
counts.heatmap(msms,fac=pData(msms)$Treat)

### Heatmap after global normalization
msms.nc <- sweep(exprs(msms),MARGIN=2,STATS=div.gbl,FUN="/")
msms.nc <- MSnSet(msms.nc,pData=pdata)
counts.heatmap(msms.nc,fac=pData(msms)$Treat)

### Heatmap after global normalization, only DEPs
msms.nc <- sweep(exprs(msms),MARGIN=2,STATS=div.gbl,FUN="/")
fnms <- rownames(nb.tbl$tres)[nb.tbl$tres$DEP]
msms.nc <- msms.nc[fnms,]
msms.nc <- MSnSet(msms.nc,pData=pdata)
counts.heatmap(msms.nc,fac=pData(msms)$Treat)
```

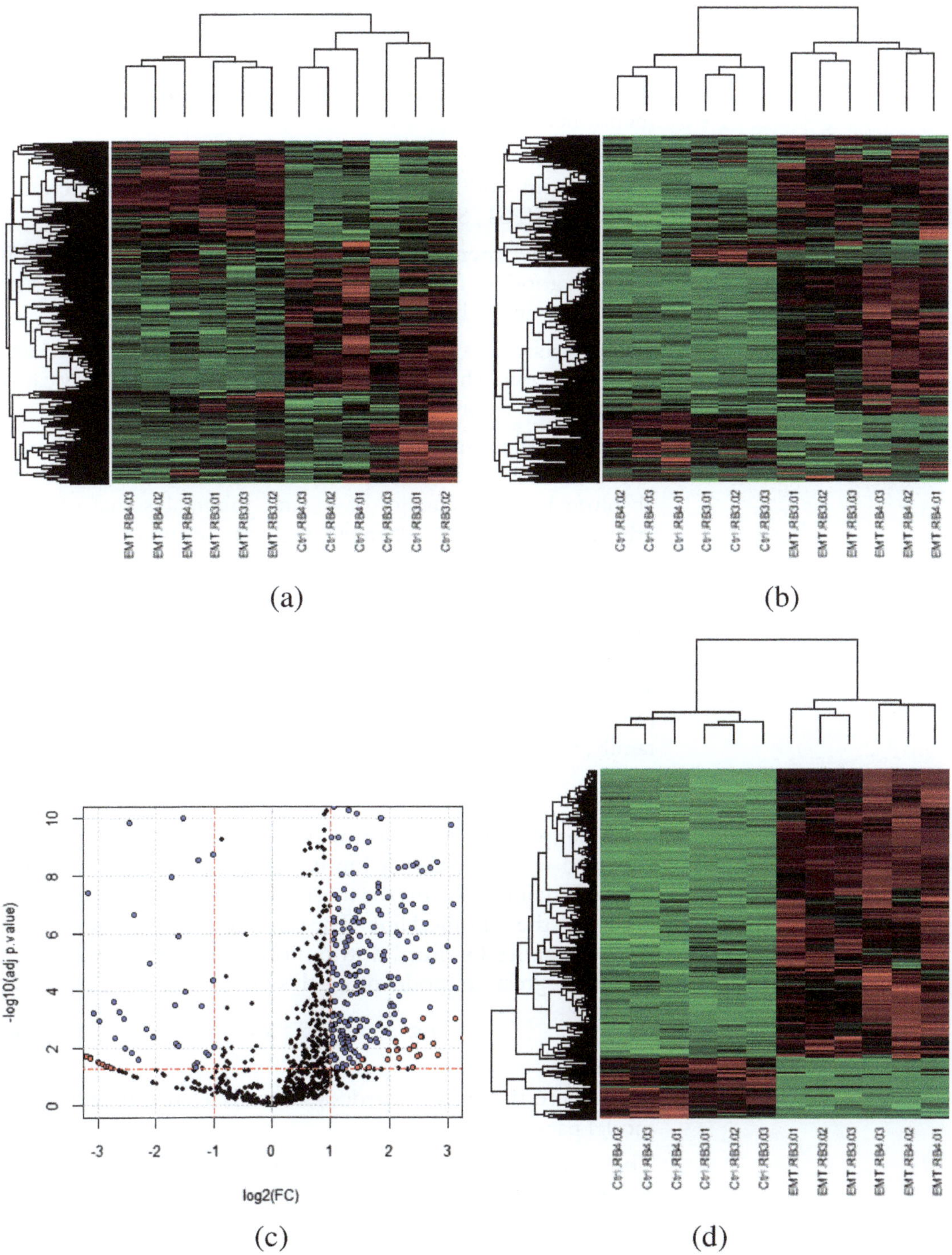

Fig. 14 Incidence of normalization and post-test filter. (**a**) Raw SpC matrix heatmap. (**b**) Bio normalized heatmap. (**c**) Volcanoplot. (**d**) Only DEPs heatmap

4 Notes

1. Both text files may be downloaded from the GitHub repository mentioned above.
2. **UPS1**: equimolar mix of 48 proteins, ranging in molecular weight from 6000 to 83,000 daltons (Universal Proteomics Standard, UPS1, Sigma-Aldrich).
3. A good **experimental design** is the best way to avoid biases. It is always good to refresh ideas in a book on experimental design and reread about blocking and randomization [32, 33].
4. Use `pp.msms.data()` on your `MSnSet` object prior to any other function. The presence of NAs or all zero rows may cause computation errors.
5. This is not the result of any test, just the comparison of means between biological conditions. The requirement of minimum absolute log-fold change of 1, and minimal signal of 2, is arbitrary and may be changed with the arguments *minLFC* and *minSpC*.
6. A **GLM** [4] is specified by three components: The response as a random variable Y and its probability distribution, a systematic component given by a linear combination of predictor variables, and a link function which relates $E(Y)$ with the linear predictor. The distribution should belong to the exponential family, which has a probability mass function factorizable as

$$f(y_i;\theta_i) = a(\theta_i)b(y_i)exp[y_i Q(\theta_i)] \tag{9}$$

   Examples are the Poisson distribution, and the negative binomial when the dispersion parameter ϕ is given. The link function that transforms the mean to the natural parameter θ_i is known as the canonical link. The canonical link for the Poisson or the negative binomial distribution is the natural logarithm, and the regression model using this link is:

$$\log \mu_i = \sum_j \beta_j x_{ij} \tag{10}$$

   where here the x_{ij} are the design matrix elements and β_j the model parameters.
7. The most basic distribution when modeling counts is the **Poisson distribution**, a distribution with just one parameter, μ, the mean.

$$Pr(X = k) = \frac{\mu^k e^{-\mu}}{k!} \tag{11}$$

   This distribution explains the uncertainty in the number of SpC positively identified as belonging to a given protein, when the expected number of SpC for this protein at a given concentration is μ. The Poisson distribution has the property that the variance equals the mean. The higher the number of expected counts, the higher is its variance.

8. The probability mass function of a **Negative Binomial** (NB) random variable with mean μ and dispersion ϕ [4, 25] is given by:

$$Pr(X = k) = \frac{\Gamma(k + \phi^{-1})}{\Gamma(\phi^{-1})\,\Gamma(k + 1)} \left(\frac{1}{1 + \mu\phi}\right)^{\phi^{-1}} \left(\frac{\mu}{\phi^{-1} + \mu}\right)^{k} \quad (12)$$

where the variance is a function of both mean and dispersion:

$$Var(X) = \mu + \phi\mu^2 \quad (13)$$

When $\phi \to 0$ the NB reduces to the Poisson distribution. Values of ϕ greater than 0 bring to overdispersed distributions. Strictly speaking any value $\phi > -\mu^{-1}$ is permitted by the model, allowing for some degree of underdispersion.

9. The **quasi-likelihood model** makes abstraction of the true distribution, and considers the mean-variance relationship:

$$Var(\Upsilon) = \psi\mu_i \quad (14)$$

for some constant ψ. The case $\psi > 1$ corresponds to overdispersion. The likelihood equations for this model are identical to the Poisson model, and the model parameter estimates are the same, but ψ is not assumed to be fixed at 1 and estimated from the data. This brings to the quasi-likelihood model which fits the Poisson model and multiplies the standard error estimates of the model parameters by the square root of $\hat{\psi}$, thus adjusting inference for overdispersion [4].

10. **Warnings in GLM outputs:** When there are two or more factors, as in this example, for some features it may happen that within a batch the SpC of all samples of both treatments to compare are 0. These situations may cause a warning. It should be of no concern as it usually happens with features of very low signal.

11. **Confusion matrix** or error matrix is a cross-tabulation between truth and prediction, in a 2 × 2 table. Each row represents a real class and each column represents a predicted class. The table cells provide the number of cases for each combination of prediction x truth, as in Table 4. True positive (TP) is a predicted positive when it is really a positive. False negative (FN) is a predicted negative when it is really a positive. True negative (TN) is a predicted negative when it is really a negative. False positive (FP) is a predicted positive when it is really a negative. In our case the significant features are the predicted positives, and the non-significant are the predicted

Table 4
Confusion matrix (correct predictions in green and errors in red)

	Predicted negative	Predicted positive
Real negative	# True negative (TN)	# False positive (FP)
Real positive	# False negative (FN)	# True positive (TP)

negatives. **Sensitivity** or True Positive Rate (TPR) is the ratio of predicted positives to true positives, in confusion matrix terms $TP/(TP+FN)$. **Specificity** or True Negative Rate (TNR) is the ratio of predicted negatives to true negatives, in confusion matrix terms $TN/(TN+FP)$.

12. **False Discovery Rate (FDR):** Procedures with control of the FDR aim to control the expected proportion of rejections of the null hypothesis that are false, that is the proportion of false positives [11]. The FDR in terms of a confusion matrix may be expressed as the expected value of the ratio $FP/(FP+TP)$, that is the expected value of the proportion of false discoveries.
13. Real life datasets have no **ground truth** associated, otherwise the experiment would be unnecessary. We have used the spiking dataset to have a means to compare methods and extract conclusions out of the results. Confusion matrices are only possible when you know the outcomes.
14. The reader is invited to check thresholds to the *p*-values other that 0.05, and to compare the results without batch correction, in which case the null model is simply $y \backsim 1$.
15. The real **effect size** of this experiment corresponds to a log-fold change of 1.The same as the threshold used in function `test.results()`, to guarantee a minimum of reproducibility. The sampling in the full LC-MS/MS process will cause that some UPS1 proteins are observed with a log-fold change over 1, and some others below 1. We may not expect to detect all UPS1 proteins in this particular case.
16. **Reproducibility:** In statistical learning terms what we do with the tests, looking for differentially expressed proteins, that is making biomarker discovery, is known as feature selection [34]. This is part of a longer process to find a model able to predict outcomes from previously unseen data with an acceptable accuracy. In this knowledge field, there is a big concern with overfitting, a term referred to a model that after being fitted on a limited dataset is unable to make good predictions from new data. To prevent overfitting, data should be splitted in training data on which to fit the model, and validation data

on which to evaluate the model. Sometimes cross-validation may be used instead of splitting in training and validation. A final step would be to get an estimate of the generalization error using and additional test dataset. Making biomarker discovery based solely on *p*-values represents a sort of overfitting which brings to results not reproducible with new data [22]. Moreover, the fewer the samples, the higher the chances of overfitting.

17. **Cell-to-cell normalization:** We aim to compare the proteins secreted by one single cell in two biological states of a cell line, even when one state is globally stimulated or depressed with respect to the other. Suppose we gather Q_j μg of total protein secreted by n_j cells in the j-th biological condition, of which q μg are digested and injected into the LC-MS/MS system to be measured. The total quantity of digested protein measured is the same for the two conditions. The ratio q/Q_j gives the proportion of total secreted protein that gets measured for condition j, hence $(q/Q_j)n_j$ is the number of cells which secreted the q μg in the j-th biological condition. Then we obtain the number of cells which produced the q μg in the j-th condition as in Eq. 15.

$$c_j = \frac{q}{Q_j} n_j \tag{15}$$

On the other hand given the expected SpC value μ of a protein at a given concentration in the total q μg, the expected value per cell is given by Eq. 16

$$E\left[\frac{y}{c_j}\right] = \frac{\mu}{c_j} \tag{16}$$

As the total protein measured for the two biological conditions is the same, the factor q contributes equally to both conditions and may be removed, bringing to Eq. 17 where the mass scale is undefined but equal for both conditions.

$$E\left[\frac{y}{c_j}\right] = \frac{\mu}{n_j/Q_j} \tag{17}$$

This allows to formulate a GLM model, as in Eq. 3, taking into account the total spectral counts by sample, *size*, the protein production rate of each condition, Q/n, the treatment factor, X, and a blocking factor to account for noncontrolled factors leading to batch effects, Z, as in Eq. 18.

$$\log(\mu) = \log\left(\frac{n}{Q}\right) + \log(size) + \alpha + \beta x + \gamma z \tag{18}$$

With this model the effect size, as logFC, is estimated as in Eq. 20.

$$FC_z = \frac{\mu_A/(size_A n_A/Q_A)}{\mu_B/(size_B n_B/Q_B)} = \frac{\exp(\alpha+\beta+\gamma z)}{\exp(\alpha+\gamma z)} = \exp(\beta) \tag{19}$$

$$log\hat{FC} = \log_2(\exp(\hat{\beta})) = 1.44\hat{\beta} \tag{20}$$

where the subindex A stands for treatment, with $x=1$, and subindex B for control, with $x=0$.

18. In this biological experiment, as in any real life experiment, there is no absolute truth. Therefore, any interpretation of the results should be done taking into account the biologic context studied.
19. **Units:** The units used for cell counts (K, M, whatever) and for secreted protein (ng, μg, whatever) to compute the secretion rate are irrelevant provided that they are consistent for both conditions. Only the secretion rates ratio between the two conditions matter. Just for the ease in interpretation, the offsets may be scaled to some useful reference. We use to scale them to the median value, so 1 is neutral, and the proportion in which any sample is off 1 gives the impact of the normalization on this sample.
20. **Secretion rate:** The offsets are defined as divisors of μ as in Eq. 3 and in Eq. 17. In this later case the divisor is the inverse of the secretion rate. Then the fold-change or ratio of expression, is the ratio of expected SpC values by condition, when a fixed and equal amount of protein in each condition is measured, μ_X, multiplied by the ratio of secretion rates Q_X/n_X.

$$\frac{\frac{\mu_A}{n_A/Q_A}}{\frac{\mu_B}{n_B/Q_B}} = \frac{\mu_A}{\mu_B}\frac{Q_A/n_A}{Q_B/n_B} \tag{21}$$

Acknowledgements

Thanks are given at our VHIO's Tumor Biomarkers wet lab colleagues, who made possible spikings, cell cultures, LC-MS/MS runs and data collection, Dr. Laura Villareal, Dr. Olga Méndez, Dr. Theodora Katsila, and Ms. Cándida Salvans.

References

1. Gregori J, Sanchez A, Villanueva J (2019) msmsTests: LC-MS/MS Differential Expression Tests. R package version 1.22.0. https://www.bioconductor.org/packages/release/bioc/html/msmsTests.html
2. Gregori J, Sanchez A, Villanueva J (2019) msmsEDA: Exploratory Data Analysis of LC-MS/MS data by spectral counts. R package version 1.22.0. https://www.bioconductor.org/packages/release/bioc/html/msmsEDA.html
3. Lundgren DH, Hwang SI, Wu L, Han DK (2010) Role of spectral counting in quantitative proteomics. Expert Rev Proteomics 7(1): 39–53. https://doi.org/10.1586/epr.09.69
4. Agresti A (2002) Categorical data analysis. Wiley Series in Probability and Statistics -

Applied Probability and Statistics Section Series. Wiley, Hoboken. https://doi.org/10.1002/0471249688

5. Scherer A (2009) Batch effects and noise in microarray experiments: sources and solutions. Wiley Series in Probability and Statistics. Wiley, Hoboken. https://doi.org/10.1002/9780470685983
6. Lazar C, Meganck S, Taminau J, Steenhoff D, Coletta A, Molter C, Weiss-Solís DY, Duque R, Bersini H, Nowé A (2013) Batch effect removal methods for microarray gene expression data integration: a survey. Briefings Bioinf 14(4):469–490. https://doi.org/10.1093/bib/bbs037
7. Gregori J, Villarreal L, Méndez O, Sánchez A, Baselga J, Villanueva J (2012) Batch effects correction improves the sensitivity of significance tests in spectral counting-based comparative discovery proteomics. J Proteomics 75(13):3938–3951. https://doi.org/10.1016/j.jprot.2012.05.005
8. Shi L, et al, MAQC-Consortium (2006) The MicroArray quality control (MAQC) project shows inter- and intraplatform reproducibility of gene expression measurements. Nat Biotechnol 24(9):1151–1161. https://doi.org/10.1038/nbt1239
9. Gregori J, Villarreal L, Sánchez A, Baselga J, Villanueva J (2013) An effect size filter improves the reproducibility in spectral counting-based comparative proteomics. J Proteomics. https://doi.org/10.1016/j.jprot.2013.05.030
10. Li M, Gray W, Zhang H, Chung CH, Billheimer D, Yarbrough WG, Liebler DC, Shyr Y, Slebos RJ (2010) Comparative shotgun proteomics using spectral count data and quasi-likelihood modeling. J Proteome Res 9(11). https://doi.org/10.1021/pr100527g
11. Benjamini Y, Hochberg Y (1995) Controlling the false discovery rate: A practical and powerful approach to multiple testing. J Roy Statist Soc Ser B (Methodological) 57(1):289–300. http://dx.doi.org/10.2307/2346101
12. Neilson KA, Ali NA, Muralidharan S, Mirzaei M, Mariani M, Assadourian G, Lee A, van Sluyter SC, Haynes PA (2011) Less label, more free: approaches in label-free quantitative mass spectrometry. Proteomics 11(4): 535–553. https://doi.org/10.1002/pmic.201000553
13. Sandin M, Krogh M, Hansson K, Levander F (2011) Generic workflow for quality assessment of quantitative label-free LC-MS analysis. Proteomics 11(6):1114–1124. https://doi.org/10.1002/pmic.201000493
14. Zhu W, Smith JW, Huang CM (2010) Mass spectrometry-based label-free quantitative proteomics. J Biomed Biotechnol 2010:840518. https://doi.org/10.1155/2010/840518
15. Patel VJ, Thalassinos K, Slade SE, Connolly JB, Crombie A, Murrell JC, Scrivens JH (2009) A comparison of labeling and label-free mass spectrometry-based proteomics approaches. J Proteome Res 8(7):3752–3759. https://doi.org/10.1021/pr900080y
16. Luo J, Schumacher M, Scherer A, Sanoudou D, Megherbi D, Davison T, Shi T, Tong W, Shi L, Hong H, Zhao C, Elloumi F, Shi W, Thomas R, Lin S, Tillinghast G, Liu G, Zhou Y, Herman D, Li Y, Deng Y, Fang H, Bushel P, Woods M, Zhang J (2010) A comparison of batch effect removal methods for enhancement of prediction performance using MAQC-II microarray gene expression data. Pharmacogenomics 10(4):278–291. https://doi.org/10.1038/tpj.2010.57
17. Knudsen S (2006) Cancer diagnostics with DNA microarrays. Wiley, Hoboken. https://doi.org/10.1002/0470041102
18. Choi H, Fermin D, Nesvizhskii AI (2008) Significance analysis of spectral count data in label-free shotgun proteomics. Mol Cell Proteom 7(12):2373–2385. https://doi.org/10.1074/mcp.M800203-MCP200
19. Li M, Gray W, Zhang H, Chung CH, Billheimer D, Yarbrough WG, Liebler DC, Shyr Y, Slebos RJC (2010) Comparative shotgun proteomics using spectral count data and quasi-likelihood modeling. J Proteom Res 9(8):4295–4305. https://doi.org/10.1021/pr100527g
20. Leitch MC, Mitra I, Sadygov RG (2012) Generalized linear and mixed models for label-free shotgun proteomics. Statist Interface 5(1):89–98. https://doi.org/10.4310/SII.2012.v5.n1.a8
21. Gregori J, Méndez O, Katsila T, Pujals M, Salvans C, Villarreal L, Arribas J, Tabernero J, Sánchez A, Villanueva J (2014) Enhancing the biological relevance of secretome-based proteomics by linking tumor cell proliferation and protein secretion. J Proteom Res 13(8): 3706–3721. https://doi.org/10.1021/pr500304g
22. Shi L, Jones WD, Jensen RV, Harris SC, Perkins RG, Goodsaid FM, Guo L, Croner LJ, Boysen C, Fang H, et al (2008) The balance of reproducibility, sensitivity, and specificity of lists of differentially expressed genes in microarray studies. BMC Bioinf 9 Suppl 9:S10. https://doi.org/10.1186/1471-2105-9-S9-S10

23. R Core Team (2019) R: A Language and environment for statistical computing. R Foundation for Statistical Computing, Vienna. https://www.R-project.org/
24. Huber W, Carey VJ, Gentleman R, Anders S, Carlson M, Carvalho BS, Bravo HC, Davis S, Gatto L, Girke T, Gottardo R, Hahne F, Hansen KD, Irizarry RA, Lawrence M, Love MI, MacDonald J, Obenchain V, Ole's AK, Pag'es H, Reyes A, Shannon P, Smyth GK, Tenenbaum D, Waldron L, Morgan M (2015) Orchestrating high-throughput genomic analysis with Bioconductor. Nat Methods 12(2): 115–121. https://doi.org/10.1038/nmeth.3252
25. Robinson MD, Smyth GK (2008) Small-sample estimation of negative binomial dispersion, with applications to sage data. Biostatistics 9(2):321–332. https://doi.org/10.1093/biostatistics/kxm030
26. Gatto L, Lilley KS (2012) MSnbase-an R/Bioconductor package for isobaric tagged mass spectrometry data visualization, processing and quantitation. Bioinformatics 28(2): 288–289. https://doi.org/10.1093/bioinformatics/btr645
27. Venables WN, Ripley BD (2002) Modern applied statistics with S, 4th edn. Springer, New York. ISBN 0-387-95457-0. https://doi.org/10.1007/978-0-387-21706-2
28. Warnes GR, Bolker B, Bonebakker L, Gentleman R, Liaw WHA, Lumley T, Maechler M, Magnusson A, Moeller S, Schwartz M, Venables B (2019) gplots: Various R Programming Tools for Plotting Data. R package version 3.0.1.1. https://CRAN.R-project.org/package=gplots
29. Neuwirth E (2014) RColorBrewer: ColorBrewer Palettes. R package version 1.1-2. https://CRAN.R-project.org/package=RColorBrewer
30. Robinson MD, McCarthy DJ, Smyth GK (2009) edgeR: a Bioconductor package for differential expression analysis of digital gene expression data. Bioinformatics 26(1): 139–140. https://doi.org/10.1093/bioinformatics/btp616
31. Storey JD, Bass AJ, Dabney A, Robinson D (2019) qvalue: Q-value estimation for false discovery rate control. R package version 2.16.0. http://github.com/jdstorey/qvalue
32. Mathews P (2005) Design of Experiments with MINITAB. ASQ. https://doi.org/10.1198/tas.2006.s46
33. Cochran WG, Cox GM (1957) Experimental designs, 2nd edn. Wiley, Hoboken
34. Hastie T, Tibshirani R, Friedman J (2009) The elements of statistical learning. Springer Series in Statistics, , 2nd edn. Springer, New York. https://doi.org/10.1007/978-0-387-84858-7

Chapter 11

Exploring Protein Interactome Data with IPinquiry: Statistical Analysis and Data Visualization by Spectral Counts

Lauriane Kuhn, Timothée Vincent, Philippe Hammann, and Hélène Zuber

Abstract

Immunoprecipitation mass spectrometry (IP-MS) is a popular method for the identification of protein-protein interactions. This approach is particularly powerful when information is collected without a priori knowledge and has been successively used as a first key step for the elucidation of many complex protein networks. IP-MS consists in the affinity purification of a protein of interest and of its interacting proteins followed by protein identification and quantification by mass spectrometry analysis. We developed an R package, named `IPinquiry`, dedicated to IP-MS analysis and based on the spectral count quantification method. The main purpose of this package is to provide a simple R pipeline with a limited number of processing steps to facilitate data exploration for biologists. This package allows to perform differential analysis of protein accumulation between two groups of IP experiments, to retrieve protein annotations, to export results, and to create different types of graphics. Here we describe the step-by-step procedure for an interactome analysis using IPinquiry from data loading to result export and plot production.

Key words Immunoprecipitation, Mass spectrometry, Data processing, Differential analysis, Volcano plots, Spectral counts, R package

1 Introduction

Affinity purification mass spectrometry (AP-MS) or immunoprecipitation mass spectrometry (IP-MS) is a popular without a priori method for the identification of protein-protein interactions that has been successfully used for resolving numerous complex protein networks [1–3]. IP-MS is a first key step of the experimental workflow for protein partner identification and usually precedes validation using alternatives approaches. IP-MS starts with the affinity purification of the protein of interest, referred to as the bait protein, and of its interacting proteins by using a resin coupled to an antibody recognizing either the bait itself or an epitope tag expressed fused to the bait. The eluted protein mixture is then

Thomas Burger (ed.), *Statistical Analysis of Proteomic Data: Methods and Tools*,
Methods in Molecular Biology, vol. 2426, https://doi.org/10.1007/978-1-0716-1967-4_11,

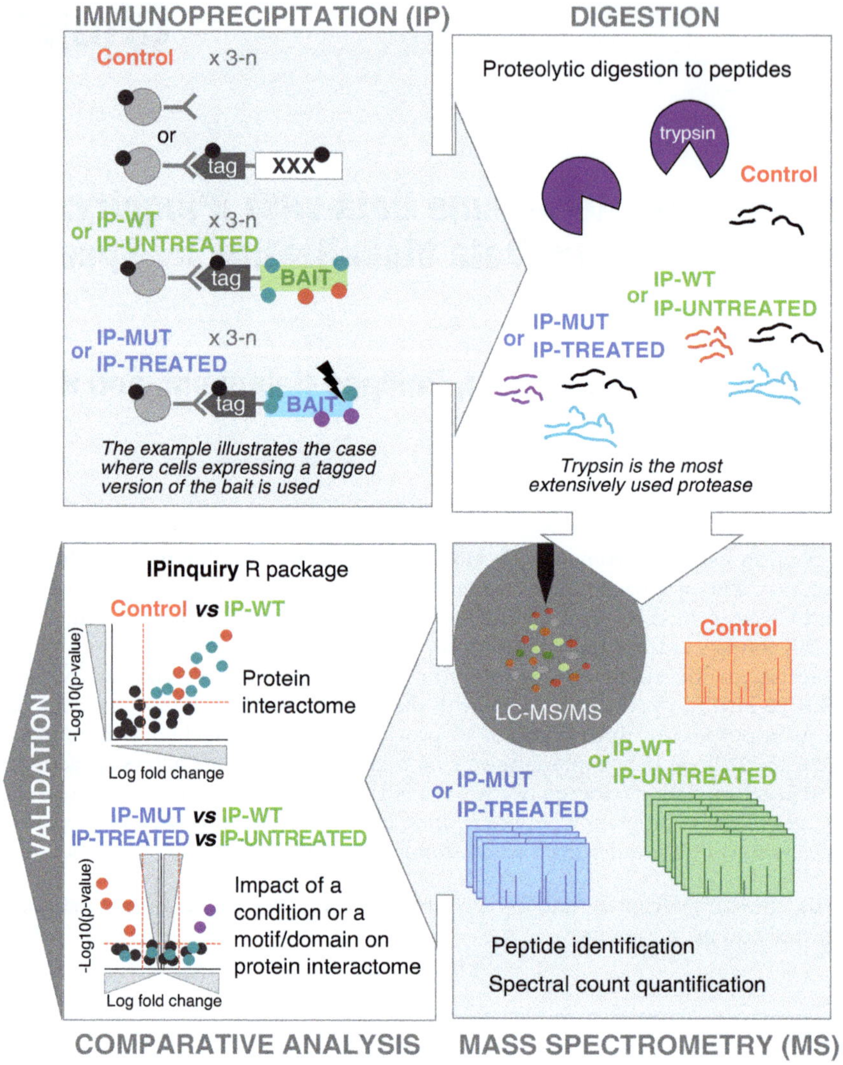

Fig. 1 Schematic overview of main steps of IP-MS approaches

subjected to proteolytic digestion and identified by MS analysis (*see* Fig. 1). The latter classically involves peptide separation by reverse-phase liquid chromatography combined with tandem mass spectrometry (LC-MS/MS). Experimental design of IP-MS approaches differs according to biological questions and available biological material. In particular, the use of appropriate controls is crucial as the eluted protein mixture contains *bona fide* protein partners but also various non-specific interactors, such as proteins binding to the epitope tag or to the resin. A good control should enable assessing protein backgrounds resulting from the various contamination sources and its choice should not be neglected. Classically, when the aim is to analyze the protein interactome of a protein of interest by using cells expressing a tagged bait protein, control IPs are performed using wild-type cells, that do not express the tagged bait protein, and/or using cells that express an unrelated tagged protein. Proteins found to be enriched in bait compared to control IPs are then considered as potential protein partners. Alternatively, when the question is to test the impact of a particular protein motif or domain on the protein interactome, IPs using the wild-type version of the bait protein are compared to the one using a mutated version. The potentially interesting proteins correspond then to those depleted in mutant IPs. Finally, a frequent question is also to test the impact of different conditions or treatments on the interactome of a protein of interest. IPs performed from samples of different conditions are then compared and both significantly enriched and depleted proteins are considered as potentially interesting. In all cases, the data analysis consists in comparing the differential accumulation of proteins between two groups of IPs. Two metrics can be used for protein quantification: the spectral count, defined as the total number of spectra identified for a protein, and the peptide abundance derived from MS1 peak area [4]. The second strategy is now often preferred notably because of its higher performance for the detection of low abundant proteins. Yet, the spectral count quantification method still represents a popular fast and simple approach that demonstrates its efficiency in IP-MS approaches to resolve protein interactomes.

Here we describe the analysis of IP-MS data based on spectral counts using the `IPinquiry` R package. The main purpose of this package is to provide a simple R pipeline with a limited number of processing steps to facilitate as much possible data exploration and plot creation for biologists. `IPinquiry` compiles several functions to: (i) identify proteins significantly enriched or depleted between two groups of IP experiments, (ii) retrieve annotations for detected proteins, (iii) export result tables, and (iv) create different graph types, such as interactive volcanoplots that display protein changes (fold changes) according to statistical significance (*p*-value). In order to calculate *p*-values associated with protein accumulation changes, the package uses the negative binomial generalized linear models, with or without quasi-likelihood tests, implemented in the `EdgeR` package [5, 6]. `EdgeR` GLM models were developed for RNA-seq analysis to assess gene differential analysis between two conditions or genotypes. RNA-seq and proteomic data share common features in a statistical point of view: both types of data are discrete, are usually linked to high biological dispersion and to a reduced number of biological replicates, often below 5. Because of these common properties, the `EdgeR` GLM model was previously proposed for analyzing MS-MS data [7] and was already successfully applied to explore protein interactome based on IP-MS experiments [8–10]. We detailed hereafter the step-by-step procedure for data analysis based on spectral counts using `IPinquiry` from data loading to result export and plot production. The package includes example datasets from [11] to help users apprehending `IPinquiry` utilization.

2 Material

2.1 Considerations for IP-MS Approaches

1. Choosing a good antibody:
   (a) When available, IP can be performed using an antibody against the protein of interest. This is the ideal situation as protein-protein interactions can be analyzed at physiological level.
   (b) When such an antibody is not available, the protein of interest fused to an epitope tag needs to be expressed in cells (*see* **Note 1**).
2. Choosing good controls for IPs, the objective being to remove as much as possible contaminant proteins.
3. Optimizing affinity purification conditions (*see* **Note 2**): optimization of the sample lysis, homogenization and grinding, choice of detergents and bead type, adjustment of wash and elution stringency, *etc.* The goal is to obtain an efficient and reproducible purification with a good balance between the number of detected proteins and the number of contaminants.
4. Optimizing and standardizing the quantity of starting material for IP and MS. Once optimized, the same number of cells or the same dry weight from plant has to be used.
5. Selecting the number of biological replicates (*see* **Note 3**). This choice depends on the expected variation and on the biological and technical variability. We recommend analyzing at least three biological replicates (*see* **Note 4**). We also encourage the analysis of different transgenic lines, when using cells expressing the tagged bait protein.
6. Selecting the method for protease digestion. Gel-free trypsin digestion is widely used and well suited in most cases as it allows to better control the reproducibility of the process compared to gel-purified protein complexes.
7. Selecting ion source. Electrospray Ionization Source (ESI) is the most widely used ion source.
8. Selecting mass analyzers. Sensitivity and high duty cycle of actual mass spectrometers allow to efficiently identify moderately complex mixtures, like in the case of an AP-MS sample.
9. Selecting and optimizing LC-MS parameters. For protein quantification with LC-MS, we favor long chromatographic gradients and a single injection under discovery mode for each sample. Several methods of data acquisition are also available such as data-dependent acquisition, targeted acquisition, and data-independent acquisition. Finally, a key parameter is the dynamic exclusion time that needs to be optimized to

obtain a good balance between the number of detected proteins and the number of spectra obtained for each protein (*see* **Note 5**).

10. Selecting protein identification method. It depends on the MS acquisition strategy. Identification should be validated according to the actual guidelines (FDR<1% on both spectral and protein level) and protein redundancy should be carefully managed.
11. Selecting the quantification method, i.e., spectral count or peptide abundance derived from MS1 peak area. This chapter is dedicated to the analysis based on spectral counts (*see* **Note 6**).

2.2 Requirements

`IPinquiry` is a package written in R [12]. If not already done, R needs to be downloaded and installed. We also recommend the use of `RStudio`, which provides a nice R user interface making life easier for R beginners. In addition, the following R packages are needed (*see* **Note 7**):

1. for statistical analysis (required for package installation): `EdgeR` [5, 6], `limma` [13], `statmod` [14].
2. for the creation of interactive volcanoplots: `plotly` [15], `htmlwidgets` [16].
3. for the creation of interactive tables: `DT` [17], `htmlwidgets` [16].
4. for the creation of others graphs: `ggplot` [18], `pheatmap` [19], `RColorBrewer` [20].
5. for protein annotation: `biomaRt` [21, 22].
6. for saving result tables as excel files: `xlsx` [23].

2.3 Software Installation

`IPinquiry` can be downloaded and installed from GitHub using devtools [24]. If needed, install devtools and load the library:

```
install.packages("devtools")
library(devtools)
```

`IPinquiry` package can then be installed (*see* **Note 8**):

```
install_github("https://github.com/hzuber67/IPinquiry4")
```

2.4 Data Format

Input data consist in two files:

1. A **count table** (text file with tab-separated values) that contains spectral counts for all proteins detected in IPs. Each row corresponds to one protein detected in IP and each column corresponds to one IP experiment (*see* Fig. 2).

	accession	Mut_L0_1	Mut_L0_2	Mut_L3_1	Mut_L3_2	URT1_L12_1_2019	URT1_L12_2_2019	URT1_L17_1_2019	URT1_L17_2_2019
1	AT1G01080.2	0	0	1	1	0	1	1	1
2	AT1G01090.1	1	2	2	2	1	2	1	2
3	AT1G01100.1	0	1	0	1	0	2	0	0
4	AT1G01300.1	11	11	10	9	8	10	9	14
5	AT1G01320.1	1	1	1	0	2	1	0	1

Fig. 2 Screenshot of the count table for the first dataset (top part)

	IP_names	sample
1	Mut_L0_1	M1
2	Mut_L0_2	M1
3	Mut_L3_1	M1
4	Mut_L3_2	M1
5	URT1_L12_1_2019	urt1
6	URT1_L12_2_2019	urt1
7	URT1_L17_1_2019	urt1
8	URT1_L17_2_2019	urt1

	IP_names	sample	batch
1	F016864_2014_S17_control_C1	control	one
2	F016865_2014_S17_control_C2	control	one
3	F016867_2014_S17_URT1_H1	urt1	one
4	F016878_2014_S25_control_C5	control	one
5	F016880_2014_S25_URT1_H5	urt1	one
6	F016882_2015_S12_CTRL_1	control	two
7	F016883_2015_S12_CTRL_2	control	two
8	F016884_2015_S12_CTRL_3	control	two
9	F016886_2015_S12_URT1_myc2	urt1	two
10	F016887_2015_S12_URT1_myc3	urt1	two

Fig. 3 Screenshots of sample tables for the first (on the left) and the second (on the right) datasets

2. A **sample table** (text file with tab-separated values) that gives information about samples. First column indicates the IP names, second column the conditions, and finally the third column is optional and allows for indicating potential batch effect, related to different experiment times, for example (*see* Fig. 3) (*see* **Note 9**).

2.5 Example Datasets

Two example datasets corresponding to IP experiments in Arabidopsis [11] (*see* **Note 10**) are included in the package:

1. The first dataset contains results for two groups of IPs performed from plants expressing a wild-type version of the URT1 TUTase fused to an epitope tag, named URT1-myc, or a mutated version, named m1URT1-myc. Each group is composed of four replicates. The goal was to test the impact of the M1 motif of URT1 on its interactome *in planta* (*see* **Note 11**). Directories for the example and sample tables are:

```
CountTable1 <- system.file("extdata",
                           "CountTable1.txt",
                           package = "IPinquiry4")
SampleTable1 <- system.file("extdata",
                            "SampleTable1.txt",
                            package = "IPinquiry4")
```

Top part of the count table can be visualized as follows (*see* Fig. 2):

```
Count_tb1 <- read.table(CountTable1, sep="\t", header=TRUE)
head(Count_tb1)
```

Sample table can be visualized as follows (*see* Fig. 3):

```
Sample_tb1 <- read.table(SampleTable1, sep="\t", header=TRUE)
print(Sample_tb1)
```

Conditions in the sample table are named "urt1" and "M1" for URT1-myc or m1URT1-myc IPs, respectively.

2. The second dataset contains results for four replicates of IPs performed from plants expressing the wild-type version of URT1 fused to an epitope tag. Control IPs were performed in parallel using wild-type plants that do not express the tagged URT1 with six biological replicates. The goal here was to identify protein partners of the URT1 TUTase *in planta*. IP experiments were performed for two different tissues at two different times inducing a batch effect that will be latter taken into account in the statistical model. Directories for the example count and sample tables are:

```
CountTable2 <- system.file("extdata", "CountTable2.txt",
                           package = "IPinquiry4")
SampleTable2 <- system.file("extdata", "SampleTable2.txt",
                            package = "IPinquiry4")
```

Top part of the count table can be visualized as follows:

```
Count_tb2 <- read.table(CountTable2, sep="\t", header=TRUE)
head(Count_tb2)
```

Sample table can be visualized as follows (*see* Fig. 3):

```
Sample_tb2 <- read.table(SampleTable2, sep="\t", header=TRUE)
print(Sample_tb2)
```

Conditions in the sample table are named "urt1" and "control" for bait and control IPs, respectively. The sample table contains a third column indicating a batch effect.

2.6 Data Loading

1. To load `IPinquiry` library in your R environment, enter in the R console:

```
library(IPinquiry4)
```

2. IP data can then be loaded using `load_IP_data` function (*see* **Note 12**). Here, the two example datasets are successively loaded by indicating their directories as defined above (*see* Subheading 2.5):

```
# Load dataset1
IP_data1 <- load_IP_Data(CountTable1, SampleTable1)
# Load dataset2
IP_data2 <- load_IP_Data(CountTable2, SampleTable2)
```

3. Arguments taken by the function are the directories for count and sample tables. When analyzing your own data, simply indicate their directories on your computer. For example:

```
my_IP_data <- load_IP_Data(
                    "/Users/me/Documents/my_count_table.txt",
                    "/Users/me/Documents/my_sample_table.txt"
                          )
```

3 Methods

3.1 Visualization of the Overall Variability Between Samples

Multidimensional scaling (MDS) plots can be used to visualize distances or dissimilarities between the different IP experiments. Here, the Euclidean distance is used to perform the MDS;

1. MDS can be plotted based on raw data (`norm="nothing"`, by default) or on normalized data either based on the total number of counts (`norm="total"`) or on the median-to-ratios method as used in DESeq2 R package [25] (`norm="DEseq"`) (*see* **Note 13**). MDS without prior normalization can be obtained as follows (*see* Fig. 4):

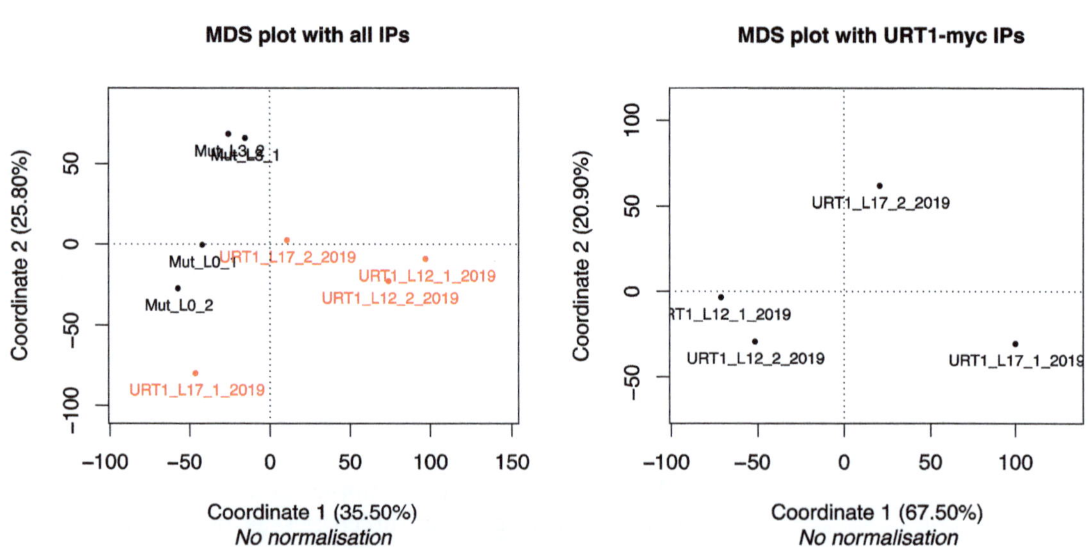

Fig. 4 MDS plot for the first dataset with all or with URT1-myc IPs

```
# MDS for the first dataset
MDSplot(IP_data1)
```

2. Functions subset_IPObj_treat and subset_IPObj_-batch can also be used to subset the dataset prior to plot the MDS. subset_IPObj_treat allows the selection of specific treatments/conditions. subset_IPObj_batch allows the selection of specific batches. Here we plot the MDS for the treatment "urt1" (*see* Fig. 4):

```
# MDS for the first dataset for the treatment ''urt1''
IP_urt1 <- subset_IPObj_treat(IP_data2, "urt1")
MDSplot(IP_urt1)
```

3.2 Statistical Analysis for Differential Analysis

The statistical analysis is based on the GLM model developed by the EdgeR package [5, 6]. This model was previously proposed to analyze MS data based on spectral counts in the msmsTests package [7]. A refined description of these statistical models is provided in **Chapter 10**. By default, IPinquiry package uses the Genewise Negative Binomial Generalized Linear Model with Quasi-likelihood Tests implemented in EdgeR (*see* **Note 14**):

1. Statistical comparison needs to be performed for each pairwise comparison. For the first dataset, we have two treatments, "urt1" and "M1", and:
   (a) Low abundance protein are filtered out before the calculation of the dispersion. Here, only proteins with a total sum of counts above 10 are used (min.disp=10). This corresponds to the 50% most abundant proteins (*see* **Note 15**).
   (b) The size correction factor (offset) is calculated using the median-to-ratio method [25] (div="DEseq"), (*see* **Note 16**):

```
test1 <- stat_test(IP_data1, "urt1", treatment = "M1",
                   div="DEseq", min.disp=10)
```

2. This function returns a data frame with six columns: Protein ID as row.names, LogFC, quasi-likelihood F-statistics, *p*-values as calculated by EdgeR, adjusted *p*-values according to Benjamini-Hochberg method and the protein rank based on adjusted *p*-values.
3. In the case of the first dataset, we are interested in identifying proteins that are significantly depleted when URT1 M1 motif is mutated. The list of proteins significantly depleted in m1URT1-myc compared URT1-myc IPs can be visualized as follows (*see* Fig. 5):

	LogFC	F	p.value	adjp	number
AT1G26110.1	-2.6756838	126.78115	9.505361e-19	1.559830e-15	1
AT5G45330.1	-6.0490263	75.81482	2.016451e-11	1.102999e-08	3
AT2G45810.1	-1.2679256	50.79735	1.550362e-08	6.360360e-06	4
AT4G00660.2	-1.0963299	41.08400	3.488022e-07	1.144769e-04	5
AT3G13300.1	-0.9810462	35.63486	1.990222e-06	4.665648e-04	7
AT3G61240.1	-1.0198458	33.62240	4.228655e-06	8.674029e-04	8
AT4G20360.1	-0.8466879	31.30745	7.568390e-06	1.379970e-03	9
AT1G27090.1	-2.1257515	30.70430	2.544579e-05	4.175654e-03	10
AT1G48410.2	-1.7462989	27.82227	6.489081e-05	9.680530e-03	11
AT3G58510.1	-0.8828883	24.12936	1.003564e-04	1.372373e-02	12
AT3G58570.1	-0.9571052	22.12210	2.273244e-04	2.664566e-02	14
AT2G42520.1	-0.9563024	20.29599	4.466130e-04	4.311129e-02	16
AT5G47010.1	-1.4477632	21.46352	4.218413e-04	4.311129e-02	17
AT4G38680.1	-1.0716476	20.39479	5.152760e-04	4.632652e-02	18
AT5G40490.1	-1.5120287	20.85829	5.363826e-04	4.632652e-02	19

Fig. 5 Data frame with statistical results for significantly depleted proteins

```
print(subset(test1, test1$adjp<0.05&test1$LogFC<0))
```

4. A batch effect can be taken into account in the statistical model by adding `batch="TRUE"` (*see* **Note 17**). If so, a third column has to be added in the sample table to indicate the batch of each IPs (*see* Subheading 2.4). For the second dataset, we take into account the batch effect as two sets of experiments were performed:

```
test2 <- stat_test(IP_data2, "control", treatment = "urt1",
                   div="DEseq", batch=TRUE)
```

5. In the case of the second dataset, we are interested in identifying proteins that are significantly enriched in URT1 IPs compared to control IPs. These proteins will be considered as potential protein partners of URT1. The list of proteins significantly enriched can be visualized as follows:

```
# Subset enriched proteins
test2_enriched <- subset(test2, test2$adjp<0.05&test2$LogFC>0)


# Print subtable
print(test2_enriched)
```

6. By adding the argument `glm="classic"`, you can use instead the `EdgeR` function based on the Genewise Negative Binomial Generalized Linear Models without Quasi-likelihood Tests (*see* **Note 18**).
7. An additional low abundance filter can be added, e.g., `filter=5` (*see* **Note 19**). If a filter value is indicated, the output data frame includes a seventh column indicating if, "YES" or "NO", proteins meet this additional criterion. This filter does not affect the statistics calculation.

3.3 Retrieve Annotations for Each Protein

Functional annotations are retrieved using the `biomaRt` package [21, 22]. Annotations are collected from the Ensembl database [26]. Active internet connection is necessary to access the remote database and query it on-line:

1. Here, annotations from *Arabidopsis thaliana* are retrieved:

```
annotated_table_At <- addBiomaRtAnnotation(
                                 test,
                                 biomart="plants_mart",
                                 dataset="athaliana_eg_gene"
                                                    )
```

2. By default, the function searches for Ensembl peptide identifiers. This argument needs to be adjusted according to the identifiers used in the row names of the count table. It can be `"ensembl_peptide_id"`, `"ensembl_transcript_id"`, `"ensembl_gene_id"`, or `"external_gene_name"`.
3. The new output data frame contains three additional columns: `ensembl ID`, `external gene name` and `description`.
4. Of course, this function can be used for all other species for which annotations are available at Ensembl. `biomart`, `dataset` and `host` arguments need to be adjusted according to the analyzed species:

   (a) For listing available databases:

```
library(biomaRt)
listMarts()
dataset_list <- listDatasets(
                    useMart("ENSEMBL_MART_ENSEMBL")
                                 )
print(dataset_list)
```

   (b) For example, for *Drosophila melanogaster*:

```
annotated_table_Dm <- addBiomaRtAnnotation(
               droso_results,
               biomart = "ENSEMBL_MART_ENSEMBL",
               dataset = "dmelanogaster_gene_ensembl",
               host = "ww{w}.ensembl.org"
                                                    )
```

(c) Another example, for human:

```
annotated_table_Hs <- addBiomaRtAnnotation(
                    human_results,
                    biomart = "ensembl",
                    dataset = "hsapiens_gene_ensembl",
                    host = "w{w}w.ensembl.org",
                    features="external_gene_name"
                                                )
```

3.4 Create and Export an Html Table

`IPinquiry` package includes a function based on `DT package` [17] to create an interactive table with results:

1. The following code creates an interactive table from the `annotated_table_At` data frame that contains statistical results and protein annotations:

```
# Interactive table for dataset 1
createTable(annotated_table_At)
```

2. This table is interactive and can be used to sort and search proteins, select and copy interesting rows, export results, *etc.* (*see* Fig. 6).
3. This interactive table can also be saved as an *html* file:

```
p <- createTable(annotated_table_At)
htmlwidgets::saveWidget(p,"interactive_table_1.html",
                        selfcontained = TRUE)
```

Copy CSV Excel Search:

	LogFC	adjp	external_gene_name	description
AT1G26110.1	-2.67568375781914	1.5598298212177e-15	DCP5	Protein decapping 5 [Source:UniProtKB/Swiss-Prot;Acc:Q9C658]
AT5G41790.1	2.07394390136945	3.6645744207509e-9	CIP1	COP1-interactive protein 1 [Source:UniProtKB/Swiss-Prot;Acc:F4JZY1]
AT5G45330.1	-6.04902632434931	1.10299859702332e-8	DCP5-L	Decapping 5-like protein [Source:UniProtKB/Swiss-Prot;Acc:Q9FH77]
AT2G45810.1	-1.26792563395301	0.00000636036009242441	RH6	DEAD-box ATP-dependent RNA helicase 6 [Source:UniProtKB/Swiss-Prot;Acc:Q94BV4]
AT4G00660.2	-1.09632992307729	0.000114476886893135	RH8	DEAD-box ATP-dependent RNA helicase 8 [Source:UniProtKB/Swiss-Prot;Acc:Q8RXK6]
AT3G45140.1	0.650623330055275	0.000237262866856185	LOX2	Lipoxygenase 2, chloroplastic [Source:UniProtKB/Swiss-Prot;Acc:P38418]
AT3G13300.1	-0.981046170327883	0.000466564823690176	VCS	Enhancer of mRNA-decapping protein 4 [Source:UniProtKB/Swiss-Prot;Acc:Q9LTT8]
AT3G61240.1	-1.01984579214868	0.000867402913446643	RH12	DEAD-box ATP-dependent RNA helicase 12 [Source:UniProtKB/Swiss-Prot;Acc:Q9M2E0]

Fig. 6 Screenshot of the interactive table with statistical results and protein annotations. The interactive table allows data sorting, section, and export

4. You can also specify the directory where the file has to be saved (*see* **Note 20**):

```
htmlwidgets::saveWidget(
    p,
    "/Users/me/Documents/My_results/interactive_table_1.html"
    selfcontained = TRUE
                   )
```

3.5 Export Result Table as Excel or Text File

Alternatively, results table can also be saved:

1. As an excel file, using the `xlsx` package with the following command line:

```
library(xlsx)
write.xlsx(annotated_table_At, "IP_results.xlsx",
           sheetName = "Statistics")
```

2. As a text file, using `write.table` R function:

```
write.table(annotated_table_At, "IP_results.txt", sep="\t",
            col.names=NA, quote=FALSE)
```

3. As previously (*see* Subheading 3.4), you can specify the directory where the file has to be saved for both functions. For example, for `write.table` function:

```
write.table(annotated_table_At,
            "/Users/me/Documents/My_results/IP_results.txt",
            sep="\t", col.names=NA, quote=FALSE)
```

3.6 Create Interactive Volcanoplot

1. The interactive volcanoplot created by `IPinquiry` is based on the `Plotly` R package [15]. The volcanoplot shows the log2 fold change according to *p*-value or to adjusted *p*-value:

   (a) To display the Volcanoplot for the first dataset (*see* Fig. 7):

```
#Dataset 1 volcanoplot
htmlPlot(annotated_table_At, sign="adjp")
```

   (b) For the second dataset, we are interested in the proteins that are enriched in URT1 IP when compared to control IP. The volcanoplot is drawn only for enriched proteins, with `LogFC>0` (*see* Fig. 7):

```
#Dataset 2 volcanoplot for enriched proteins
htmlPlot(subset(test2, test2$LogFC>0), sign="adjp")
```

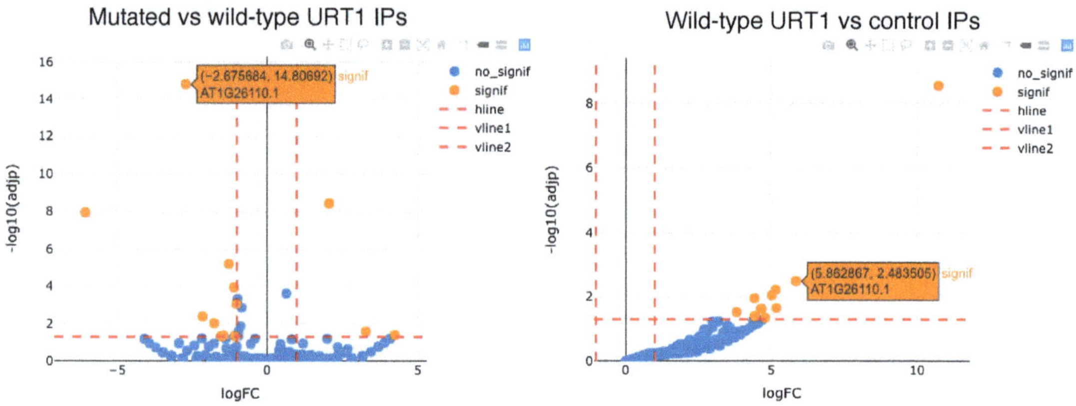

Fig. 7 Screenshots of interactive volcanoplots obtained, for example, datasets 1 and 2 shown on the left and on the right, respectively. Text labels of points appear when the cursor is moved over them

2. By default, point labels correspond to row names of the input table accompanied with the point coordinates. Custom texts can also be used instead using the `custom_text` argument. For example, the R code below allows using the 30 first letters of protein annotation found in the description column of the result table:

```
# extract the 30 first letters of the description column
my_test = paste(row.names(annotated_table_At) , "_",
               substr(annotated_table_At$description,1,30))
# add annotation as point label for the dataset 1 volcanoplot
htmlPlot(annotated_table_At,  custom_text=my_test)
```

3. By default, point colors are set according to the significance and dotted lines are set both according to *p*-value and LogFC. *P*-value and LogFC cut-offs can be adjusted using `max.pval` and `min.LFC` arguments. Default values are 0.05 and 1, respectively. Point colors can also be used to highlight specific proteins. For example, the R code below is used to pinpoint three proteins linked to decapping (*see* **Note 21**):

```
# List of interesting proteins
smallTrueList <- c("AT1G26110.1", "AT5G45330.1", "AT3G13300")
# Create interactive plot
htmlPlot(annotated_table_At, listGenes = smallTrueList,
         custom_text=my_test)
```

4. There is also the possibility to set colors according to a supplemental column. This column can contain additional information, for example, concerning gene ontology. IPinquiry package contains a text file with supplemental information related to the first dataset. This file is composed of two columns, a first one with protein identifiers and a second one with protein classifications based on their molecular function:
   (a) Directory for this information table is:

```
Supplemental <- system.file("extdata",
                            "Supplemental_information.txt",


                            package = "IPinquiry4")
```

   (b) When analyzing your own data, simply indicate the directory on your computer of the text table containing the information of interest. For example:

```
Supplemental_me <-
   "/Users/me/Documents/Interesting_information.txt"
```

   (c) The add_suppl_information function of IPinquiry can be used to combine your result table with another table containing classification criteria. The code below combines the result table for the first dataset with the information table:

```
# Add suppl column with criteria for color classification
annotated_table_At2 <- add_suppl_information(
                              annotated_table_At,
                             Supplemental
                                            )
head(annotated_table_At2)
```

   (d) This new column can then be used to set point colors:

```
# Create the volcanoplot with colors
# according to this new column
htmlPlot(annotated_table_At2,
         colforcolor = annotated_table_At2$Classification)
```

5. The interactive volcanoplot can be directly saved under html format:

```
p <- htmlPlot(
        annotated_table_At2,
        colforcolor = annotated_table_At2$Classification
             )
htmlwidgets::saveWidget(p,"interactive_volcanoplot_plot.html"
                        selfcontained = TRUE)
```

3.7 Create ggplot2 Based Volcanoplot

1. IPinquiry also includes a function, named PDF_Plot, to create a volcanoplot based on the ggplot2 package [18]. As previously, the volcanoplot shows the log2 fold change according to *p*-value or to adjusted *p*-value. The advantage of using ggplot2 is that the volcanoplot can then be saved as a vector image, using pdf or eps format (*see* **Note 22**):

```
# Volcanoplot for example dataset1
PDF_Plot(annotated_table_At2)
# Volcanoplot for example dataset2
PDF_Plot(subset(test2, test2$LogFC>0), sign="adjp")
```

2. PDF_Plot function contains many arguments that can be adjusted (*see* Fig. 8):
   (a) Point colors and sizes.
   (b) Axis limits.
   (c) *p*-value and LogFC cut-offs. By default, max.pval = 0.05 and min.LFC = 1.
   (d) Text labels and font size. Text labels are added only for proteins with a significant *p*-value.
   (e) Text labels and cut-off red lines can also be removed.

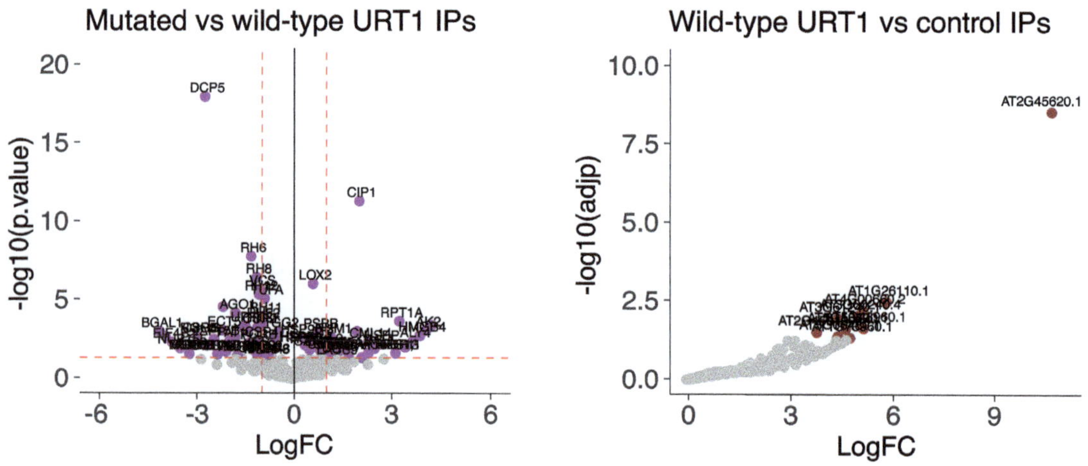

Fig. 8 Volcanoplots obtained using PDF_Plot function, for example, datasets 1 and 2 shown on the left and on the right, respectively. These graphs can be saved as pdf file using the ggsave function

```
# Custom volcanoplot for example dataset 1
graph1 <- PDF_Plot(
           annotated_table_At2,
           sign="p.value", max.pval = 0.05,
           min.LFC = 1, line=TRUE,
           point_color= c("gray", "purple"),
           min_x=-6, max_x=6, min_y=0, max_y=20,
           point_size=3, label=TRUE, label_size=2,
           custom_text=annotated_table_At2$external_gene_name
           title="Mutated vs wild-type URT1 IPs"
                 )
graph1

# Custom volcanoplot for example dataset 2
graph2 <- PDF_Plot(subset(test2, test2$LogFC>0),
                   sign="adjp",
                   max.pval = 0.05, line=FALSE,
                   point_color= c("gray", "darkred"),
                   min_x=-0.5, max_x=11, min_y=-0.5,


                   max_y=10, point_size=3,
                   label=TRUE, label_size=2,
                   title="Wild-type URT1 vs control IPs")
graph2
```

3. Volcanoplots can be saved as a pdf file using the ggsave function of `ggplot2` R package:

```
library(ggplot2)
ggsave("Volcanoplot1.pdf", graph1, height=3, width=5)
ggsave("Volcanoplot2.pdf", graph2, height=3, width=5)
```

3.8 Create Heatmap

1. The heatmap allows the visualization of protein expression pattern between samples. It can be useful when you have multiple groups and you want to sort your interesting proteins based on their abundance in the different groups of IPs. The function `IP_heatmap` creates a heatmap for a list of selected proteins. This heatmap is performed based on the `pheatmap` package [19].
2. Heatmap can be drawn for all detected proteins or for a subset of interesting proteins, for example, proteins that show differential accumulation according to the conditions. For the example dataset 1, the heatmap can be drawn for proteins related to RNA metabolism based on the functional classification in the last column of the result table (*see* Subheading 3.6):

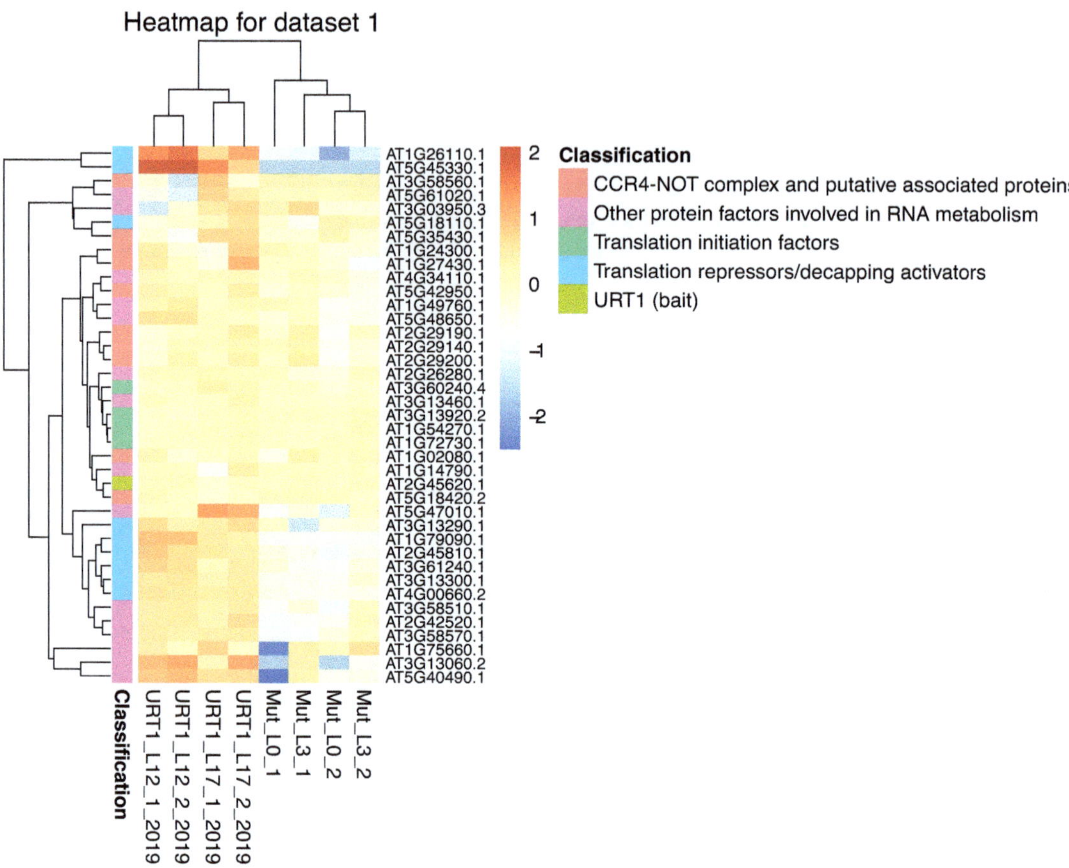

Fig. 9 Heatmap for the example dataset 1 drawn for proteins related to RNA metabolism. Color code on the left of the heatmap indicates the different classifications. These graphs can be saved as pdf file using `pdf` and `dev.off` functions

```
# Make a table with only proteins with classification linked
# to RNA metabolism.
# Remove proteins with empty classification (NA)
class <- annotated_table_At2[
          !is.na(annotated_table_At2$Classification),]
```

3. Nicely, `pheatmap` package allows also to add a color code based annotation for columns or rows. For example, for the dataset 1, row color code can be added to indicate classification of the proteins used for the heatmap (*see* Fig. 9). This color code can be added by specifying, as `annotation_row` argument, a dataframe with protein identifiers as row names and their corresponding classification as first column:

```
# Creation of a one-column dataframe with
# the classification for each selected protein
# and the protein ID as row.names
class2 <- class[,"Classification", drop=F]
```

4. Heatmap can be plotted based on raw data (`norm="nothing"`, by default), or on normalized data either based on the total number of count (`norm="total"`) or on the median-to-ratios method as used in DESeq2 R package [25] (`norm="DEseq"`). Here, the median-to-ratio method (DEseq2) was used to normalize data (*see* **Note 23**):

```
IP_pheatmap(IP_data1, GeneList=row.names(class), norm="DEseq"
            annotation_row = class2, fontsize_row=8)
```

5. Several other arguments of `IP_pheatmap` can also be adjusted:
   (a) font sizes.
   (b) title and its font size.
   (c) hierarchical clustering of columns or rows can be removed.

   For argument usages enter:

```
help(IP_heatmap)
```

6. The heatmap can also be saved as a pdf file using `pdf` and `dev.off` R functions:

```
pdf("Heatmap_dataset1.pdf", width=8, height=6)
IP_pheatmap(IP_data, GeneList=row.names(class), norm="DEseq",
            annotation_row = class2, fontsize_row=8,
            title="Heatmap for dataset 1")
dev.off()
```

4 Notes

1. Stable expression system should be favored and, when possible, the expression level of the tagged protein should be as close as possible of the endogenous level of the protein of interest in order to better reflect physiological protein-protein interactions.
2. In this chapter, we focus on IP-MS approach but other affinity approaches are often used in interactomics, as, for example, the Tandem Affinity Purification tag system (TAP-tag) [27].
3. Biological replicates are parallel measurements of biologically distinct samples that reflect random biological variation, whereas technical replicates are repeated measurements of the same biological sample that reflect random technical variability [28]. In the context of IP-MS, technical variability can be linked to sample preparation or affinity purification.

4. In addition to biological replicates, we also encourage analyzing affinity replicates in a first approach to evaluate technical variability related to affinity purification.
5. Dynamic exclusion can be enabled or not for spectral count based quantification. Yet, [29] showed that enabling dynamic exclusion leads to higher peptide counts and better reproducibility for the detection of relatively low abundant proteins. They found that the optimal duration of this exclusion depends on the average width of the chromatographic peak, mass spectrometry parameters and sample complexity.
6. The quantification method based on spectral counts is a fast and simple approach to resolve a short list of potential protein interactors. One shortcoming of the spectral count approach is its limitation towards the detection of low abundant proteins that can lead to an underestimation of differentially accumulated proteins. Alternatively, or as a complement, MS1 peak area quantification can be performed on the same MS raw files. Of note, the statistical model implemented in `IPinquiry` is not appropriate for statistical analysis based on MS1 peak area quantification.
7. These packages are not automatically installed when installing `IPinquiry` and have to be installed from CRAN [12] or Bioconductor [30].
8. `IPinquiry` is meant to evolve in order to allow for bug fixes and/or improvements. Please update `IPinquiry` regularly.
9. Sample names in count and sample tables have to be identical and must not start with numbers.
10. In this study [11], Scheer, de Almeida et al. performed interactomic and functional analyses of the TUTase URT1, the main enzyme responsible for mRNA uridylation in Arabidopsis. Their data supports that URT1 participates in a molecular network connecting several translational repressors/decapping activators.
11. M1 motif is a short linear motif in the N-terminal region of URT1. In [11], M1 was shown to mediate direct interaction between URT1 and DCP5, a decapping activator.
12. Documentation can be accessed by using the R function `help` for each function of the `IPinquiry` package.
13. `IPinquiry` includes three methods of data normalization. When `norm="nothing"` is used, scale factor is set to 1. When `norm="total"` is used, spectral counts are divided by the total number of counts and scale factors are calculating using the R code:

```
div<- apply(data,2,sum)
```

Finally, when `norm="DEseq"` is used, counts are divided by sample-specific size factors determined by median ratio of spectral counts relative to geometric mean per protein. The geometric mean is calculated using the following R code:

```
n <- length(x)
prod(x)\^{ }(1/n)
```

The scale factor is calculated using the following R code:

```
div <- apply((data+1)/ apply(data + 1,1,gmean),2, median)
```

By default, `norm="total"`.

14. When `glm=QL`, the `stat_test` function applies the three following `EdgeR` functions: `estimateDisp`, `glmQLFit`, `glmQLFTest`.
15. Low abundance proteins can adversely affect the dispersion estimation. The `min.disp` argument allows users to set an appropriate cut-off value for the calculation of the dispersion. Only proteins with total sum of counts above this value are used. By default, the cut-off value used by the `EdgeR` `estimateDisp` function is 5.
16. GLM models implemented in `EdgeR` and `msmsTest` packages normalize data with the help of an offset term in the model. `IPinquiry` includes three alternative ways for the offset calculation: no normalization (`norm="nothing"`), normalization using the total number of counts (`norm="total"`) and normalization based on the median-to-ratio method (`norm="DEseq"`) (*see* **Note 13** for details about scale factor calculation).
17. If `batch=TRUE`, the batch variable is added as a blocking factor in the GLM model.
18. When `glm=classic`, the stat_test function applies the three following `EdgeR` functions: `estimateDisp`, `glmFit`, `glmLRT`. Output includes the same elements except that quasi-likelihood F-statistic values are replaced by likelihood ratio statistic values. This model is the one included in `msmsTest` package [7].
19. "YES" or "NO" tags indicate proteins with sum of counts across all IPs higher or lower than this filter value, respectively.
20. The directory where data are saved can be specified for all functions allowing data export, for example, in this pipeline for `saveWidget`, `write.xlsx`, `write.table`, `ggsave`, and `pdf`.
21. Decapping is a critical step of mRNA degradation and consists in the hydrolysis of the 5' cap structure of mRNA. Data in

Scheer, de Almeida et al. suggest that URT1 connects decapping activators.

22. The aim of `PDF_Plot` is to facilitate the creation of graphs that are suitable for pdf or eps saving. Of course, if you are familiar with the `ggplot2` package, you can skip the `PDF_Plot` function and use the `ggplot2` suite of functions. You will then have the possibility to control more graphical parameters.
23. Counts are log2 transformed using the $y = log2(x + 1)$ formula.

Acknowledgements

The authors gratefully acknowledge Dominique Gagliardi for proofreading of the manuscript, Johana Chicher and all persons involved in `IPinquiry` testing at IBMP. Activities in our groups are currently supported by the Centre National de la Recherche Scientifique (CNRS) and research grants from the French National Research Agency as part of the "Investments for the Future" program under the framework of the LABEX: ANR-10-LABX-0036_NETRNA and ANR-17-EURE-0023. The work was also supported by an IdEx grant from the University of Strasbourg to HZ and by an IdEx grant from the University of Strasbourg for the funding of a QExactive Plus mass spectrometer (ThermoScientific).

References

1. Dunham WH, Mullin M, Gingras AC (2012) Affinity-purification coupled to mass spectrometry: basic principles and strategies. Proteomics 12(10):1576–1590. https://doi.org/10.1002/pmic.201100523
2. Smits AH, Vermeulen M (2016) Characterizing protein–protein interactions using mass spectrometry: challenges and opportunities. Trends Biotechnol 34(10):825–834. https://doi.org/10.1016/j.tibtech.2016.02.014
3. Yugandhar K, Gupta S, Yu H (2019) Inferring protein-protein interaction networks from mass spectrometry-based proteomic approaches: a mini-review. Comput Struct Biotechnol J 17:805–811. https://doi.org/10.1016/j.csbj.2019.05.007
4. Bubis JA, Levitsky LI, Ivanov MV, Tarasova IA, Gorshkov MV (2017) Comparative evaluation of label-free quantification methods for shotgun proteomics. Rapid Commun Mass Spectrom 31(7):606–612. https://doi.org/10.1002/rcm.7829
5. Robinson MD, McCarthy DJ, Smyth GK (2010) Edger: a bioconductor package for differential expression analysis of digital gene expression data. Bioinformatics 26(1):139–140. https://doi.org/10.1093/bioinformatics/btp616
6. McCarthy DJ, Chen Y, Smyth GK (2012) Differential expression analysis of multifactor RNA-seq experiments with respect to biological variation. Nucl Acids Res 40(10):4288–4297. https://doi.org/10.1093/nar/gks042
7. Gregori J, Sanchez A, Villanueva J (2019) msmsTests: LC-MS/MS differential expression tests. R package version 1.22.0. https://www.bioconductor.org/packages/release/bioc/html/msmsTests.html
8. Chicois C, Scheer H, Garcia S, Zuber H, Mutterer J, Chicher J, Hammann P, Gagliardi D, Garcia D (2018) The upf1 interactome reveals interaction networks between RNA degradation and translation repression factors in arabidopsis. Plant J 96(1):119–132. https://doi.org/10.1111/tpj.14022
9. Lange H, Ndecky SY, Gomez-Diaz C, Pflieger D, Butel N, Zumsteg J, Kuhn L, Piermaria C, Chicher J, Christie M, et al. (2019) RST1 and RIPR connect the cytosolic

RNA exosome to the Ski complex in arabidopsis. Nat Commun 10(1):1–12. https://doi.org/10.1038/s41467-019-11807-4
10. Bouchoucha A, Waltz F, Bonnard G, Arrivé M, Hammann P, Kuhn L, Schelcher C, Zuber H, Gobert A, Giegé P (2019) Determination of protein-only RNase P interactome in arabidopsis mitochondria and chloroplasts identifies a complex between PRORP1 and another NYN domain nuclease. Plant J 100(3):549–561. https://doi.org/10.1111/tpj.14458
11. Scheer H, de Almeida C, Ferrier E, Simonnot Q, Poirier L, Pflieger D, Sement FM, Koechler S, Piermaria C, Krawczyk P, et al. (2021) The TUTase urt1 connects decapping activators and prevents the accumulation of excessively deadenylated mRNAs to avoid siRNA biogenesis. Nat Commun 12(1):1–17. https://doi.org/10.1038/s41467-021-21382-2
12. R Core Team (2017) R: A language and environment for statistical computing. R Foundation for Statistical Computing, Vienna, Austria. https://www.R-project.org/
13. Ritchie ME, Phipson B, Wu D, Hu Y, Law CW, Shi W, Smyth GK (2015) Limma powers differential expression analyses for RNA-sequencing and microarray studies. Nucl Acids Res 43(7):e47–e47. https://doi.org/10.1101/2020.05.26.114322
14. Giner G, Smyth GK (2016) Statmod: probability calculations for the inverse gaussian distribution. R J 8(1):339–351. https://journal.r-project.org/archive/2016-1/giner-smyth.pdf
15. Sievert C (2020) Interactive web-based data visualization with R, plotly, and shiny. Chapman and Hall/CRC, London. https://plotly-r.com
16. Vaidyanathan R, Xie Y, Allaire J, Cheng J, Russell K (2019) htmlwidgets: HTML Widgets for R. R package version 1.5.1. https://CRAN.R-project.org/package=htmlwidgets
17. Xie Y, Cheng J, Tan X (2020) DT: a wrapper of the JavaScript library 'DataTables'. R package version 0.13. https://CRAN.R-project.org/package=DT
18. Wickham H (2016) ggplot2: elegant graphics for data analysis. Springer, New York. https://ggplot2.tidyverse.org
19. Kolde R (2019) pheatmap: pretty heatmaps. R package version 1.0.12. https://CRAN.R-project.org/package=pheatmap
20. Neuwirth E (2014) RColorBrewer: colorBrewer palettes. R package version 1.1-2. https://CRAN.R-project.org/package=RColorBrewer
21. Durinck S, Moreau Y, Kasprzyk A, Davis S, De Moor B, Brazma A, Huber W (2005) Biomart and bioconductor: a powerful link between biological databases and microarray data analysis. Bioinformatics 21:3439–3440. https://doi.org/10.1093/bioinformatics/bti525
22. Durinck S, Spellman PT, Birney E, Huber W (2009) Mapping identifiers for the integration of genomic datasets with the r/bioconductor package biomart. Nat Protocols 4:1184–1191. https://doi.org/10.1038/nprot.2009.97
23. Dragulescu A, Arendt C (2020) xlsx: read, write, format excel 2007 and excel 97/2000/XP/2003 files. R package version 0.6.3. https://CRAN.R-project.org/package=xlsx
24. Wickham H, Hester J, Chang W (2020) devtools: tools to make developing R packages easier. R package version 2.3.1. https://CRAN.R-project.org/package=devtools
25. Love MI, Huber W, Anders S (2014) Moderated estimation of fold change and dispersion for RNA-Seq data with deseq2. Genom Biol 15:550. https://doi.org/10.1186/s13059-014-0550-8
26. Hubbard T, Barker D, Birney E, Cameron G, Chen Y, Clark L, Cox T, Cuff J, Curwen V, Down T, et al. (2002) The ensembl genome database project. Nucl Acids Res 30(1):38–41. https://doi.org/10.1093/nar/30.1.38
27. Gerace E, Moazed D (2015) Affinity purification of protein complexes using tap tags. In: Methods in enzymology, vol 559. Elsevier, Amsterdam, pp 37–52. https://doi.org/10.1016/bs.mie.2014.11.007
28. Blainey P, Krzywinski M, Altman N (2014) Replication: quality is often more important than quantity. Nat Methods 11(9):879–881. https://doi.org/10.1038/nmeth.3091
29. Zhang Y, Wen Z, Washburn MP, Florens L (2009) Effect of dynamic exclusion duration on spectral count based quantitative proteomics. Anal Chem 81(15):6317–6326. https://doi.org/10.1021/ac9004887
30. Gentleman RC, Carey VJ, Bates DM, Bolstad B, Dettling M, Dudoit S, Ellis B, Gautier L, Ge Y, Gentry J, et al. (2004) Bioconductor: open software development for computational biology and bioinformatics. Genom Biol 5(10):R80. https://doi.org/10.1186/gb-2004-5-10-r80

Chapter 12

Statistical Analysis of Post-Translational Modifications Quantified by Label-Free Proteomics Across Multiple Biological Conditions with R: Illustration from SARS-CoV-2 Infected Cells

Quentin Giai Gianetto

Abstract

Protein post-translational modifications (PTMs) are essential elements of cellular communication. Their variations in abundance can affect cellular pathways, leading to cellular disorders and diseases. A widely used method for revealing PTM-mediated regulatory networks is their label-free quantitation (LFQ) by high-resolution mass spectrometry. The raw data resulting from such experiments are generally interpreted using specific software, such as MaxQuant, MassChroQ, or Proline for instance. They provide data matrices containing quantified intensities for each modified peptide identified. Statistical analyses are then necessary (1) to ensure that the quantified data are of good enough quality and sufficiently reproducible, (2) to highlight the modified peptides that are differentially abundant between the biological conditions under study. The objective of this chapter is therefore to provide a complete data analysis pipeline for analyzing the quantified values of modified peptides in presence of two or more biological conditions using the R software. We illustrate our pipeline starting from MaxQuant outputs dealing with the analysis of A549-ACE2 cells infected by SARS-CoV-2 at different time stamps, freely available on PRIDE (PXD020019).

Key words Statistics, R, Data quality control, Clustering, Post-translational modifications, Label-free proteomics, Relative quantification

1 Introduction

Mass spectrometry (MS)-based proteomics allow the identification and quantification of a large number of post-translational protein modifications [1, 2]. A general workflow consists of (a) a proteolytic digestion of proteins into peptides, (b) an enrichment step to increase the concentration of modified peptides in the samples, and (c) the analysis of the samples by liquid chromatography-tandem mass spectrometry (LC-MS/MS) [3]. It is often used to study protein phosphorylations [4, 5], which is the most commonly studied post-translational modification. This

Thomas Burger (ed.), *Statistical Analysis of Proteomic Data: Methods and Tools*,
Methods in Molecular Biology, vol. 2426, https://doi.org/10.1007/978-1-0716-1967-4_12,

workflow can also be adapted to study other post-translational modifications such as ubiquitinations [6], methylations [7], acetylations [8], or several kind of modifications in unified pipelines [9].

Label-free quantification enables large-scale analyzes and can be applied to experiments composed of many samples and biological conditions, as it can be the case in the field of clinical screening for instance. It allows avoiding drawbacks of labeling methods, whether metabolic (SILAC) or chemical (iTRAQ), which are limited by the cost of labeling reagents, some ineffective labeling, or the limited number of samples that can be analyzed. However, label-free quantification of post-translational modifications requires careful experimental designs to achieve good reproducibility of quantified values. Indeed, the enrichment step introduces additional variations to the ones traditionally produced by the fluctuation of the liquid chromatography and the ionization conditions in MS-based proteomics. This is the reason why a special care must be taken during this step to carry it out as rigorously as possible in order to maximize the reproducibility of experiments.

The MS/MS spectra obtained after the LC-MS/MS analysis have to be associated with peptides. The identification of spectra is generally carried out by searching the measured MS/MS spectra in a theoretical database, freely downloadable from UniProt website for instance (https://www.uniprot.org/). The localization of a modification on a specific amino acid of the identified peptide is assessed using a score or a probability, depending on the method applied. It is usually calculated by comparing the measured spectrum with theoretical spectra where the modification is placed on each possible amino acid of the considered peptide [10]. Currently, this research is generally done thanks to specific software, such as MaxQuant [11], MassChroQ [12], or Proline [13], but can also be performed from R (*see* **Note 1**). Depending on the quantification method, either the MS spectra or the MS/MS spectra are next used to quantify the abundance of each modified peptides identified. In the end, such bioinformatics analyses result in data matrices composed of the values quantified in each of the samples for all the modified peptides identified.

It is common to identify and quantify several tens of thousands peptides in usual experiments. To extract useful information from these large datasets, statistical analyzes of quantified data matrices are required, to check the reproducibility of the analyses and to determine the modified peptides of interest. The objective of this chapter is therefore to provide a data analysis pipeline for analyzing the quantified values of modified peptides, in presence of two or more biological conditions using the R software. This analysis pipeline is provided with examples of R codes adapted to a case study. The informed reader can easily change these codes to adapt them to his/her problematic and to what seems relevant to put forward in his/her analysis. We start from outputs of the MaxQuant software [11], which is one of the most popular software tool for this kind of analysis, but the proposed data analysis pipeline can easily be applied to the outputs of various software, as long as they provide the quantified values of the modified peptides and the proteins they belong to.

2 Material

2.1 R and RStudio Installation

R can be freely installed on most operating systems (Windows, Linux, Mac OS) from https://cran.r-project.org/. The basic R installation includes the language itself as well as packages, most of which are geared toward statistical analysis. One of the strengths of R is the large number of packages that can be downloaded, mainly from the CRAN repository (https://cran.r-project.org/) and, as far as bioinformatics is concerned, from Bioconductor (http://bioconductor.org/). In addition to R, we recommend that the reader install RStudio (https://rstudio.com/products/rstudio/download/) which is a development environment to facilitate the development of R codes and packages. However, all the codes presented herein can be executed from a basic R installation.

Throughout this chapter we assume that the reader has some prior knowledge of R, and we give code examples without dealing with the basics of R, as for example the different types of R objects. For more details on R and its programming language, the reader may refer to the many courses available for free on the R and Bioconductor websites as well as many good books on R and Bioconductor such as [14–16].

2.2 R Packages

In addition to the functions provided in the standard version of R, we will use functions from other R packages in the sample codes provided. These packages are the next: `VennDiagram` [17], `UpSetR` [18], `ggplot2` [19], `grid`, `ggplotify`, `jaccard`, `ggdendro` [20], `ggridges` [21], `limma` [22], `cp4p` [23], `ssize.fdr` [24], `imp4p` [25], `car` [26], `multivariance` [27], `factoextra` [28], `cluster`, `reshape2` [29], `ggpubr` [30].

It is up to the reader to download these packages and to check that they are properly installed on his/her machine.

2.3 Data Type

Throughout this chapter we assume that the data are outputs from software used to identify and quantify the abundance of modified peptides and their proteins, such as MaxQuant [11], MassChroQ [12], or Proline [13]. The quantitative data should fit into two matrices: the first contains quantified values for PTMs identified on peptides; the second contains quantified values for the non-modified proteins to which the modified peptides of the first matrix are associated (i.e. values deduced from non-modified peptides of associated proteins).

For instance, when using MaxQuant [11], the first dataset can be extracted from the "Phospho (STY)Sites.txt" (phosphorylation) or "GlyGly (K)Sites.txt" (ubiquitination), while the second dataset is in the "proteinGroups.txt" file.

The first matrix thus makes it possible to determine the dynamics of the abundance of a PTM between several biological conditions; while the second makes it possible to compare this dynamic with that of the unmodified protein across multiple biological conditions. This comparison is important to ensure that a difference in the abundance of a modified peptide is actually due to the abundance of the modification and not to the dynamic of the associated unmodified protein between compared conditions.

2.4 Case Study Dataset

To illustrate this protocol, we used a dataset that the reader can freely download from PRIDE (PXD020019). On this server, the reader will find MaxQuant outputs as well as the raw data from the mass spectrometers, such as he/she can reanalyze them using their own workflow and software of interest. This dataset corresponds to the analysis of phosphorylated and ubiquitinated peptides in A549-ACE2 cells infected by SARS-CoV-2 at different time stamps [31]. Hereafter, we will focus on the analysis of phosphorylated peptides, but the same analyses can be performed from the ubiquitinated peptides. To reproduce the results presented along this protocol, the careful reader has to download the MaxQuant outputs "proteinGroups.txt" and "Phospho (STY)Sites.txt" from the PRIDE website.

3 Methods

3.1 Data Preprocessing Steps

1. Importing data in R: First, the software outputs used to identify and quantify the proteins as well as the peptides have to be saved into a format easily loadable in R. Generally, this can be performed by saving the software outputs into a "txt" or "csv" file. Then, these files can be imported in the R session into a data frame object (*see* **Note 2**). For instance, using MaxQuant outputs, the `read.csv` function can be used for this purpose:

```
data.prot=read.csv("(path_to_your_file)/proteinGroups.txt",
                   header=TRUE,sep="\t",quote="")
data.ptm=read.csv(
      "(path_to_your_file)/Phospho (STY)Sites.txt",
      header=TRUE,sep="\t",quote=""
          )
```

2. Filtering peptides which are associated with "Reverse" or "Potential contaminant" peptides: In MaxQuant, "reverse" peptides are artifactual peptides whose amino-acid sequence corresponds to the one of peptides of the input database used for the identification, yet in reverse order (*see* **Note 3**). Additionally, "Potential contaminant" peptides are associated with proteins that commonly contaminate samples. Thus, these two kinds of identified peptides can be deleted before subsequent analysis. They are indicated by "+ " in the reverse and potential contaminant columns:

```
data.ptm=data.ptm[
      which((data.ptm$Reverse!="+")&
      (data.ptm$Potential.contaminant!="+")),
                 ]
```

3. Filtering peptides which PTM location is not sufficiently well identified: In MaxQuant outputs, the maximum localization probability that has been estimated in one of the analyzed samples is in the "Localization.prob" column. In literature, it is common to keep only the modified peptides for which this probability is greater than 75% (*see* **Note 4**):

```
data.ptm=data.ptm[which(data.ptm$Localization.prob>0.75),]
```

4. Separating intensities from metadata: The metadata generally contains multiple information, notably related to the identification of the modified peptides and the proteins they belong to. To separate the quantified intensities from the metadata, columns containing quantified intensities can be searched using the `grep` function, and the data can be converted into a matrix object in R with the `as.matrix` function:

```
int.prot=as.matrix(data.prot[,grep("Intensity.",
                   names(data.prot), value=TRUE)])
```

For the proteins, only the columns relative to the quantification of the unmodified proteins can be kept using (*see* **Note 5**):

```
 int.prot=int.prot[,!colnames(int.prot)%in%grep("Phospho",
                   colnames(int.prot), value=TRUE)]
int.prot=int.prot[,!colnames(int.prot)%in%grep("Ubi",
                   colnames(int.prot), value=TRUE)]
```

Similarly, the intensities of the modified peptides can be extracted using (*see* **Note 6**):

```
int.ptm=as.matrix(data.ptm[,grep("Intensity.Phospho",
                  names(data.ptm), value=TRUE)])
```

5. Replacing 0 by NA: This step is useful to easily detect missing values in subsequent analyses:

```
int.prot[int.prot==0]=NA
int.ptm[int.ptm==0]=NA
```

6. Distinguishing between the different versions of a modified peptide: In MaxQuant outputs, the quantified intensities derived from one, two, or three and higher PTMs are in columns whose name ends with ___1, ___2, ___3. Thus three lines must be created for each modified peptide of the file:

```
int.ptm.mod=rbind(int.ptm[,grep("___1",colnames(int.ptm))],
                  int.ptm[,grep("___2",colnames(int.ptm))],
                  int.ptm[,grep("___3",colnames(int.ptm))])
colnames(int.ptm.mod)=unlist(
            strsplit(colnames(int.ptm)[
            grep("___1",colnames(int.ptm))],
            split ="___1")
                             )
```

Each modified peptide is next characterized by its identification number, the proteins to which it may belong (quantified in the "proteinGroups.txt" file) and the number of PTMs used to quantify ("multiplicity" column):

```
id.ptm=rep(data.ptm$id,3)
id.pg=rep(data.ptm$Protein.group.IDs,3)
multiplicity=c(rep("___1",nrow(int.ptm)),
               rep("___2", nrow(int.ptm)),
               rep("___3",nrow(int.ptm)))
```

7. Removing modified peptide with no quantification values: The modified peptides associated with no quantification value in any sample are useless for performing subsequent statistical analysis (*see* **Note 7**):

```
# Compute the number of observed values on each row
sum_notna=apply(int.ptm.mod,1,function(x)sum(!is.na(x)))

# Extract peptides with at least one observed value
# on each row
int.ptm.mod=int.ptm.mod[which(sum_notna>0),]
id.ptm=id.ptm[which(sum_notna>0)]
id.pg=id.pg[which(sum_notna>0)]
multiplicity=multiplicity[which(sum_notna>0)]
```

8. Renaming the columns of the dataset: It can generally be useful to rename the columns with shorter names for further analyses, in particular to make them appear in graphics:

```
short.name=unlist(strsplit(colnames(int.ptm.mod),
                  split="Intensity.Phospho_"))
short.name=short.name[short.name!=""]
colnames(int.ptm.mod)=short.name
data.ptm.mod=data.frame(id.ptm,id.pg,multiplicity,
                        int.ptm.mod)
```

9. Defining vectors containing the design of experiments: The experimental design is specific to each experiment. It is extremely important, especially for the subsequent statistical analysis. The vectors containing the experimental design must correspond to the columns containing the intensity values. It is advisable to do one for modified peptides, another for proteins, and one containing both. In our case study, we have two factors, the time (6 h or 24 h) and the biological conditions (mock or SARS-Cov-2 infected samples), three replicated samples have been performed in each condition for studying the modifications and four replicates have been used for the unmodified proteins. Experimental design corresponding to the columns of `int.ptm.mod` is defined by:

```
# colnames(int.ptm.mod) can be used to check the column names
# of int.ptm.mod
Cond.ptm=factor(c(rep("mock",6),rep("SARS_COV_2",6)),
                levels=c("mock","SARS_COV_2"))
Time.ptm=factor(rep(c(rep("24h",3),rep("6h",3)),2),
                levels=c("6h","24h"))
CondTime.ptm=as.factor(paste(Cond.ptm,Time.ptm,sep="."))
```

Experimental design corresponding to the columns of `int.prot` is defined by:

```
Cond.prot=factor(c(rep("mock",8),rep("SARS_COV_2",8)),
                 levels=c("mock","SARS_COV_2"))
Time.prot=factor(rep(c(rep("24h",4), rep("6h",4)),2),
                 levels=c("6h","24h"))
CondTime.prot=as.factor(paste(Cond.prot,Time.prot,sep="."))
```

Both experimental designs can be combined using:

```
Cond.ptmprot=unlist(list(Cond.ptm,Cond.prot))
Time.ptmprot=unlist(list(Time.ptm,Time.prot))
```

Additionally, a vector indicating whether the combined design is for the peptide or protein will be useful in subsequent statistical analysis (*see* Subheading 3.9, **Step 1**):

```
# Comp.ptmprot is equal to 1 if the coordinate is related
# to the modified peptide and 0 otherwise
Comp.ptmprot=c(rep(1,length(Cond.ptm)),
               rep(0,length(Cond.prot)))
```

3.2 Checking the Reproducibility of Identifications

When the samples are highly reproducible, one can expect that the same modified peptides will be identified in each of the samples. The intersections between the sets of modified peptides identified in the samples can be visualized using Venn diagrams or UpSet graphs to highlight any aberrant sample:

1. Creating a matrix containing the row number of an identified peptide if it is quantified:

```
tr.int.ptm=int.ptm.mod;
tr.int.ptm[tr.int.ptm>0]=1;
vecto=as.matrix(1:nrow(tr.int.ptm));
row.nb.id=apply(tr.int.ptm,2,function(x) x*vecto);
```

2. Calculating the size of each intersection between samples using the `intersect` function:

```
peptid=row.nb.id[,grep("SARS_COV2_6h",colnames(row.nb.id))]
nb.sample=ncol(peptid)
# n2 contains the sizes of intersections between 2 samples
n2=matrix(0,nb.sample,nb.sample);
# n3 contains the sizes of intersections between 3 samples
n3=array(0,dim=rep(nb.sample,3));
for (i in 1:nb.sample){
  for (j in i:nb.sample){
    inter=intersect(peptid[,i],peptid[,j]);
    n2[i,j]=length(inter[inter!=0]);
    for (k in j:nb.sample){
      inter=intersect(peptid[,i],
                      intersect(peptid[,j],peptid[,k]));
      n3[i,j,k]=length(inter[inter!=0]);
    }
  }
}
```

3. Using the `draw.triple.venn()` function of R package `VennDiagram` to plot a Venn diagram (*see* **Note 8**). The following code displays the numbers of modified peptides found in common or specific of each SARS-Cov-2 samples at 6 h (*see* Fig. 1a):

```
require(VennDiagram)
venn.plot = draw.triple.venn(
  area1=n2[1,1], area2=n2[2,2],
  area3=n2[3,3], n12=n2[1,2],
  n13=n2[1,3], n23=n2[2,3], n123=n3[1,2,3],
  category = colnames(peptid),
  fill = c("dodgerblue", "goldenrod1", "seagreen3"),
  cat.col = c("dodgerblue", "goldenrod4", "seagreen4"),
  cat.cex = 1.5,cat.dist=0.1,
  margin = 0.1,
  cex = c(1.5,1.5,1.5,1.5,2,1.5,1.5), ind=TRUE
                              );
```

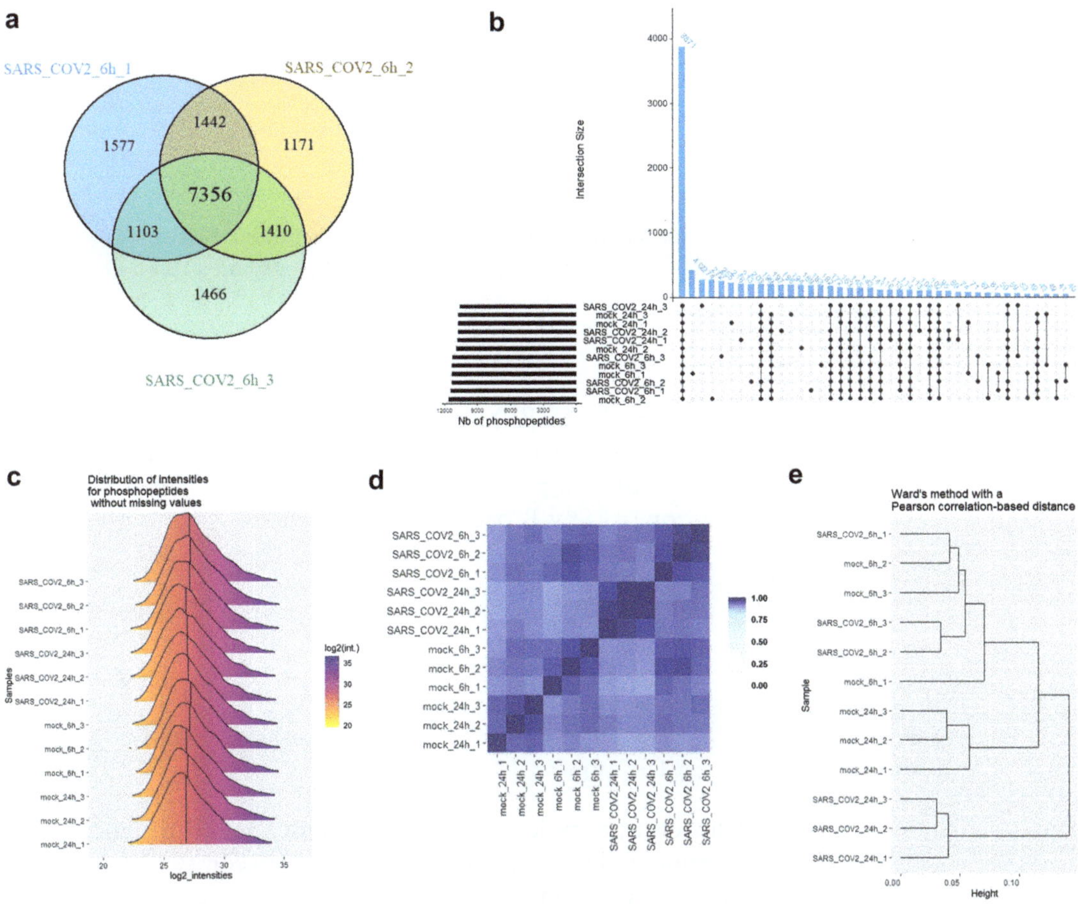

Fig. 1 Graphs to check the reproducibility of experiments. (**a**) Venn diagram checking the reproducibility of modified peptides identified in SARS-CoV-2 samples at 6 h. (**b**) UpSet graph checking the reproducibility of modified peptides identified in all samples (3871 are found in common). (**c**) Estimated distributions of log2 (intensities) of modified peptides found in all samples. This makes it possible to detect shifts in all the intensity values of one or more samples. (**d**) Pearson correlation matrix between all samples using shades of blue. Darker blues (higher correlation values) should appear between replicates of the same condition. (**e**) Hierarchical clustering of samples from Pearson correlation values. Samples of a same condition should cluster together, and not with the one of another condition if the conditions proteomes are not close. If they are, clustering may have a hard time distinguishing them, but they will appear close (with darker blue) in the correlation matrix plotted in (**d**)

4. Visualizing the intersections using UpSet graphs: Venn diagrams represent the sets of identified peptides using circles or ellipses. However, the larger the number of samples, the more difficult the Venn diagrams are readable. This is the reason why it may be preferable to use UpSet graphs (*see* Fig. 1b) when the number of samples becomes large (i.e., superior to 5):

```
require(UpSetR)

data.upset=as.data.frame(tr.int.ptm)
names(data.upset)=colnames(data.upset)
data.upset[is.na(data.upset)]=0

upset.plot = upset(data.upset , order.by = "freq",
                   number.angles=315, main.bar.color=4,
                   nsets=ncol(data.upset), sets.x.label="Nb",
                   text.scale = c(1.5,1.5,1.5,1.5,1.5,1));
# To visualize
upset.plot
```

5. (OPTIONAL STEP) Converting the plots into ggplot objects: It can be useful to convert the obtained UpSet plot or the Venn diagram into `ggplot` objects (*see* **Note 9**). For this, the `grid` and `ggplotify` R packages can be used:

```
require(ggplot2)
require(ggplotify)
require(grid)

grid.newpage();
p = grobTree(venn.plot);
venn.plot = as.ggplot(p);
upset.plot = as.ggplot(upset.plot);
```

6. Clustering samples using Jaccard index-based distances: The reproducibility of the identifications between two samples can be evaluated using Jaccard index. This index can be computed using the `jaccard()` function of the R package `jaccard`:

```
require(jaccard)
mat.jaccard=matrix(NA,ncol(data.upset),ncol(data.upset))
for (i in 1:ncol(data.upset)){
  for (j in 1:ncol(data.upset)){
     mat.jaccard[i,j]=jaccard(data.upset[,i],data.upset[,j]);
  }
}
colnames(mat.jaccard)=colnames(data.upset)
rownames(mat.jaccard)=colnames(data.upset)
```

This matrix can be visualized using shades of blue (*see* **Note 10**). Additionally, a distance matrix based on this index can be computed with the `as.dist()` function, and a hierarchical clustering of samples can be performed with the `hclust()` function:

```
hv = hclust(as.dist(1-mat.jaccard),method="ward.D2")
```

The clustering can be visualized with the ggdendrogram function of the R package ggdendro:

```
require(ggdendro)

ggd = ggdendrogram(hv, rotate = TRUE, theme_dendro = FALSE)
ggd = ggd + theme(axis.text.y = element_text(hjust = 1))
ggd = ggd + xlab("Sample")
ggd = ggd + ylab("Height")
ggd = ggd + ggtitle("Ward's method with a \n
                    Jaccard index-based distance")
```

Additionally, the same approach can be used to evaluate the replication of the non-identifications between samples (*see* **Note 11**).

3.3 Checking the Reproducibility of Quantified Values

When the samples are highly reproducible, one can expect the quantified values of a same modified peptides to be similar across the biological replicates.

1. Log2-transforming the intensity values: Before analyzing quantified values, a log2 transformation is usually applied, in particular to decorrelate the intensity levels of peptides from their variances (*see* [32] and its supplementary material):

```
log2.int.ptm=log2(int.ptm.mod);
```

2. Plotting the distributions of observed values in all samples: The differences between the distributions of the observed values can be used to see whether samples have globally lower or higher quantified values than in the others (*see* Fig. 1c). To check if the distribution of quantified values are the same in different samples, it is important to focus exclusively on modified peptides which are quantified in all samples you want to compare (*see* **Note 12**):

```
# compute number of observed values on each row
sum_notna=apply(log2.int.ptm,1,function(x){sum(!is.na(x));})

# keeping log2 intensities of modified peptides with no
# missing values
log2.int.ptm.notNA=log2.int.ptm[
                      sum_notna==ncol(log2.int.ptm),
                             ]

# formatting into a dataframe to use ggplot
df=stack(data.frame(log2.int.ptm.notNA));
colnames(df)=c("log2_intensities","Samples");

# plotting using stat_density_ridges function
require(ggridges)

dis.obs = ggplot(df, aes(x=log2_intensities,
                        y=Samples,
                        fill=stat(x)))
dis.obs=dis.obs+
  stat_density_ridges(geom="density_ridges_gradient",
                      scale=3,size=0.3,rel_min_height = 0.01,
                      quantile_lines=TRUE,
                      quantiles = 0.5, alpha = 0.7)
dis.obs = dis.obs+
  scale_fill_viridis_c(name="log2(int.)",
                       option="C", direction=-1)
dis.obs = dis.obs+
  labs(title = "Distribution of intensities
      for phosphopeptides\n without missing values")
```

3. Clustering samples using the Pearson correlation matrix: A Pearson correlation matrix using only complete pairs of observations between two samples to compute each Pearson correlation coefficient can be obtained by using:

```
mat.cor=cor(log2.int.ptm, use = "pairwise.complete.obs");
```

The correlation matrix can next be visualized using shades of blue using the qplot function as in Fig. 1d (*see* **Note 13**). Of note, Principal Component Analysis is strongly related to this matrix (*see* **Note 14**) and can also be used to check the clustering of samples by condition. Clustering methods can be applied from the obtained Pearson correlation-based distance matrix to evaluate if the intensities measured in different samples are close or not:

```
hv = hclust(as.dist(1-mat.cor),method="ward.D2")

require(ggdendro)
ggd = ggdendrogram(hv, rotate = TRUE, theme_dendro = FALSE)
ggd = ggd + theme(axis.text.y = element_text(hjust = 1))
ggd = ggd + xlab("Sample")
ggd = ggd + ylab("Height")
ggd = ggd + ggtitle("Ward's method with a \n Pearson
                    correlation-based distance")
```

Such clustering can be useful for determining which biological conditions are similarly distributed, and which are less so (*see* Fig. 1e).

3.4 Is the Number of Replicates Sufficient in Each Condition?

A sufficient number of samples to confidently trust the results of a statistical test can be determined by seeking the minimum number of samples reaching a satisfactory statistical power. The calculation of this statistical power depends on several factors. In case of a standard *t*-test comparing the average intensities between two conditions, it depends on the minimal average difference intensity between conditions that is expected to be detectable, the variance of the intensities, the type 1 error (threshold on the *p*-values), and the number of samples in each condition (sample size). That is why it is logical to conduct a power analysis from a first MS-based experiment to assess how many samples are sufficient to obtain robust statistical results in a subsequent analysis (*see* **Note 15**). This analysis can also be performed after the final data have been acquired to evaluate what is the confidence in the results.

The R package `ssize.fdr` contains several functions that can be used to calculate minimum sample sizes [24]. The purpose of each of these functions is to estimate an average statistical power over all the tests carried out on all the peptides in function of the sample size. In case of more than two conditions, the functions `ssize.F` and `ssize.Fvary` can be used by defining a level of false discovery rate, a user-specified proportion of non-differentially abundant peptides, a design matrix, and a standard deviation (or the parameters of inverse gamma distribution followed by variances of peptides for `ssize.Fvary`). We detailed hereafter how to use the latter one.

1. Creating a design matrix from the vectors defining the design of experiment: Additional explanations on this matrix and the contrast matrix can be found below (*see* Subheading 3.9, **Steps 1** and **2**).

```
dat.design=data.frame(Cond.ptm,Time.ptm)
design=model.matrix(~Cond.ptm+Time.ptm,data=dat.design)
```

2. Creating a contrast matrix specific to a statistical test from which average statistical power will be computed:

```
# test if the second coefficient related to biological
# condition (mock or SARS-CoV-2) is null
ct=cbind(c(0,1,0))
# test if the second coefficient related to biological
# condition (mock or SARS-CoV-2)) is null
# and the third coefficient related to the time is also null
ct=cbind(c(0,1,-1),c(0,0,1))
```

3. Using `limma` [22] to estimate the model and determine prior variance and degrees of freedom related to the test:

```
require(limma)
fit=eBayes(contrasts.fit(lmFit(log2.int.ptm.notNA,design),ct)
           robust=TRUE)
s02=fit$s2.prior;
d0=mean(fit$df.prior);
```

4. Estimating the proportion of non-differentially abundant peptides: Histogram and calibration plot of *p*-values [23, 32] are displayed in Fig. 2a, b:

```
pval_ct=fit$F.p.value
# histogram of pvalues
hist(pval_ct,50,main="Histogram of pvalues",col=4)

require(cp4p)
# Calibration plot of pvalues to choose a method
calibration.plot(pval_ct,"ALL")

#The proportion of non-differentially abundant peptides is
#estimated using the distribution of the p-values related
#to the contrasted test with default method ("pounds")
prop=estim.pi0(pval_ct)
```

5. Using the `ssize.Fvary` function, additional explanations can be found in [24]:

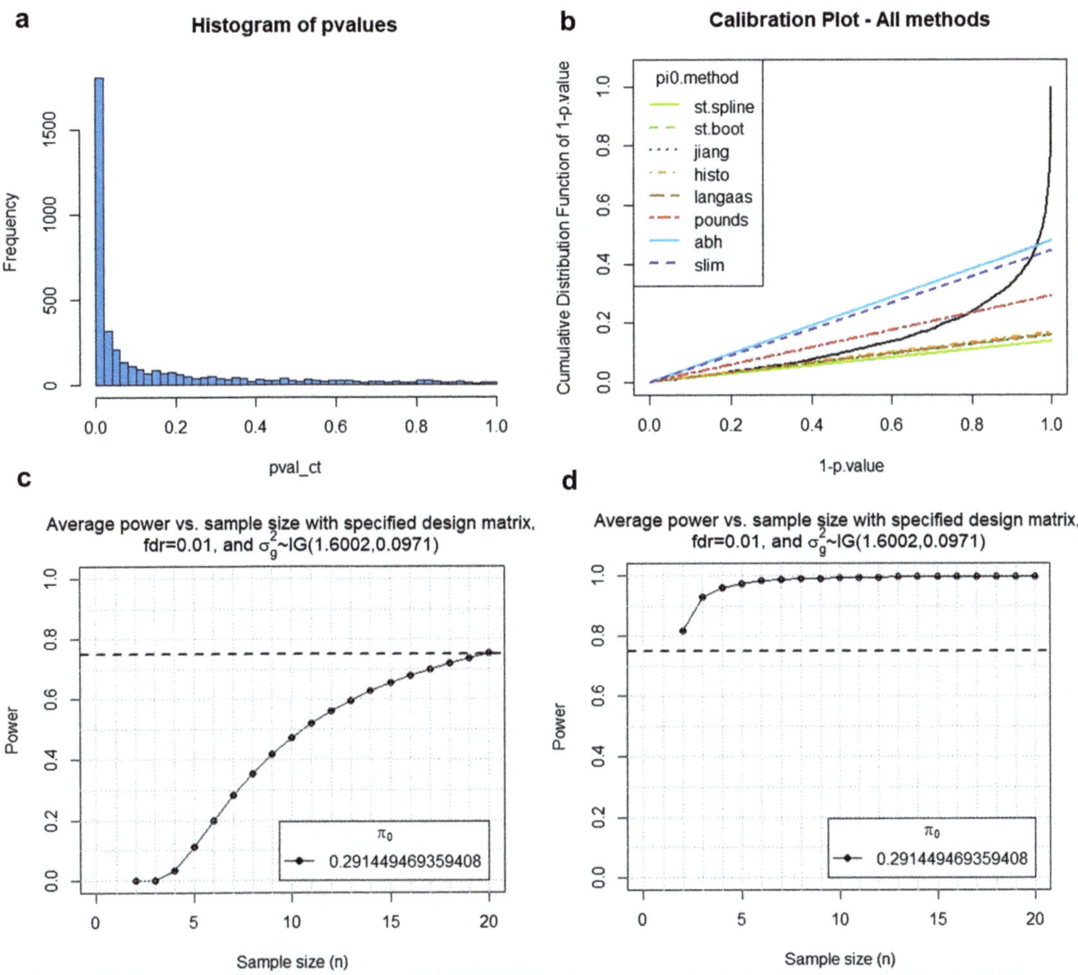

Fig. 2 (**a**) Histogram of *p*-values related to the used test. (**b**) Calibration plot to choose a method to estimate the proportion of true null hypotheses. (**c**) Average power in function of the sample size with eps $= 0.1$ and the proportion of true null hypotheses estimated by the method of Pounds and Cheng [33]. (**d**) Average power in function of the sample size with eps $= 0.5$ and the proportion of true null hypotheses estimated by the method of Pounds and Cheng [33]

```
# FDR level required
fdr=0.01

# Minimum statistical power required
pwr=0.75

# Function computing the degrees of freedom related to the
# design of experiment: here we have 4 groups (two conditions
# at two time points) and two parameters estimated in the model
df=function(n){4*n-2;}

#parameters of the inverse gamma distribution of the variances
alph=d0/2
beta=d0*s02/2

#eps represents values for the model parameters related to
#biological conditions and the time. Because we test the
#nullity of the parameters, more this value is close to 0
#and the harder it will be on the statistical test to know
#if the coefficients are indeed zero or not from data:
#the average statistical power will logically be lower.
eps=0.1
require(ssize.fdr)
ftv=ssize.Fvary(X=design,beta=c(1,eps,-eps),L=ct,dn=df,a=alph,
                b=beta,fdr=fdr,power=pwr,
                pi0= prop$pi0.est$pi0.Pounds,maxN=20)
```

Figure 2c, d represent the average statistical power in function of the sample size for different values of `eps`. The closer it is to 0, and lower the statistical power will be.

3.5 Normalization Step

As explained in the introduction, the enrichment step induces a lower reproducibility among the reported intensities. That is why a normalization step is of prime importance to correct the variations of intensities. In literature, several methods have been proposed [34, 35]. Several normalization methods can be used from the `wrapper.normalizeD` function of the R package `DAPAR` [36] after creating an object of class MSnSet. A classical approach consists to apply a median-centering function in the samples which are a replication of a same biological condition:

1. Creating a median-centering function:

```
median.norm=function(intensities,condition){
  int.norm = intensities
  lev=levels(condition)
  medianes=rep(0,ncol(intensities))
  for (i in 1:length(lev)){
    # column of the condition
    col_cond=which(condition == lev[i])
    # nb N.A. in the condition for proteins/peptides
    nbna_cond=apply(intensities[,col_cond],1,
                function(x){sum(is.na(x));})
    #medianes from proteins/peptides without missing values
    medianes[col_cond]=apply(intensities[which(nbna_cond==0),
                col_cond],2,median,na.rm=T);
    moy=mean(medianes[col_cond]);
    for (j in col_cond){
      #computing differences between each median and
      #the average median in each condition
      int.norm[,j] = intensities[,j] - medianes[j] + moy
    }
  }
  return(int.norm)
}
```

2. Performing the normalization on intensities of modified peptides:

```
log2.int.ptm.norm=median.norm(intensities=log2.int.ptm,
                               condition= CondTime.ptm)
```

3. Performing the normalization method on the protein intensities:

```
log2.int.prot=log2(int.prot)
log2.int.prot.norm= median.norm(intensities=log2.int.prot,
                                 condition=CondTime.prot)
```

3.6 Dealing with Missing Values

In case of multiple biological conditions, modified peptides can have similar detection profiles: they are detected, or not detected, under the same biological conditions. These detection profiles may be of greatest interest. Indeed, non-detection is generally synonymous either with an absence of the modified peptide in the biological condition; or with its low abundance, as missing values mainly result of the limit of detection of the mass spectrometer. The detection profile of the modified peptide has to be compared with those of their belonging protein in order to highlight over- or under-abundance of the modification relatively to the one of the unmodified protein they belong to (*see* **Note 16**):

1. Creating a function returning detection profiles (1 if the modified peptide is detected and 0 if not):

```
detect.prof=function(mat,condition){
  lev=levels(condition)
  detect=matrix(0,nrow(mat),length(lev))
  for (k in 1:nrow(mat)){
      for (i in 1:length(lev)){
          for (j in which(condition ==lev[i])){
              if (!is.na(mat[k,j])){
                  detect[k,i]=1
              }
          }
      }
  }
  colnames(detect)=levels(condition)
  return(detect)
}
```

2. Computing detection profiles for modified peptides and their proteins:

```
detect.ptm=detect.prof(log2.int.ptm.norm,CondTime.ptm)
detect.prot=detect.prof(log2.int.prot.norm,CondTime.prot)
```

Missing values can be replaced within a same condition using imputation algorithms. If no value has been measured along replications of a same experiment in a condition, then either a value can be inferred assuming that there is a relationship to the values measured in the other conditions, or no value can be inferred (*see* **Note 17**). Two main kinds of missing values arise in MS-based proteomics: either missing not at random (MNAR) values or missing completely at random (MCAR) values [37]. The R package imp4p proposes a multiple imputation strategy to deal with these two kinds of missing values and functions to analyze missing value mechanisms [25].

3. Analyzing missing value mechanisms:

```
require(imp4p)
log2.int.ptm.impMCAR=impute.mle(log2.int.ptm.norm,
                                conditions=CondTime.ptm)
resmod=estim.mix(tab=log2.int.ptm.norm,
                 tab.imp=log2.int.ptm.impMCAR,
                 conditions= CondTime.ptm)
resmod$pi.na
resmod$pi.mcar
```

The proportion of missing values is almost 30% in each sample. The proportion of MCAR values is estimated between 0 and 30% depending on the considered sample.

4. Imputing missing values of peptides using a multiple imputation strategy combining MCAR and MNAR values with the `impute.mi()` function:

```
log2.int.ptm.imp=impute.mi(tab = log2.int.ptm.norm,
                           conditions = CondTime.ptm)
colnames(log2.int.ptm.imp)=colnames(log2.int.ptm)
```

5. Imputing missing values of proteins using a MCAR-devoted imputation algorithm:

```
log2.int.prot.impMCAR=impute.mle(log2.int.prot.norm,
                                 conditions=CondTime.prot)
colnames(log2.int.prot.impMCAR)=colnames(log2.int.prot.norm)
```

3.7 Mapping the Quantified Values of the Unmodified Protein to the Ones of Modified Peptides

After the steps of data pre-processing, normalization, and imputation, it is useful to create a dataset containing all the modified peptides retained on rows with their values by columns, as well as the values of the potential proteins to which they belong for subsequent statistical analysis. For each modified peptide, these intensities can be retrieved using the identification numbers of their protein groups in MaxQuant: they are in the column "id" of the "proteinGroups.txt" file.

The following R code creates a dataset containing all the information necessary for subsequent analysis:

```
int.ptm.prot=NULL;id.ptm.mod=NULL;
id.pg.mod=NULL;multiplicity.mod=NULL;
prof.ptm=NULL;prof.prot=NULL;id.prot.mod=NULL;
for (i in 1:nrow(data.ptm.mod)){
  id_prot=unlist(strsplit(data.ptm.mod$id.pg[i],split =";"))
  for (j in 1:length(id_prot)){
    id.ptm.mod=c(id.ptm.mod,data.ptm.mod$id.ptm[i])
    id.prot.mod=c(id.prot.mod,id_prot[j])
    id.pg.mod=c(id.pg.mod,data.ptm.mod$id.pg[i])
    prof.ptm=rbind(prof.ptm,detect.ptm[i,])
    multiplicity.mod=c(multiplicity.mod,
                       data.ptm.mod$multiplicity[i])
    if (length(data.prot$id%in%id_prot[j])!=0){
     prof.prot=rbind(
            prof.prot,
            detect.prot[which(data.prot$id%in%id_prot[j]),]
                    )
     int.ptm.prot=rbind(int.ptm.prot,c(log2.int.ptm.imp[i,],
     log2.int.prot.impMCAR[which(data.prot$id%in%id_prot[j]),]))
    }else{
     prof.prot=rbind(prof.prot,rep(NA,ncol(detect.prot)))
     int.ptm.prot=rbind(int.ptm.prot,c(log2.int.ptm.imp[i,],
                        rep(NA,ncol(log2.int.prot.impMCAR))))
    }
}
data.ptm.prot=data.frame(id.ptm.mod,id.pg.mod,id.prot.mod,
                         multiplicity.mod,
                         prof.ptm,prof.prot,int.ptm.prot)
```

3.8 Extracting Modified Peptides with Specific Detection Profiles Relatively to Their Unmodified Protein

Before analyzing the quantified intensities, specific detection profiles can be extracted to highlight modified peptides that are detected while their unmodified protein is not quantified. For instance, this can be performed with the following steps for the "mock" condition of our case study dataset:

1. Putting aside peptides with no quantified values in the considered condition:

```
nb.mock.ptm=apply(
         prof.ptm[,grep("mock",colnames(prof.ptm))],
         1,function(x){sum(x==1)}
                 )
ind_noval=which(nb.mock.ptm==0)
```

2. Putting aside peptides with associated proteins having no quantified values in the considered condition:

```
nb.mock.prot=apply(
         prof.prot[,grep("mock",colnames(prof.prot))],
         1,function(x){sum(x==1)}
                  )
ind_noprot=which(nb.mock.prot==0)
```

3. Putting aside peptides quantified at time stamps where their protein is not quantified:

```
# peptides quantified at time stamps where their protein is
# not quantified.
ind=NULL
for (i in 1:nrow(prof.ptm)){
    if (length(which(
        prof.ptm[i,grep("mock",colnames(prof.ptm))]==
        prof.prot[i,grep("mock",colnames(prof.prot))]
        ))==0){
       ind=c(ind,i)
    }
}
ind_diffprof=ind[!ind%in%c(which(nb.mock.prot==0),
                 which(nb.mock.ptm==0))]
```

4. (OPTIONAL STEP) Exporting data in a txt file: Next, peptides of interest can be exported outside R in "txt" files using the `write.table` function for instance. For peptides quantified at time stamps where their protein is not quantified, they can be exported using:

```
write.table(data.ptm.prot[ind_diffprof,],
            "(path_to_your_file)/pept_wo_prot.txt",sep="\t",
            row.names=F)
```

3.9 Extracting Modified Peptides Evolving Significantly Differently from Their Unmodified Protein

A common strategy aims to focus on the evolution of the intensities of modified peptides at different time stamps and under different biological conditions, notably to highlight PTM-mediated pathways involved in diseases or cellular disorders. For example, this can happen if one has samples related to patients belonging to different categories, and whose modified proteomes are measured at different time stamps. In this framework, the measured intensity of a modified peptide in a sample can be explained by three factors: the abundance of its membership protein, the patient's category, and the time stamps.

When no value is missing in the dataset, the following linear model can easily be estimated in each condition using R:

$$y_j = \alpha + \beta \mathbb{1}_{comp(j)=pept} + \sum_{t \in [t_1,\ldots,t_n]} \theta_t \mathbb{1}_{time(j)=t} + \delta_t \mathbb{1}_{time(j)=t} \mathbb{1}_{comp(j)=pept} + \epsilon_j \tag{1}$$

where ϵ_j is a Gaussian white noise.

In this equation, $Y=(y_j)_{j \in [1,J]}$ is a vector containing the intensities of both the modified peptide and its unmodified protein, and the function $\mathbb{1}_{comp(j)=pept}$ is equal to 1 if the jth coordinate of the vector Y is an intensity related to the modified peptide and 0 if it is related to the protein. Similarly, the function $\mathbb{1}_{time(j)=t}$ is equal to 1 if the jth coordinate of the vector y is an intensity related to the time stamp t and 0 otherwise. Here, the measured intensity y_j is explained in function of four parameters: the intercept α which represents the average intensity level of the protein to which the peptide belongs at time t_0; the β parameter which represents the gap between this average at t_0 and the one of the intensities of the modified peptide; the list of θ_t which are the effects of each other time point regardless of whether the intensity is that of the peptide or the protein; and the list of δ_t that correspond to the interaction effects between the peptide intensities and the time stamps.

Testing the nullity of a linear combination of the model coefficients allows to select peptides of interest. For this, a design matrix (defining the linear model) and a contrast matrix (defining the linear combination of model coefficients that have to be tested) have to be specified in R.

1. Defining the design matrix: The model (1) can also be represented as a product between a design matrix X, composed of ones and zeros, and a vector of parameters $B=(\alpha, \beta, \theta_1, \ldots, \theta_T, \delta_1, \ldots, \delta_T)^T$, such as $Y=XB+\epsilon$, where $\epsilon=(\epsilon_j)_{j \in [1,J]}$. In R, the design matrix can be defined by using the `model.matrix` function (*see* **Note 18**):

```
design = model.matrix(~Comp.ptmprot*Time.ptmprot)
```

2. Defining the contrast matrix: This matrix (referred to as C) is used to test if a linear combination of coefficients composing the estimated model is null. In this framework, the null hypothesis of the test is defined by $H_0: C^T B=0$ and the alternative hypothesis by $H_1: C^T B \neq 0$.

   To test if the modified peptide has a dynamic similar to that of the protein over time, we have to test that all the interaction parameters δ_t are equal to 0, i.e., $\delta_1=0$ and $\delta_1-\delta_2=0$, $\delta_2-\delta_3=0$, etc. For instance, this contrast matrix will test $\delta_1=0$ and $\delta_1-\delta_2=0$:

$$C^T B = 0 \Leftrightarrow \begin{pmatrix} 0 & 0 & 0 & \cdots & 0 & 1 & 0 & 0 & \cdots & 0 \\ 0 & 0 & 0 & \cdots & 0 & 1 & -1 & 0 & \cdots & 0 \end{pmatrix} \begin{pmatrix} \alpha \\ \beta \\ \theta_1 \\ \vdots \\ \theta_T \\ \delta_1 \\ \delta_2 \\ \vdots \\ \delta_T \end{pmatrix} = \begin{pmatrix} \delta_1 \\ \delta_1 - \delta_2 \end{pmatrix} = 0 \tag{2}$$

   In our case study dataset of two time stamps, only one interaction parameter is in the model defined by the design matrix. Using this model, the contrast matrix testing the nullity of this parameter is defined by only one column:

```
ct=cbind(c(0,0,0,1))
```

   Alternatively, if it is desired to remove all the peptides which are constant over time, it is advisable to test the nullity of all the coefficients in relation to time, i.e., the nullity of the (θ_t) and (δ_t) parameters. In our case study, the contrast matrix to be used is:

```
ct=cbind(c(0,0,1,0),c(0,0,1,-1))
```

3. For the following steps, we consider a dataset without missing values in order to be able to estimate linear models and perform statistical tests:

```
nbna=apply(int.ptm.prot,1,function(x)sum(is.na(x)))
data.ptm.prot=data.ptm.prot[which(nbna==0),]
int.ptm.prot=int.ptm.prot[which(nbna==0),]
```

4. Apply the chosen statistical test:
   (a) Classical F test (Fisher's): Once the design matrix and the contrast matrix have been defined, a classical Fisher's F test can be applied in a condition (here, "mock") for all the modified peptides of our dataset using the `linearHypothesis` function of the `car` R package [26]:

```
require(car)
condition="mock"
# condition="SARS_COV_2"
p.value_ANOVA=rep(NA, nrow(int.ptm.prot))
for (i in 1:nrow(int.ptm.prot)){
  Int=int.ptm.prot[i,Cond.ptmprot==condition]
  Comp=Comp.ptmprot[Cond.ptmprot==condition]
  Time=Time.ptmprot[Cond.ptmprot==condition]
  mod <- lm(Int ~ Comp * Time)
  p.value_ANOVA[i]=linearHypothesis(mod,t(ct))$'Pr(>F)'[2]
}
```

   (b) Limma F test: The `limma` R package can alternatively be used to test the nullity of the linear combination of coefficients with Fisher's F test. It uses a regularization of the variances based on the assumption that they follow an inverse Gamma distribution [22]:

```
require(limma)
design=model.matrix(~Comp.ptmprot[Cond.ptmprot==condition]
                    *Time.ptmprot[Cond.ptmprot== condition])
fit=eBayes(contrasts.fit(lmFit(int.ptm.prot[,Cond.ptmprot==
    condition], design),ct),robust=TRUE)
p.value_LIMMA=fit$F.p.value
```

5. Selecting peptides with a significantly different dynamic than their parent unmodified protein with a chosen FDR threshold: For this, an adaptive FDR control procedure can be applied to get adjusted pvalues using the `cp4p` R package [23]:

```
require(cp4p)
a=adjust.p(p.value_ANOVA,pi0.method = "pounds")
a.pvalue_ANOVA=a$adjp$adjusted.p
ptm.sel.mock=which(a.pvalue_ANOVA<0.01)
#ptm.sel.sarscov2=which(a.pvalue_ANOVA<0.01)
```

Here, a 1% FDR threshold has been used and the proportion of true null hypotheses has been estimated using the method of Pounds and Cheng [33]. Likewise, modified peptides which have a significantly different dynamic from their unmodified protein in the SARS-CoV-2 condition can be recovered by using `condition="SARS_COV_2"` and listing them in a vector `ptm.sel.sarscov2` from previous step (*see* Subheading 3.9, **Step 4**).

6. Extracting peptides of interest: Finally, a dataset containing all the modified peptides evolving significantly differently than their unmodified proteins in at least one condition can be extracted from `ptm.sel.mock` and `ptm.sel.sarscov2` with:

```
data.ptm.prot.sel=data.ptm.prot[unique(c(ptm.sel.mock,
                                ptm.sel.sarscov2)),]
int.ptm.prot.sel=int.ptm.prot[unique(c(ptm.sel.mock,
                              ptm.sel.sarscov2)),]
```

They can next be exported outside R using the `write.table` function for instance (*see* Subheading 3.8, **Step 4**).

3.10 Clustering of Modified Peptides Using Their Dynamics Relatively to Their Unmodified Protein

When prior information is not available to perform groupings of peptides, unsupervised clustering methods can be applied to highlight clusters of peptides behaving similarly for subsequent biological interpretation.

In label-free proteomics, many physico-chemical properties can lead to higher or lower intensity levels for different peptides with similar abundance in different conditions. Consequently, the cluster analysis of peptides has to be performed without considering the global intensity levels of the peptide, or the one of its protein, but rather their deviation from a reference intensity level. With this purpose, the intensities of the modified peptide and of its unmodified protein can be modeled by a linear model with interaction parameters using the three factors that are time stamps, biological conditions, and the studied component (protein or peptide):

$$
\begin{aligned}
y_j = \alpha &+ \beta_0 \mathbb{1}_{comp(j)=pept} \\
&+ \sum_{k\in[1,K]} \beta_k \mathbb{1}_{cond(j)=k} + \gamma_k \mathbb{1}_{cond(j)=k} \mathbb{1}_{comp(j)=pept} \\
&+ \sum_{t\in[t_1,t_n]} \theta_t \mathbb{1}_{time(j)=t} + \vartheta_t \mathbb{1}_{time(j)=t} \mathbb{1}_{comp(j)=pept} \\
&+ \sum_{k\in[1,K]} \sum_{t\in[t_1,t_n]} \delta_{tk} \mathbb{1}_{time(j)=t} \mathbb{1}_{cond(j)=k} \\
&+ \varphi_{tk} \mathbb{1}_{time(j)=t} \mathbb{1}_{cond(j)=k} \mathbb{1}_{comp(j)=pept} \\
&+ \epsilon_j
\end{aligned}
\qquad (3)
$$

where ϵ_j is a Gaussian white noise.

Thus, the dynamics of the modified peptide compared to the one of its unmodified protein is characterized by all the parameters multiplied by $\mathbb{1}_{comp(j)=pept}$, with the exception of the β_0 parameter which represents the overall difference between the intensity level of the modified peptide and that of its unmodified protein. Therefore, the dynamic of the modified peptide relative to that of its protein can be characterized by the parameter vector $(\gamma_1, \ldots, \gamma_K, \vartheta_{t_1}, \ldots, \vartheta_{t_n}, \varphi_{t_1 1}, \ldots, \varphi_{t_n K})^T$, while the dynamic of its protein is characterized by the parameter vector $(\beta_1, \ldots, \beta_K, \theta_{t_1}, \ldots, \theta_{t_n}, \delta_{t_1 1}, \ldots, \delta_{t_n K})^T$. A clustering of these vectors can be performed to cluster peptides with similar dynamics.

1. Extracting model parameters using a classical approach:

```
param=matrix(NA, nrow(int.ptm.prot.sel),6)
for (i in 1:nrow(int.ptm.prot.sel)){
  Int=int.ptm.prot.sel[i,]
  mod <- lm(Int ~ Cond.ptmprot*Time.ptmprot*Comp.ptmprot)
  param[i,]=mod$coefficients[unique(c(grep("Cond",
  names(mod$coefficients)),
  grep("Time",names(mod$coefficients))))]
}
colnames(param)=names(
          mod$coefficients[unique(
          c(grep("Cond",names(mod$coefficients)),
          grep("Time",names(mod$coefficients))))]
                     )

# Extract coefficients only related to the dynamics of the
# modified peptides relatively to their protein
 param_pept=param[,grep(":Comp.ptmprot",colnames(param))]
```

2. Extracting model parameters using `limma` R package (alternative to previous step):

```
require(limma)
design=model.matrix(~Cond.ptmprot*Time.ptmprot*Comp.ptmprot)
fit=eBayes(lmFit(int.ptm.prot.sel, design),robust=TRUE)
param_LIMMA=fit$
   coefficients[,unique(c(grep("Cond",
                             names(mod$coefficients)),
                         grep("Time",
                             names(mod$coefficients))
                         )
                     )
          ]
```

3. Choosing the clustering algorithm: Many unsupervised clustering algorithms proposed in the literature could be applied on our dataset. We propose in the next to use the classical

"hard" version of the k-means algorithm (*see* **Note 19**). This algorithm depends on two main parameters that have to be shortlisted by the user: the distance measure and the number of clusters.

4. Distance measure and dissimilarity matrix: Generally, Euclidean distances or Pearson correlation-based distances between observed values are used (*see* **Note 20**). The Euclidean distance aims to measure the absolute deviations between the vector coordinates, while the Pearson correlation-based distance characterizes the linear evolution between the vector coordinates. In our case, we search to cluster parameter vectors and no dynamic is expected between these parameters. Thus, a classical Euclidean distance seems to be an appropriate choice. Once the distance chosen, a dissimilarity matrix can be computed. To this end, the `multivariance` R package [27] provides a function `fastdist` allowing fast computation of Euclidean distance matrix, which is of definite interest with respect to the dataset size (several tens of thousands of peptides):

```
require(multivariance)
diss.mat=fastdist(param)
```

5. Determining the optimal number of clusters: Many indices can be used to determine an optimal number of clusters [38]. A classical method consists in using a Gap statistic [39]. For this, the `fviz_nbclust` function of the R package `factoextra` can be used:

```
require(factoextra)
gap.stat=fviz_nbclust(param,diss = diss.mat, kmeans,
                      nstart = 25, method = "gap_stat",
                      nboot = 10, k.max = 30)
gap.stat=gap.stat+labs(subtitle = "Gap statistic method")
gap.stat
```

Figure 3a, b display the Gap statistics in function of the number of clusters for `param` and `param_pept` vectors (*see* Subheading 3.10, **Step 1**).

6. Performing the clustering: The `eclust` function of `factoextra` R package can be used for enhancing the workflow of clustering analyses and `ggplot2`-based elegant data visualization:

```
# Using a k-means clustering with 20 clusters.
res.km=eclust(param, hc_metric = "euclidean", "kmeans",
              k=20, nstart = 10, graph=TRUE)
```

Alternatively, the k-medioids algorithm can be used using the `pam` R function from `cluster` R package, while a visualization of this clustering using Principal Component Analysis is obtained with the `fviz_cluster` function:

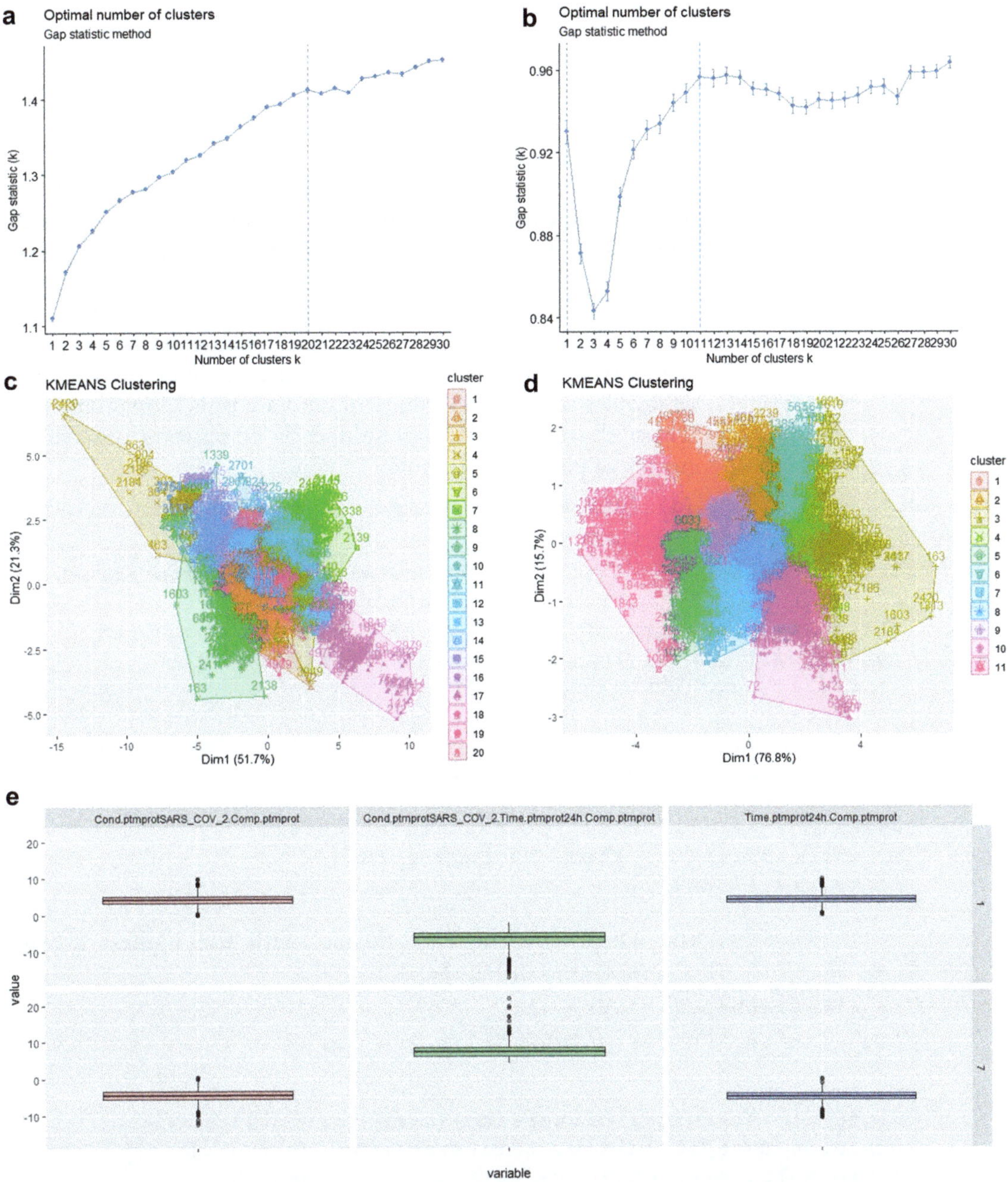

Fig. 3 (**a**) Gap statistic in function of the number of clusters for param vectors (*see* Subheading 3.10, **Step 1**). (**b**) Gap statistic in function of the number of clusters for param_pept vectors (*see* Subheading 3.10, **Step 1**). (**c**) Visualization using PCA of 20 clusters of param vectors found with k-means. (**d**) Visualization using PCA of 11 clusters of param_pept vectors found with k-means. (**e**) Profiles of parameters for the clusters with the highest average norms for param_pept vectors. Here, param_pept contains coefficients only related to the dynamics of the modified peptides relatively to their protein. The cluster 1 is characterized by peptides with positive coefficients for Time 24 h and SARS-Cov-2 conditions, as well as negative coefficients for their interaction

```
require(cluster)
res.pam=pam(x=diss.mat,k=20,diss=TRUE)
res.pam$data=param
fviz_cluster(res.pam,main = "K MEDIOIDS Clustering")
```

7. Visualization of results: There are two main ways to visualize cluster results. One way is to use dimension reduction methods as Principal Component Analysis or Multidimensional Scaling Methods. The `graph=TRUE` option of the `eclust` function or the `fviz_cluster` function can be used with this purpose. A second way is to plot the profiles of the model parameters used to cluster with boxplots or heatmaps. The clusters associated with strong deviations from 0 of these parameters are synonym of profiles strongly discriminant, and are thus of main interest. These clusters can be highlighted by computing the average norms of parameter vectors:

   (a) Estimating the average norms of parameter vectors by cluster:

```
dataplot=data.frame(param_pept,cluster=res.pam$clustering,
                        peptID=1:nrow(param))
# compute the norms of parameter vectors for param_pept
normv=apply(param_pept,1,function(x)sqrt(sum(x^2)))
# average norm for each cluster
mc=NULL
for (i in 1:length(levels(as.factor(dataplot$cluster)))){
    mc=c(mc,mean(normv[dataplot$cluster==i]))
}
names(mc)=levels(as.factor(dataplot$cluster))
# rank clusters by their average norms
mc=mc[order(-mc)]
```

   (b) Displaying the two clusters with the highest average norms as in Fig. 3e:

```
require(reshape2) #for melt
require(ggpubr) #for facet_grid

nb_to_plot=2
cl_to_plot=as.numeric(names(mc)[1:nb_to_plot])
dataplot=dataplot[dataplot$cluster%in%cl_to_plot,]
datap=melt(dataplot,id.var=c("cluster","peptID"))

gg= ggplot(datap, aes(x = variable, y = value) )
gg=gg+geom_boxplot(aes(fill = variable), alpha = 0.5)
gg=gg+facet_grid(cluster~variable,scales="free_x")
gg=gg+scale_x_discrete(labels = "")
gg=gg+theme(legend.position = "none")
gg
```

3.11 Subsequent Functional Analysis

1. Motif enrichment analysis: Motif analysis can be used to know the main sequence of amino acids surrounding modification sites of interest. For instance, this kind of analysis can be useful to find out which kinases are involved in phosphorylation. In R, motif analysis can be done using the `rmotifx` R package [40] and the `ggseqlogo` R package can be used to visualize the results [41] (*see* **Note 21**).
2. Enrichment analysis of Gene Ontology terms and Pathways: An analysis of the enrichment of Gene Ontology terms or biological pathways referenced in databases such as KEGG or Reactome is often carried out to highlight biological processes or specific signaling pathways enriched in the lists of modified peptides of interest coming from statistical analysis. Two important points have to be kept in mind when performing these analyses in our framework:

   (a) This enrichment is generally carried out in relation to a "background" using hypergeometric tests. In our context, it is very important to choose the set of modified proteins identified by mass spectrometry as background. Indeed, the enrichment protocols can display a bias in favor of enrichment for peptides with a single modification versus peptides with several modifications, or toward enrichment for hydrophobic versus hydrophilic modified peptides. Such biases will not be taken into account if the whole proteome or genome of the organism studied is used as background.

   (b) Most enrichment tools are based on gene-centric databases. However, two modified peptides of the same protein can sometimes evolve differently between two conditions, for example if the abundance of one proteoform of the protein decreases while that of another increases. In such a case, using the annotations of the reference protein without taking the modification into account may lead to falsely highlighting certain biological processes that are associated with a particular proteoform and not with another. Fortunately, recent resources have been developed to take this aspect into account [42].
3. Visualizing networks of PTMomics data: This step can be insightful to find modified proteins with known interactions. To do so, the reader can start from the statistical analyzes carried out from R, export them, and then use the Cytoscape software [43] to visualize interaction networks. Two Cytoscape applications freely downloadable from the Cytoscape App Store are particularly interesting for visualizing PTMomics data: stringApp [44] and Omics Visualizer

[45]. stringApp is an application allowing to query the protein–protein interaction database of STRING [46] from Cytoscape, while Omics Visualizer allows to optimize the visualization of the obtained network. The latter is particularly suitable for viewing information on each node of the network, for instance the number of modified peptides regulated for each protein. Specific tutorials are freely available online (https://jensenlab.org/training/).

4 Notes

1. A number of recently proposed packages allow one to consider pipelines of analyzes performed entirely from R: for example, the `rawR` package can read RAW files from mass spectrometers [47], while the Bioconductor packages `rTandem` [48] or `MSGFplus` [49, 50] can be used to perform identification of mass spectra from these RAW files. Additionally, the `RforProteomics` package offer useful functionalities to manipulate and visualize mass spectra [51].
2. The first line of the input file has to contain the column names of the data table. Decimal separator for quantitative values has to be the dot ".". Other R functions, such as `read.xlsx`, from the R package `openxlsx`; or `read_excel` from the R package `readxl` can also be used if the file is exported from Excel. Of note, the fread function of the R package `data.table` can be used to read large datasets.
3. These peptides are used to calculate a false discovery rate linked to the identification of peptides [52].
4. The higher the probability of localization, the more certain it is that the measured spectrum is associated with the peptide containing the modification.
5. The "proteinGroups.txt" contains intensities for the proteins using either their unmodified peptides or their modified peptides in different columns. In the case study dataset, samples enriched in phosphorylation contain "Phospho" while the ones enriched in ubiquitination contain "Ubi". However, these names can be different with other datasets depending on how they have been defined in MaxQuant or in the used software. Thus, the reader has to adapt these names in such cases. The `colnames` function can be used to check these names.
6. Similarly to previous note, the names of enriched samples can be different with other datasets depending on how they have been defined in MaxQuant or in the used software: the reader has to adapt these names in such cases.

7. These peptides are generally present because they have been identified. However, no quantified values are available because of the quantification algorithm used in MaxQuant.
8. The same approach can be used with 2, 4, or 5 samples using the `draw.pairwise.venn`, `draw.quad.venn`, and `draw.quintuple.venn` functions of the `VennDiagram` package.
9. It is generally useful to convert a plot into a `ggplot` object, for example to easily include it in a pdf or powerpoint document created automatically with R Markdown or the `officer` R package.
10. This can be performed using the `qplot` function of R package `ggplot2` with the following R code:

```
require(reshape2)
gradientRate=1.2;
plot.j=qplot(x=Var1,y=Var2,data=melt(mat.jaccard),
             fill=value,geom="tile")
plot.j=plot.j+theme(
    axis.text=element_text(size=10),
    axis.title=element_text(size=10,face="bold"),
    axis.text.x=element_text(angle=90,vjust=1,hjust=1),
    legend.text=element_text(colour="black",size=8,face="bold");
                   )
plot.j=plot.j+theme(legend.title=element_blank())
plot.j=plot.j+labs(x="",y="")
plot.j=plot.j+scale_fill_gradientn(
    colours=colorRampPalette(
      c("white","lightblue","darkblue"))(101),
    values=c(pexp(seq(0,1,0.01),
    rate=gradientRate),1),limits=c(0,1)
                                    )
```

11. For this, `data.upset` has to contain 1 is the modified peptide is not identified and 0 if it is identified. Of note, the Jaccard index does not give the same value when evaluating mutual presence or mutual absence, such that the result of this second approach will be different from the first. An alternative to the Jaccard index consists to use the Rand index, this one will lead to a single value measuring at the same time mutual presence and absence.
12. In this way, the differences observed between the various distributions will highlight whether if the intensities measured in one sample are lower or higher than in others, without confounding effects related to different sets of peptides used to plot the distributions.
13. This can be performed similarly as in **Note 10**, except that `data=melt(mat.cor)` in the `qplot` function. Also, we advise to choose `gradientRate=` 3.5 for correlation matrix

based on quantified values to see more clearly clusters of samples in the matrix.

14. Because the Principal Component Analysis (PCA) is based on the eigendecomposition of the covariance matrix, it will summarize the same information that appears in the correlation matrix. The `prcomp` function can be used to perform PCA in R.
15. In general, it is necessary to carry out a first MS-based experiment with a minimum number of replicates per biological condition (at least 3 to be able to estimate the variance of intensities for each modified peptide in each condition). This kind of test experiment is generally intended to check that the enrichment step has worked correctly but can also be used to assess how many samples are sufficient to obtain robust statistical results in a subsequent analysis.
16. In each condition, there are therefore two possibilities: either the modified peptide is detected in one of the samples of the condition, or it is not. In the case of N conditions, $2^N - 1$ detection profiles are possible. These detection profiles should be compared with the ones associated with the corresponding proteins, which further increases the number of possible cases. Hopefully, it can generally be expected that many detection profiles will not be encountered in the experiment. It should be noted that some cases lead to not being able to conclude on the over- or under-abundance of the modified peptide compared to the unmodified protein. For example, if they are detected in the same condition and not in another, nothing can be concluded by comparing these conditions.
17. For example, with conditions representing different time stamps, a value can be deduced by applying methods developed for time series data [53]. In the case where no value can be inferred in the condition from observed ones, then an analysis of the detection profiles can be conducted to highlight interesting modified peptides.
18. Here, the "*" operator is used to specify that interaction parameters have to be included in the model. If one does not want interactor parameters, one uses only the "+ " operator.
19. Either hierarchical clustering methods or partitional clustering methods (such as the k-means algorithm) can be used [54]. Moreover, either hard clustering versions of these methods, assigning a unique cluster for each peptide, or fuzzy clustering versions, assigning probabilities of belonging to each cluster for each peptide could be used [55]. The main advantage of hierarchical methods is that they does not need to specify a number of clusters. Their results are generally represented in the form of a dendrogram. However, in our context,

it is preferable to cluster the peptides into a finite number of clusters in order to facilitate the subsequent biological interpretation of the results. Partitional clustering methods seem thus more appropriate. The most famous one is the k-means algorithm. The fuzzy version of this algorithm has serious limitations in high dimensional spaces, as it is generally the case in peptidomics datasets [56].

20. However, alternative distances can also be investigated as, in case of time series, Dynamic Time Warping distances [57] or Short time series distances [58]. The parallelDist package can also be used to compute a broad variety of distance for large datasets.

21. Pre-alignment of the modified peptides around the detected sites of modification has to be performed before to use these packages.

Acknowledgements

The author wants to acknowledge Mariette Matondo, Thibaut Douché, Thibault Chaze, and Magalie Duchateau of the Proteomics platform of the Institut Pasteur for fruitful discussions about proteomics and phosphoproteomics.

References

1. Witze ES, Old WM, Resing KA, Ahn NG (2007) Mapping protein post-translational modifications with mass spectrometry. Nat Methods 4(10):798–806. https://doi.org/10.1038/nmeth1100
2. Zhao Y, Jensen ON (2009) Modification-specific proteomics: strategies for characterization of post-translational modifications using enrichment techniques. Proteomics 9(20):4632–4641. https://doi.org/10.1002/pmic.200900398
3. Pieroni L, Iavarone F, Olianas A, Greco V, Desiderio C, Martelli C, Manconi B, Sanna MT, Messana I, Castagnola M et al. (2020) Enrichments of post-translational modifications in proteomic studies. J Sep Sci 43(1):313–336. https://doi.org/10.1002/jssc.201900804
4. Ficarro SB, Adelmant G, Tomar MN, Zhang Y, Cheng VJ, Marto JA (2009) Magnetic bead processor for rapid evaluation and optimization of parameters for phosphopeptide enrichment. Anal Chem 81(11):4566–4575. https://doi.org/10.1021/ac9004452
5. Pinkse MW, Lemeer S, Heck AJ (2011) A protocol on the use of titanium dioxide chromatography for phosphoproteomics. In: Gel-free proteomics. Springer, pp 215–228, https://doi.org/10.1007/978-1-61779-148-2_14
6. Udeshi ND, Svinkina T, Mertins P, Kuhn E, Mani D, Qiao JW, Carr SA (2013) Refined preparation and use of anti-diglycine remnant (k-ε-gg) antibody enables routine quantification of 10,000 s of ubiquitination sites in single proteomics experiments. Mol Cell Proteomics 12(3):825–831. https://doi.org/10.1074/mcp.O112.027094
7. Carlson SM, Moore KE, Green EM, Martín GM, Gozani O (2014) Proteome-wide enrichment of proteins modified by lysine methylation. Nat Protoc 9(1):37–50. https://doi.org/10.1038/nprot.2013.164
8. Kim SC, Sprung R, Chen Y, Xu Y, Ball H, Pei J, Cheng T, Kho Y, Xiao H, Xiao L et al. (2006) Substrate and functional diversity of lysine acetylation revealed by a proteomics survey. Mol Cell 23(4):607–618. https://doi.org/10.1016/j.molcel.2006.06.026
9. Mertins P, Qiao JW, Patel J, Udeshi ND, Clauser KR, Mani D, Burgess MW, Gillette MA, Jaffe JD, Carr SA (2013) Integrated proteomic analysis of post-translational modifications by serial enrichment. Nat Methods 10(7):634. https://doi.org/10.1038/nmeth.2518

10. Chalkley RJ, Clauser KR (2012) Modification site localization scoring: strategies and performance. Mol Cell Proteomics 11(5):3–14. https://doi.org/10.1074/mcp.R111.015305
11. Tyanova S, Temu T, Cox J (2016) The maxquant computational platform for mass spectrometry-based shotgun proteomics. Nat Protoc 11(12):2301. https://doi.org/10.1038/nprot.2016.136
12. Valot B, Langella O, Nano E, Zivy M (2011) Masschroq: a versatile tool for mass spectrometry quantification. Proteomics 11(17): 3572–3577. https://doi.org/10.1002/pmic.201100120
13. Bouyssié D, Hesse AM, Mouton-Barbosa E, Rompais M, Macron C, Carapito C, Gonzalez de Peredo A, Couté Y, Dupierris V, Burel A et al. (2020) Proline: an efficient and user-friendly software suite for large-scale proteomics. Bioinformatics 36(10):3148–3155. https://doi.org/10.1093/bioinformatics/btaa118
14. Gentleman R (2008) R programming for bioinformatics. CRC Press. https://doi.org/10.18637/jss.v029.b08
15. Chambers J (2008) Software for data analysis: programming with R. Springer Science & Business Media. https://doi.org/10.1007/978-0-387-75936-4
16. Millot G (2011) Comprendre et réaliser les tests statistiques à l'aide de r. De Boeck, Brussels. ISBN 978-2-8073-0291-4
17. Chen H, Boutros PC (2011) Venndiagram: a package for the generation of highly-customizable Venn and Euler diagrams in R. BMC Bioinformatics 12(1):1–7. https://doi.org/10.1186/1471-2105-12-35
18. Conway JR, Lex A, Gehlenborg N (2017) UpSetR: an R package for the visualization of intersecting sets and their properties. Bioinformatics 33(18):2938–2940. https://doi.org/10.1093/bioinformatics/btx364
19. Wickham H (2016) ggplot2: elegant graphics for data analysis. Springer. https://doi.org/10.1080/15366367.2019.1565254
20. de Vries A, Ripley BD (2020) Ggdendro: create dendrograms and tree diagrams using 'ggplot2'. R package version 0122. https://cran.r-project.org/web/packages/ggdendro/index.html
21. Wilke CO (2021) ggridges: ridgeline plots in 'ggplot2'. R package version 053. https://cran.r-project.org/web/packages/ggridges/index.html
22. Ritchie ME, Phipson B, Wu D, Hu Y, Law CW, Shi W, Smyth GK (2015) limma powers differential expression analyses for RNA-sequencing and microarray studies. Nucleic Acids Res 43(7):e47–e47. https://doi.org/10.1093/nar/gkv007
23. Giai Gianetto Q, Combes F, Ramus C, Bruley C, Couté Y, Burger T (2016) Calibration plot for proteomics: a graphical tool to visually check the assumptions underlying FDR control in quantitative experiments. Proteomics 16(1):29–32. https://doi.org/10.1002/pmic.201500189
24. Liu P, Hwang JG (2007) Quick calculation for sample size while controlling false discovery rate with application to microarray analysis. Bioinformatics 23(6):739–746. https://doi.org/10.1093/bioinformatics/btl664
25. Gianetto QG, Wieczorek S, Couté Y, Burger T (2020) A peptide-level multiple imputation strategy accounting for the different natures of missing values in proteomics data. bioRxiv. https://doi.org/10.1101/2020.05.29.122770
26. Fox J, Weisberg S, Adler D, Bates D, Baud-Bovy G, Ellison S, Firth D, Friendly M, Gorjanc G, Graves S et al. (2020) car: companion to applied regression. R package version 30-10. https://cran.r-project.org/web/packages/car/index.html
27. Böttcher B (2020) Copula versions of distance multivariance and dHSIC via the distributional transform–a general approach to construct invariant dependence measures. Statistics 1–18. https://doi.org/10.1080/02331888.2020.1748029
28. Kassambara A, Mundt F (2020) factoextra: extract and visualize the results of multivariate data analyses. R package version 107. https://cran.r-project.org/web/packages/factoextra/index.html
29. Wickham H (2020) reshape2: flexibly reshape data: a reboot of the reshape package. R package version 144. https://cran.r-project.org/web/packages/reshape2/index.html
30. Kassambara A (2020) ggpubr: "ggplot2" based publication ready plots. https://cran.r-project.org/web/packages/ggpubr/index.html
31. Stukalov A, Girault V, Grass V, Bergant V, Karayel O, Urban C, Haas DA, Huang Y, Oubraham L, Wang A et al. (2020) Multi-level proteomics reveals host-perturbation strategies of SARS-CoV-2 and SARS-CoV. BioRxiv https://doi.org/10.1101/2020.06.17.156455
32. Wieczorek S, Gianetto QG, Burger T (2019) Five simple yet essential steps to correctly estimate the rate of false differentially abundant proteins in mass spectrometry analyses. J

Proteomics 207:103441. https://doi.org/10.1016/j.jprot.2019.103441
33. Pounds S, Cheng C (2006) Robust estimation of the false discovery rate. Bioinformatics 22(16):1979–1987. https://doi.org/10.1093/bioinformatics/btl328
34. Kauko O, Laajala TD, Jumppanen M, Hintsanen P, Suni V, Haapaniemi P, Corthals G, Aittokallio T, Westermarck J, Imanishi SY (2015) Label-free quantitative phosphoproteomics with novel pairwise abundance normalization reveals synergistic RAS and CIP2A signaling. Sci Rep 5:13099. https://doi.org/10.1038/srep13099
35. Saraei S, Suomi T, Kauko O, Elo LL (2018) Phosphonormalizer: an R package for normalization of MS-based label-free phosphoproteomics. Bioinformatics 34(4):693–694. https://doi.org/10.1093/bioinformatics/btx573
36. Wieczorek S, Combes F, Lazar C, Giai Gianetto Q, Gatto L, Dorffer A, Hesse AM, Coute Y, Ferro M, Bruley C, Burger T (2017) DAPAR & ProStaR: software to perform statistical analyses in quantitative discovery proteomics. Bioinformatics 33(1):135–136. https://doi.org/10.1093/bioinformatics/btw580
37. Lazar C, Gatto L, Ferro M, Bruley C, Burger T (2016) Accounting for the multiple natures of missing values in label-free quantitative proteomics data sets to compare imputation strategies. J Proteome Res 15(4):1116–1125. https://doi.org/10.1021/acs.jproteome.5b00981
38. Charrad M, Ghazzali N, Boiteau V, Niknafs A (2014) NbClust: an R package for determining the relevant number of clusters in a data set. J Stat Softw 61:1–36. https://doi.org/10.18637/jss.v061.i06
39. Tibshirani R, Walther G, Hastie T (2001) Estimating the number of clusters in a data set via the gap statistic. J R Stat Soc Ser B Stat Methodol 63(2):411–423. https://rss.onlinelibrary.wiley.com/doi/abs/10.1111/1467-9868.00293
40. Wagih O, Sugiyama N, Ishihama Y, Beltrao P (2016) Uncovering phosphorylation-based specificities through functional interaction networks. Mol Cell Proteomics 15(1):236–245. https://doi.org/10.1074/mcp.M115.052357
41. Wagih O (2017) ggseqlogo: a versatile r package for drawing sequence logos. Bioinformatics 33(22):3645–3647. https://doi.org/10.1093/bioinformatics/btx469
42. Krug K, Mertins P, Zhang B, Hornbeck P, Raju R, Ahmad R, Szucs M, Mundt F, Forestier D, Jane-Valbuena J et al. (2019) A curated resource for phosphosite-specific signature analysis. Mol Cell Proteomics 18(3):576–593. https://doi.org/10.1074/mcp.TIR118.000943
43. Shannon P, Markiel A, Ozier O, Baliga NS, Wang JT, Ramage D, Amin N, Schwikowski B, Ideker T (2003) Cytoscape: a software environment for integrated models of biomolecular interaction networks. Genome Res 13(11):2498–2504. https://doi.org/10.1101/gr.1239303
44. Doncheva NT, Morris JH, Gorodkin J, Jensen LJ (2018) Cytoscape stringApp: network analysis and visualization of proteomics data. J Proteome Res 18(2):623–632. https://doi.org/10.1021/acs.jproteome.8b00702
45. Legeay M, Doncheva NT, Morris JH, Jensen LJ (2020) Visualize omics data on networks with omics visualizer, a cytoscape app. F1000Research 9. https://doi.org/10.12688/f1000research.22280.2
46. Szklarczyk D, Gable AL, Lyon D, Junge A, Wyder S, Huerta-Cepas J, Simonovic M, Doncheva NT, Morris JH, Bork P et al. (2019) String v11: protein–protein association networks with increased coverage, supporting functional discovery in genome-wide experimental datasets. Nucleic Acids Res 47(D1):D607–D613. https://doi.org/10.1093/nar/gky1131
47. Kockmann T, Panse C (2020) rawR-direct access to raw mass spectrometry data in r. bioRxiv. https://doi.org/10.1101/2020.10.30.362533
48. Fournier F, Joly Beauparlant C, Paradis R, Droit A (2014) rTANDEM, an R/Bioconductor package for MS/MS protein identification. Bioinformatics 30(15):2233–2234. https://doi.org/10.1093/bioinformatics/btu178
49. Kim S, Pevzner PA (2014) MS-GF+ makes progress towards a universal database search tool for proteomics. Nat Commun 5:5277. https://doi.org/10.1038/ncomms6277
50. Pedersen TL (2020) Msgfplus: an interface between R and MS-GF+. R package version 1240. http://www.bioconductor.org/packages/release/bioc/html/MSGFplus.html
51. Gatto L, Breckels LM, Naake T, Gibb S (2015) Visualization of proteomics data using R and bioconductor. Proteomics 15(8):1375–1389. https://doi.org/10.1002/pmic.201400392
52. Couté Y, Bruley C, Burger T (2020) Beyond target–decoy competition: Stable validation of peptide and protein identifications in mass spectrometry-based discovery proteomics.

Anal Chem 92(22):14898–14906. https://doi.org/10.1021/acs.analchem.0c00328

53. Pratama I, Permanasari AE, Ardiyanto I, Indrayani R (2016) A review of missing values handling methods on time-series data. In 2016 International conference on information technology systems and innovation (ICITSI). IEEE, pp 1–6. https://doi.org/10.1109/ICITSI.2016.7858189
54. Gan G, Ma C, Wu J (2020) Data clustering: theory, algorithms, and applications. SIAM. https://doi.org/10.1137/1.9780898718348
55. Schwämmle V, Jensen ON (2018) VSClust: feature-based variance-sensitive clustering of omics data. Bioinformatics 34(17): 2965–2972. https://doi.org/10.1093/bioinformatics/bty224
56. Winkler R, Klawonn F, Kruse R (2011) Fuzzy C-means in high dimensional spaces. Int J Fuzzy Syst Appl 1(1):1–16. https://doi.org/10.4018/IJFSA.2011010101
57. Giorgino T et al. (2009) Computing and visualizing dynamic time warping alignments in R: the dtw package. J Stat Softw 31(7):1–24. https://doi.org/10.18637/jss.v031.i07
58. Mori U, Mendiburu A, Lozano JA (2016) Distance measures for time series in R: the TSdist package. R J 8(2):451. https://doi.org/10.32614/RJ-2016-058

Chapter 13

Fast, Free, and Flexible Peptide and Protein Quantification with FlashLFQ

Robert J. Millikin, Michael R. Shortreed, Mark Scalf, and Lloyd M. Smith

Abstract

The rapid and accurate quantification of peptides is a critical element of modern proteomics that has become increasingly challenging as proteomic data sets grow in size and complexity. We present here FlashLFQ, a computer program for high-speed label-free quantification of peptides and proteins following a search of bottom-up mass spectrometry data. FlashLFQ is approximately an order of magnitude faster than established label-free quantification methods and can quantify data-dependent analysis (DDA) search results from any proteomics search program. It is available as a graphical user interface program, a command line tool, a Docker image, and integrated into the MetaMorpheus search software.

Key words Label-free quantification, Post-translational modifications, Quantitative proteomics, Software

1 Introduction

Modern bottom-up proteomics workflows involve digesting a protein sample with a protease, followed by online separation of the resultant peptides and subsequent analysis by tandem mass spectrometry (MS/MS). In data-dependent acquisition (DDA), peptides are first observed in a survey (MS1) scan and then identified by their fragmentation (MS2) spectra. The MS1 scan contains isotopic envelopes of intact peptide ions, the signal intensity of which is a proxy for abundance of that ion in the mass spectrometer. This intensity can be used for quantification of the peptide species. There are many complex factors that determine how well a peptide ionizes. Accordingly, a relative quantification strategy is typically employed in which the signal intensity of a peptide is compared to the same peptide's signal in another sample, rather than attempting absolute quantification. The trace of MS1 signal across time for a particular ion is called the extracted ion chromatogram (XIC). The

Thomas Burger (ed.), *Statistical Analysis of Proteomic Data: Methods and Tools*,
Methods in Molecular Biology, vol. 2426, https://doi.org/10.1007/978-1-0716-1967-4_13,

absence of an added chemical label to assist in quantification gives this strategy its name: label-free quantification (LFQ).

Given the large amount of raw data afforded by these experiments (typically on the order of thousands of peptides in a single run), bioinformatics software is used for automatic data interpretation to ease the burden of manual annotation and quantification. FlashLFQ [1] is a software program that was created to quickly quantify peptides and proteins in LC-MS/MS data through XIC tracing. It was specifically designed to be a free, open-source LFQ tool that is agnostic of upstream and downstream software. One important motivation for creating FlashLFQ was the desire to quantify post-translational modification (PTM)-containing peptides discovered by MetaMorpheus and its global PTM discovery (G-PTM-D) engine (*see* **Chapter 3** and [2]).

At its core, FlashLFQ is an algorithm for rapid XIC extraction/peakfinding. Several features have been added since its initial debut as a peptide quantification program, including match-between-runs, intensity normalization, and Bayesian protein quantification and hypothesis testing [3].

The Bayesian hypothesis testing option is intended to replace the typical workflow of importing peptide or protein intensities into a separate program to perform the Student's t-test and multiple testing correction. While the t-test is a perfectly valid option for the interpretation of proteomics data, FlashLFQ's Bayesian solution aggregates intensity-dependent uncertainty for each peptide into a protein-level uncertainty, increasing the selectivity of the results (*see* **Note 1**).

FlashLFQ (https://github.com/smith-chem-wisc/FlashLFQ) is available as a command-line interface (CLI) and a graphical user interface (GUI) program. It is also integrated into the MetaMorpheus search software program. The CLI version can be run in Microsoft Windows, Apple macOS, and Linux; a CLI FlashLFQ/Linux Docker image is also available on Docker Hub (https://hub.docker.com/r/smithchemwisc/flashlfq). The GUI version of FlashLFQ is currently only available on Microsoft Windows.

FlashLFQ requires spectral data in .mzML or .raw file format, along with a list of peptide identifications from a proteomics search program (e.g., MetaMorpheus [2], Morpheus [4], Andromeda [5], SearchGUI [6], etc.).

2 Material

2.1 Data Inputs

FlashLFQ requires a list of peptide identifications (peptide spectral matches [PSMs]) as well as LC-MS/MS data. Each peptide identification is required to have a(n):

1. spectra file name,
2. amino acid sequence,
3. string representation of the peptide containing any potential modifications (such as phosphorylation),
4. monoisotopic mass,
5. retention time,
6. charge,
7. protein identifier (e.g., protein accession).

The required data values are provided to FlashLFQ as a tab-delimited text file(s).

2.2 Accepted Data Formats

Several data formats from common search engines are recognized in their native format without modification:

1. MetaMorpheus .psmtsv,
2. Morpheus .tsv,
3. MaxQuant msms.txt,
4. PeptideShaker .tabular.

Output from other search engines can be modified to be compatible with FlashLFQ. The columns:

1. "File Name"
2. "Base Sequence"
3. "Full Sequence"
4. "Peptide Monoisotopic Mass"
5. "Scan Retention Time"
6. "Precursor Charge"
7. "Protein Accession"

must be defined if this generic format is used, as described in the Data Inputs section above (*see* Subheading 2.1, as well as https://github.com/smith-chem-wisc/FlashLFQ/wiki/Identification-Input-Formats and Fig. 1 for an example of the generic PSM input format).

The .mzML and Thermo .raw spectra file formats are valid inputs for the LC-MS/MS data. Other file formats can be converted to .mzML using ProteoWizard's free MSConvert software [7] (*see* https://github.com/smith-chem-wisc/FlashLFQ/wiki/Converting-spectral-data-files-with-MSConvert).

2.3 Hardware Requirements

There are no formal requirements for CPU speed or number of cores, but faster processors with additional cores will result in a faster processing time. FlashLFQ is a multithreaded program and

	A	B	C	D	E	F	G
1	File Name	Scan Retention Time	Precursor Charge	Base Sequence	Full Sequence	Peptide Monoisotopic Mass	Protein Accession
2	Small_Yeast	24.06874	2	EKAEAEAEK	EKAEAEAEK	1003.4822	P40212;Q12690
3	Small_Yeast	24.39968	2	ALKQEGAANK	ALKQEGAANK	1028.56146	P0CX49;P0CX50
4	Small_Yeast	24.67303	2	SKDVTDSATTK	SKDVTDSATTK	1151.567	P39015
5	Small_Yeast	24.45053	2	KLEDHPK	KLEDHPK	865.46577	P02994
6	Small_Yeast	24.77398	1	HIDAGAK	HIDAGAK	710.37114	P00358;P00359
7	Small_Yeast	24.9022	2	YLAKEEEKK	YLAKEEEKK	1136.60774	P32589
8	Small_Yeast	24.76278	2	YAGEVSHDDK	YAGEVSHDDK	1119.48327	P00358;P00359
9	Small_Yeast	24.06107	2	RGNVCGDAK	RGNVC[Mod:Carbamidomethyl]GDAK	975.45561	P02994
10	Small_Yeast	24.89485	2	SGTHNMYK	SGTHNMYK	936.41235	P0CX23;P0CX24
11	Small_Yeast	24.8133	2	KQAIETANK	KQAIETANK	1001.55056	P0CS90
12	Small_Yeast	24.36436	2	AAAHSSLK	AAAHSSLK	783.4239	Q12118

Fig. 1 Example of a "generic" identification input file. The columns are tab-delimited. FlashLFQ automatically reads the output of MetaMorpheus, Morpheus, MaxQuant, and PeptideShaker, but output from other search programs can be adapted to this generic format for use by FlashLFQ

uses all but one core by default. We advise that the user have at least 4 GB of RAM, plus 500 MB for each spectra file if match-between-runs is enabled.

2.4 Software Requirements

1. The operating system (OS) must be Microsoft Windows, Apple macOS, or Linux for the command-line version. For the GUI version of FlashLFQ, the OS must be Microsoft Windows.
2. .NET Core 3.1 must be installed (*see* **Note 2**).
3. To use the Docker image (*see* **Note 3**) of FlashLFQ, the software requirements above can be ignored; only a Docker installation is required, as the Docker image includes Alpine Linux and .NET Core 3.1.

2.5 Installation

1. **GUI** (Microsoft Windows only). The GUI (graphical user interface) version of FlashLFQ can be downloaded from GitHub by navigating to the releases page (https://github.com/smith-chem-wisc/FlashLFQ/releases/latest) and clicking FlashLFQ.zip. After extracting the archive to a folder, open FlashLFQ.exe.
2. **Command-line**:
   (a) Microsoft Windows: Download FlashLFQ.zip and extract it to a folder following the same steps as the GUI version. Running CMD.exe in the terminal will start FlashLFQ and display valid parameters.
   (b) Linux and Apple macOS: Download FlashLFQ_DotNetCore.zip and extract it to a folder following the same steps as the GUI version. Running the command `dotnet CMD.dll` in the terminal will start FlashLFQ and display valid parameters.
3. **Docker**. A Docker image of FlashLFQ is hosted on DockerHub (https://github.com/smith-chem-wisc/FlashLFQ/wiki/Docker-Image). It can be pulled with the `docker pull` command; for example:

```
docker pull smithchemwisc/flashlfq:1.2.0
```

will download the Docker image containing FlashLFQ version 1.2.0. The most recent version of FlashLFQ is always tagged with `latest` (i.e., `smithchemwisc/flashlfq:latest`).

3 Methods

3.1 Adding Identification Files

Identification (PSM) files can be added to FlashLFQ's GUI interface by drag-and-drop or by clicking the Add Identifications button in the Identifications tab. In the command-line version, the `--idt` flag is used to specify an identification file. An example of a MetaMorpheus PSM file added in the GUI is shown in Fig. 2.

3.2 Adding Spectra Files

Spectra files can be added by drag-and-drop or by clicking the Add Spectra button in the GUI in the Spectra tab. In the command-line version, the `--rep` flag is used to specify a folder containing the spectra files. Each spectra file is associated with metadata containing:

1. the run's experimental condition (e.g., normal or treatment),
2. sample number,
3. fraction number,
4. replicate number.

This collection of metadata is called the experimental design, which is saved to a file called ExperimentalDesign.tsv. This information is used to normalize intensities, perform statistical analysis, and improve match-between-runs. If you simply want to quantify peptides without performing these extra functions, the experimental design can be ignored. The Condition can be any text whereas the Sample, the Fraction, and the Replicate must be an integer value. The Sample, Fraction, and Replicate values must begin

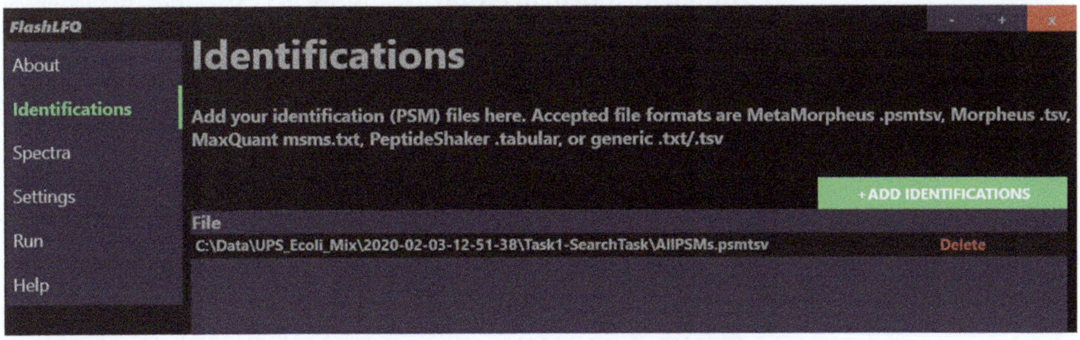

Fig. 2 Example of the "Identifications" page in the FlashLFQ GUI. A MetaMorpheus .psmtsv output file has been added

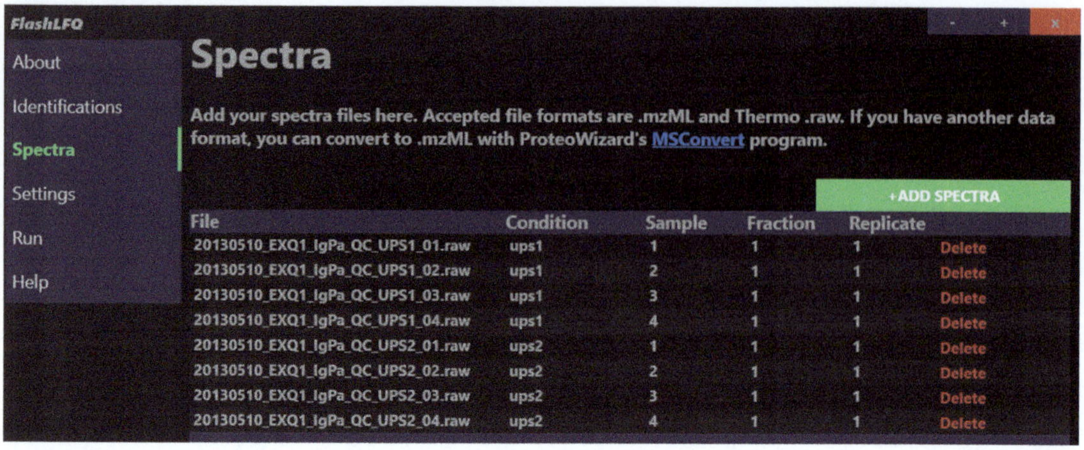

Fig. 3 Example of the "Spectra" page in the FlashLFQ GUI. Several Thermo .raw data files containing the Universal Proteomics Standard (UPS) proteins have been added, with UPS1 and UPS2 files divided between two experimental conditions, 4 samples per condition. The samples are unfractionated

with 1 and there can be no missing values. An example from the GUI is shown in Fig. 3, where the experimental design has been defined for an 8-file run.

Command-line users must build the ExperimentalDesign.tsv file manually. You can find a detailed description of the contents of the file at https://github.com/smith-chem-wisc/FlashLFQ/wiki/Experimental-Design, along with an example file available for download.

3.3 Settings

Settings are specified in the Settings tab in the GUI. Most settings can be enabled or disabled via a checkbox. In the command-line version, each setting has its own flag. An example in the GUI is shown in Fig. 4:

1. **PPM tolerance** (*see* **Note 4**). Specifies the mass-error tolerance for the XIC peakfinding algorithm, in parts per million (ppm). The default is 10ppm.
2. **Intensity normalization** (*see* **Note 5**). Specifies whether or not intensity normalization should be performed. FlashLFQ's intensity normalization algorithm uses a median normalization, which means that the median peptide intensity difference between samples is assumed to be zero (i.e., not changing). This normalization process rests on the assumption that most proteins are not changing in abundance between samples.
3. **Match-between-runs** (*see* **Note 6**). Specifies whether or not match-between-runs (MBR) should be performed. MBR attempts to identify peptides that were not fragmented and identified in some runs, using retention-time alignment between runs.

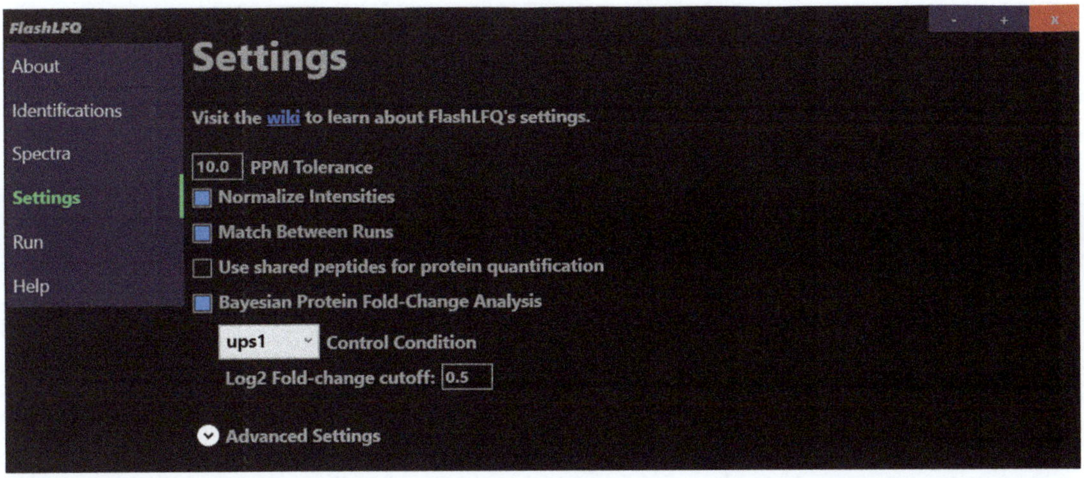

Fig. 4 Example of the "Settings" page in the FlashLFQ GUI. Normalization, match-between-runs, and the Bayesian protein quantification have been enabled

4. **Using shared peptides for protein quantification** (*see* **Note 7**). Specifies whether or not shared peptides should be used for protein quantification. If this is disabled, only peptides unique to a protein will be used for quantification (*see* **Note 8**).
5. **Bayesian protein quantification** (*see* **Note 9**). Specifies whether or not FlashLFQ's Bayesian protein quantification system should be used. A control condition (*see* **Note 10**) must be specified, along with a fold-change cutoff (*see* **Note 11**).

3.4 Run

On the Run tab, you can specify an output directory. A default directory will already be populated for you; this default directory is where your spectra files are located. The field `$DATETIME` is included in this name, which will be automatically replaced with the date and time of the FlashLFQ run. After you are satisfied with your settings, click Run FlashLFQ. The command-line flag to (optionally) specify an output directory is `--out`. An example in the GUI is shown in Fig. 5, with the output folder being left as default.

3.5 Output

FlashLFQ writes several files as output: QuantifiedPeaks.tsv, QuantifiedPeptides.tsv, QuantifiedProteins.tsv, FlashLfqSettings.toml, and (optionally) BayesianFoldChangeAnalysis.tsv, which are hereafter described:

1. **QuantifiedPeaks.tsv** reports each chromatographic peak along with information about that peak, such as its begin and end time, apex intensity, etc. Some peaks list 0 for their intensity, which indicates an unquantifiable peak (i.e., an identification was passed in to FlashLFQ, but that identification could not be quantified). Some peaks list "—" as part of their base

Fig. 5 Example of the "Run" page in the FlashLFQ GUI. The default output directory is shown

sequence or full sequence fields, which indicates that multiple identifications were associated with the same chromatographic peak (i.e., the identity of the chromatographic peak is ambiguous).

2. **QuantifiedPeptides.tsv** reports all peptide sequences along with their intensity in all spectra files and their method of quantification (by MS/MS, by match-between-runs, etc.) in all files. For fractionated data, each fraction's intensity is reported separately; these fraction intensities can be summed to get a sample intensity.
3. **QuantifiedProteins.tsv** reports protein intensities per spectra file, similar to QuantifiedPeptides.tsv. The median polish algorithm is used to calculate protein quantities in each sample (*see* **Note 12**).
4. **FlashLfqSettings.toml** reports the settings used for the FlashLFQ run, so that results can be reproduced easily.
5. **BayesianFoldChangeAnalysis.tsv** reports output from the Bayesian protein quantification and hypothesis testing. It lists each protein, along with a fold-change (relative to the control condition specified in the settings) and the probability that the protein's fold change is less than the fold-change cutoff (the posterior error probability [PEP]).

3.6 Using the Docker Image

Once a Docker image has been pulled (downloaded) from Docker Hub, you can run FlashLFQ within the Docker container with the `docker run` command. For example:

```
docker run [Docker args]
       smithchemwisc/flashlfq:latest
       [FlashLFQ args]
```

Typical usage will look something like this:

```
docker run --rm -v C:/MyDataFolder:/mnt/
          smithchemwisc/flashlfq:latest
          --idt ./mnt/AllPSMs.psmtsv
         --rep ./mnt/
         --out ./mnt/FlashLFQ _Output
```

The `--rm` flag tells Docker to clean up or remove the image after FlashLFQ is done executing. The `-v C:/MyDataFolder:/mnt/` part "mounts" the 🗁`C:/MyDataFolder` directory to the Docker container, i.e., lets the container access your hard drive and gives that directory an alias of 🗁 `/mnt/`. The rest of the line runs FlashLFQ with the passed arguments. To ease the use of the Docker image, we recommend the following (*see* **Note 13**):

1. Sometimes Docker has trouble getting permission from your computer to access the hard disk; for Windows users, the C: drive is usually easiest to access and you might get a popup asking for permission.
2. Try not to use spaces in your path names if you can help it (in the Dockerfile, terminal commands, input or output files, etc.). It usually makes things harder than they need to be.

4 Notes

1. There were a few motivations for creating this system. Most protein quantification algorithms report only one protein-level intensity per sample, which is then used in downstream statistics. However, this throws out quite a lot of information; each peptide is an independent measurement of a protein's abundance in a sample, and we wanted to use this information. In addition, uncertainty in peptide measurements increases with lower intensity, and these peptide-level uncertainty measurements can be aggregated into a protein-level uncertainty; again, this information would be lost in most statistical pipelines. FlashLFQ takes a peptide-centric approach to bottom-up protein quantification and measures peptide fold changes across experimental conditions, as well as peptide and protein uncertainty within a condition to estimate quantitative false discovery rates; see [3] for details.
2. To use the Windows GUI, the .NET Core *Desktop* Runtime must be installed. This framework allows both GUI and CLI .NET Core programs to be run. This is in contrast to the .NET Core Runtime, which only runs CLI .NET Core programs.

3. Docker images are "containers," which encapsulate an operating system and pre-installed software. These containers can be run independent of one's own operating system or other software requirements. The purpose of using a Docker container over a more "traditional" installation is usually to avoid installing software prerequisites, either out of convenience or inability, such as on a server in which the user does not have permission to install software.
4. CMD flag: `--ppm`.
5. CMD flag: `--nor`.
6. CMD flag: `--mbr`.
7. CMD flag: `--sha`.
8. Currently, if shared peptides are used for protein quantification, shared peptides are treated the same as unique peptides for protein quantification.
9. CMD flag: `--bay`.
10. Currently, one condition must be specified as the control to compare the other conditions to. For example, if you were to have 3 conditions of samples, "Normal," "Tumor1," and "Tumor2," with "Normal" specified as the control, then the "Tumor1" and "Tumor2" conditions would be quantified relative to the "Normal" condition.
11. A change in a given protein's relative abundance below which you are not interested.
12. See https://mgimond.github.io/ES218/Week11a.html for a tutorial on how the median polish algorithm works.
13. More documentation is given on the FlashLFQ wiki: https://github.com/smith-chem-wisc/FlashLFQ/wiki/Docker-Image.

Acknowledgments

This work was supported by grant R35GM126914 from the National Institute of General Medical Sciences. R.J.M. was supported by an NHGRI training grant to the Genomic Sciences Training Program 5T32HG002760. We thank the software development team of the Smith Lab (Stefan K. Solntsev, Anthony J. Cesnik, Khairina Ibrahim, Lei Lu, Rachel M. Miller, Zach Rolfs, and Leah V. Schaffer), who contributed daily input and guidance for FlashLFQ's improvement.

References

1. Millikin RJ, Solntsev SK, Shortreed MR, Smith LM (2018) Ultrafast peptide label-free quantification with FlashLFQ. J Proteome Res 17(1): 386–391. https://doi.org/10.1021/acs.jproteome.7b00608
2. Li Q, Shortreed MR, Wenger CD, Frey BL, Schaffer LV, Scalf M, Smith LM (2017) Global post-translational modification discovery. J Proteome Res 16(4):1383–1390. https://doi.org/10.1021/acs.jproteome.6b00034
3. Millikin RJ, Shortreed MR, Scalf M, Smith LM (2020) A Bayesian null interval hypothesis test controls false discovery rates and improves sensitivity in label-free quantitative proteomics. J Proteome Res 19(5):1975–1981. https://doi.org/10.1021/acs.jproteome.9b00796
4. Wenger CD, Coon JJ (2013) A proteomics search algorithm specifically designed for high-resolution tandem mass spectra. J Proteome Res 12(3):1377–1386. https://doi.org/10.1021/pr301024c
5. Cox J, Neuhauser N, Michalski A, Scheltema RA, Olsen JV, Mann M (2011) Andromeda: a peptide search engine integrated into the MaxQuant environment. J Proteome Res 10(4): 1794–1805. https://doi.org/10.1021/pr101065j
6. Barsnes H, Vaudel M (2018) SearchGUI: A highly adaptable common interface for proteomics search and de novo engines. J Proteome Res 17(7):2552–2555. https://doi.org/10.1021/acs.jproteome.8b00175
7. Kessner D, Chambers M, Burke R, Agus D, Mallick P (2008) ProteoWizard: open source software for rapid proteomics tools development. Bioinformatics 24(21):2534–2536. https://doi.org/10.1093/bioinformatics/btn323

Chapter 14

Robust Prediction and Protein Selection with Adaptive PENSE

David Kepplinger and Gabriela V. Cohen Freue

Abstract

Adaptive PENSE is a method that can be used to build models for predicting clinical outcomes from a small subset of a potentially large number of candidate proteins. Adaptive PENSE is designed to give reliable results under two common challenges often encountered in these kinds of studies: (1) the number of samples with known clinical outcome and proteomic data is small, while the number of candidate proteins is large and/or (2) proteomic data and the clinical outcome measurements suffer from data quality issues in a small fraction of samples. Even in the presence of these challenges, adaptive PENSE reliably identifies proteins relevant for prediction and estimates accurate predictive models. Adaptive PENSE is designed to be resilient to data quality issues in up to 50% of samples. Almost half of the samples could have aberrant values in the measured protein levels and clinical outcome values without causing severe detrimental effects to the estimated predictive model. The method is implemented as an R package and supports the user in the model selection process by automating most steps and providing diagnostic visualizations to guide the user. Users can choose among several predictive models to select the model with high prediction accuracy and an appropriate number of selected proteins.

Key words Prediction, Protein selection, Robust estimation, Linear regression, High-dimensional data

1 Introduction

Accurate prediction of a clinical outcome from a small panel of protein levels is a common goal of many proteomic studies. To achieve this goal, it is necessary to (1) identify a small number of proteins relevant for prediction from a large number of candidate proteins and (2) find a suitable combination of these protein levels to achieve high prediction accuracy. Linear predictive models are a useful tool in these studies, especially if the number of samples is limited, while at the same time many candidate proteins are available to choose from. Common methods to build such predictive models include the LASSO, univariate screening, or prior knowledge about the relation between the available proteins and the

Thomas Burger (ed.), *Statistical Analysis of Proteomic Data: Methods and Tools*,
Methods in Molecular Biology, vol. 2426, https://doi.org/10.1007/978-1-0716-1967-4_14,

clinical outcome. However, in addition to the challenge of few samples and many candidate proteins, it is difficult to guarantee that all of the measured values are of good quality and trustworthy, i.e., the data set is free of outliers. In the presence of outliers, most methods do not yield reliable results.

In particular, in studies with many candidate proteins or response values that are difficult to measure, anomalous protein levels or outcome values are a common issue. If these data quality issues (summarized under the broad term "contamination") are not properly addressed, they can lead to predictive models with poor prediction accuracy when applied to future samples. Therefore, if a small fraction of samples is contaminated and not properly handled, generalizability of the estimated predictive model may be questionable, jeopardizing the translation of biomarkers into clinical utility.

In data sets with many candidate proteins, contamination is typically difficult to identify prior to knowing the relevant proteins. Even if some values in the data set are suspected to be anomalous, excluding the affected samples from the analysis is counterproductive. Anomalous values may only affect candidate proteins not relevant for prediction, and excluding these samples may remove valuable information. This is particularly detrimental in studies with a small number of samples.

In this chapter, we demonstrate the use of the adaptive PENSE method, proposed in [1] and extended in [2]. The method identifies relevant proteins for prediction and estimates the predictive model based on samples comprising protein levels and values of the response variable measured on the same individuals. Adaptive PENSE affords up to 50% of aberrant contamination in the data set at hand. This means adaptive PENSE gives reliable results even if up to 50% of samples in the data have anomalous response values or one or more aberrant measurements of protein levels.

Adaptive PENSE simultaneously identifies a panel of proteins relevant for prediction and estimates the coefficients in the linear model to yield accurate predictions of the clinical outcome. The method leverages as much information as possible from the samples by focusing on those samples without contamination in the relevant proteins. Furthermore, adaptive PENSE is not affected by low prediction accuracy for a small fraction of samples if they are affected by contamination. Adaptive PENSE is thus more reliable for biomarker discovery than other methods commonly used in studies with a limited number of samples and a large number of candidate proteins (e.g., univariate screening, LASSO, or elastic net regression). Adaptive PENSE leads to predictive models with better chances of being generalizable, even if some samples may be affected by contamination.

The adaptive PENSE method is showcased in this chapter for identifying a panel of proteins for predicting an artificial clinical outcome measured between 0 and 10 on 41 fictional individuals. The synthetic proteomic data comprises abundance of 40 proteins measured on the same 41 individuals. The synthetic data follows a similar structure as the proteomic data in the biomarker discovery study in [1]. After applying the adaptive PENSE method, the practitioner must choose appropriate values for hyper-parameters to select the predictive model with highly accurate predictions and an appropriately sized panel of relevant proteins. This chapter provides guidance for choosing these hyper-parameters based on the samples at hand and explains how to tune the method for a variety of challenges faced in practice.

2 Material

2.1 Data Format

1. The quantitative data should be stored in two separate files, one for the proteomic data and one for the response values. The proteomic data should be formatted in a tabulated text file where the first line contains the column names (*see* Fig. 1 (left) for an example). It is recommended that the column names contain only alphanumeric characters ("a" to "z" and "0" to "9") and underscores ("_"). Most other characters are automatically removed and should be avoided. Ensure that the dot character (".") is used as decimal separator for the quantitative values.
2. The columns in the proteomic data file correspond to one protein each, and the lines (except for the first line) correspond to individual samples. The (i-th, j-th) cell of the tabulated file represents the measured abundance of the j-th protein in the i-th sample.
3. The numeric response values should also be stored as a tabulated text file, with the first line containing the column name (*see* Fig. 1 (right) for an example). The response data file must contain only a single column, where the i-th line (excluding the first line) corresponds to the known response value for the i-th sample.
4. It is important that the lines in the two files line up, i.e., the i-th sample in the proteomic data is for the same individual as the i-th value in the response data.

2.2 Data Size

Adaptive PENSE requires at least two proteins (columns in the proteomic data file) and 10 samples (lines in the proteomic data file and the response file). Ten samples are the absolute minimum, and the more samples, the better the predictive ability and protein selection, with 30 samples as a strongly recommended lower limit (*see* **Note 1**). While there is no upper limit to either the number of proteins or the number of samples, computing times on a regular

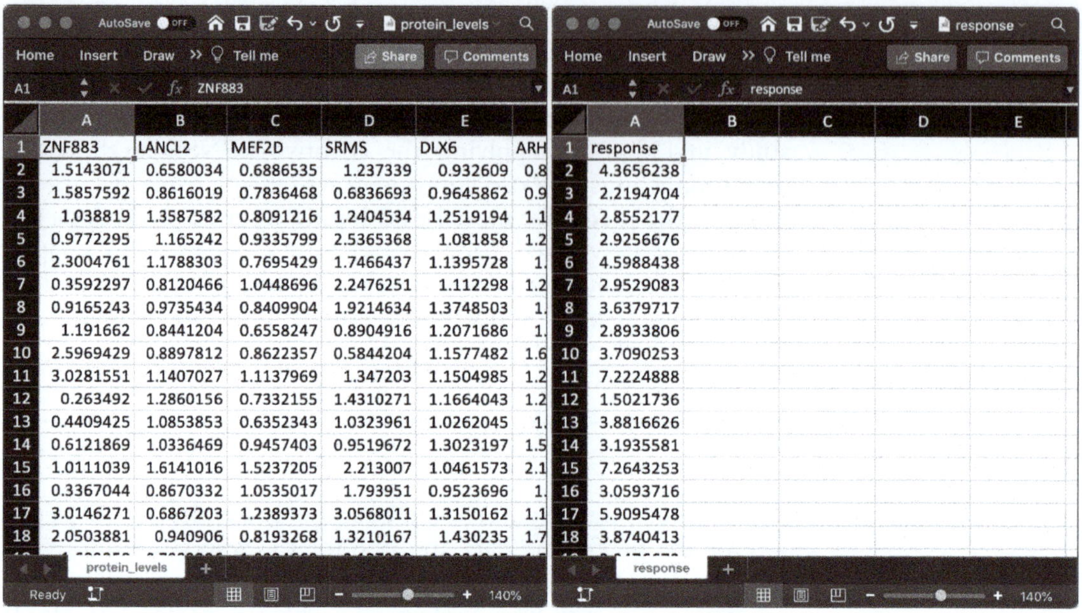

	A	B	C	D	E	
1	ZNF883	LANCL2	MEF2D	SRMS	DLX6	ARH
2	1.5143071	0.6580034	0.6886535	1.237339	0.932609	0.8
3	1.5857592	0.8616019	0.7836468	0.6836693	0.9645862	0.9
4	1.038819	1.3587582	0.8091216	1.2404534	1.2519194	1.1
5	0.9772295	1.165242	0.9335799	2.5365368	1.081858	1.2
6	2.3004761	1.1788303	0.7695429	1.7466437	1.1395728	1.
7	0.3592297	0.8120466	1.0448696	2.2476251	1.112298	1.2
8	0.9165243	0.9735434	0.8409904	1.9214634	1.3748503	1.
9	1.191662	0.8441204	0.6558247	0.8904916	1.2071686	1.
10	2.5969429	0.8897812	0.8622357	0.5844204	1.1577482	1.6
11	3.0281551	1.1407027	1.1137969	1.347203	1.1504985	1.2
12	0.263492	1.2860156	0.7332155	1.4310271	1.1664043	1.2
13	0.4409425	1.0853853	0.6352343	1.0323961	1.0262045	1.
14	0.6121869	1.0336469	0.9457403	0.9519672	1.3023197	1.5
15	1.0111039	1.6141016	1.5237205	2.213007	1.0461573	2.1
16	0.3367044	0.8670332	1.0535017	1.793951	0.9523696	1.
17	3.0146271	0.6867203	1.2389373	3.0568011	1.3150162	1.1
18	2.0503881	0.940906	0.8193268	1.3210167	1.430235	1.7

	A
1	response
2	4.3656238
3	2.2194704
4	2.8552177
5	2.9256676
6	4.5988438
7	2.9529083
8	3.6379717
9	2.8933806
10	3.7090253
11	7.2224888
12	1.5021736
13	3.8816626
14	3.1935581
15	7.2643253
16	3.0593716
17	5.9095478
18	3.8740413

Fig. 1 Format of the tabulated text files with proteomic data (left) and response data (right)

desktop computer may be unreasonable for data sets with more than 1000 proteins or more than 500 samples. The default algorithm settings afford up to 300 samples and 200 proteins on a recent desktop machine, but these settings can be adjusted to make larger data sets amenable to adaptive PENSE.

2.3 Hardware and Software Requirements

Adaptive PENSE can be used on a local desktop computer or on a server. A recent desktop computer is recommended. Depending on the data size, at least 4GB of memory is advised. To keep computation time reasonably short, at least 4 CPU cores are suggested, but this is not a requirement and adaptive PENSE works also on desktop computers with a single CPU core.

Adaptive PENSE requires a recent version of the R software, at least version 3.5.0 (*see* **Note 2**).

2.4 Installation

To install the adaptive PENSE method, enter the following line in the R console on the computer where the analysis is to be performed (local desktop machine or server):

```
install.packages("pense")
```

This installs the pense package and a number of dependencies required to run the adaptive PENSE method.

3 Methods

3.1 Starting PENSE

To get started with adaptive PENSE, start a new R session. Then, to load the pense package, enter in the R console:

```
library(pense)
```

3.2 Loading Data

1. The data needs to be in the format described above (*see* Subheading 2.1). For this example, the proteomic data file and the response data file are assumed to be available as *protein_levels.tsv* and *response.tsv*, respectively. Both of these files are assumed to exist in a directory called proteomic-study/data/. These two data files can be loaded into R with the following commands (*see* **Note 3**):

```
setwd("proteomic-study/data/")
protein_levels <- as.matrix(read.table("protein_levels.tsv",
                                       header = TRUE))
response <- as.matrix(read.table("response.tsv",
                                 header = TRUE))
```

   This creates two R objects: the object `protein_levels` is a matrix with the measured protein levels from file *protein_levels.tsv* and the object `response` is a matrix with the measured response values from file *response.tsv*.

2. It is prudent to verify the files are read correctly as numeric data, for example, using the function `str()`:

```
str(protein_levels)
```

```
 num [1:41, 1:40] 1.514 1.586 1.039 0.977 ...
 - attr(*, "dimnames")=List of 2
  ..$ : NULL
  ..$ : chr [1:40] "ZNF883" "LANCL2" "MEF2D" ...
```

```
str(response)
```

```
 num [1:41, 1] 4.37 2.22 2.86 2.93 ...
 - attr(*, "dimnames")=List of 2
  ..$ : NULL
  ..$ : chr "response"
```

   The matrices `protein_levels` and `response` both must contain numeric data (indicated above by `num`) and have the same number of rows (41 in the output above). The number of columns of `protein_levels` must match the expected number of proteins (40 in this example), while `response` must have only a single column.

3. Missing values in the proteomic or response data are not supported by adaptive PENSE. The user has two options for handling missing values prior to computing adaptive PENSE estimates.

   (a) Removing samples with missing values from the input data. To omit samples with missing values either in the proteomic data or the response values, run the following commands in the R console:

```
complete_samples <- complete.cases(protein_levels,response)
protein_levels <- protein_levels[complete_samples, ]
response <- response[complete_samples]
```

   The user should check if the input data still contains enough samples (*see* Subheading 2.2) after filtering samples with missing values, e.g., using the following R commands:

```
dim(protein_levels)
```

```
[1] 37 40
```

```
length(response)
```

```
[1] 37
```

   In this case, a total of 4 samples are omitted because they have missing values in one or more measurements. The remaining number of 37 samples is greater than the suggested minimum of 30 and should be enough to build a predictive model with sufficient prediction accuracy. If the input data set does not have enough samples after excluding samples with missing values, the user may need to exclude proteins with a high number of missing values prior to omitting samples with missing values (*see* **Note 4**).

   (b) Imputing missing values in the proteomic data on the peptide level (*see* **Chapter 6**, as well as, e.g., [3] for details). For proteins with a small proportion of missing values, imputation can help retain a larger number of samples in the input data and hence more information for estimating a predictive model with adaptive PENSE. Imputing missing values in the response data is not advised as it may severely bias the predictive model and decrease generalizability.

3.3 Fitting a Predictive Model

1. To harness all available computing resources, set up a "cluster" of R processes to leverage more than one CPU core (*see* **Note 5**) by running the following commands in the R console:

```
library(parallel)
cluster <- makeCluster(3)
```

2. The predictive model is fitted to the observed data (*see* **Note 6**) by issuing the following commands in the R console (*see* **Note 7**):

```
set.seed(123)
fit <- adapense_cv(protein_levels, response,
                   alpha = 0.75, cv_k = 5,
                   cv_repl = 50, cl = cluster)
```

This fits the linear model using the adaptive PENSE method using many different penalization levels (*see* **Note 8**) and assigns the results to object `fit`. The `alpha` hyper-parameter controls the balance between aggressiveness and stability of protein selection and needs to be fixed for fitting the model, with 0.75 the suggested default value (*see* **Note 9**). For selecting an `alpha` value based on the samples at hand *see* Subheading 3.4. To estimate the prediction performance on future samples for each of the considered penalization levels, `adapense_cv()` performs cross-validation (CV) [4, p. 176 *ff.*]. Arguments `cv_k` and `cv_repl` are control parameters for cross-validation: `cv_k` determines the number of disjoint chunks for splitting the data set in a single CV run, whereas argument `cv_repl` controls the number of CV runs (*see* **Note 10**). By default, adaptive PENSE can tolerate up to 25% of contaminated samples, but this can be adjusted with argument `bdp` (*see* **Note 11**). Setting the seed with `set.seed()` is technically not necessary for adaptive PENSE, but it is essential for reproducible results (*see* **Note 12**).

3. Evaluate stability of protein selection by plotting the cross-validated prediction performance for different penalization levels. This is done by running the following R command:

```
plot(fit)
```

The resulting plot (*see* Fig. 2, top left) guides the selection of the appropriate penalization level. For each penalization level, the plot shows the average prediction accuracy of the model as well as the associated standard error, across the `cv_repl` CV runs. The dark blue dot marks the penalization level achieving the best average prediction performance. The light blue dot farther right marks the penalization level achieving almost as good prediction performance, but with higher penalization level and potentially fewer proteins. If the curve of the estimated prediction performance is non-smooth as in the top right panel of Fig. 2 (*see* **Note 13**) or the grid is not well

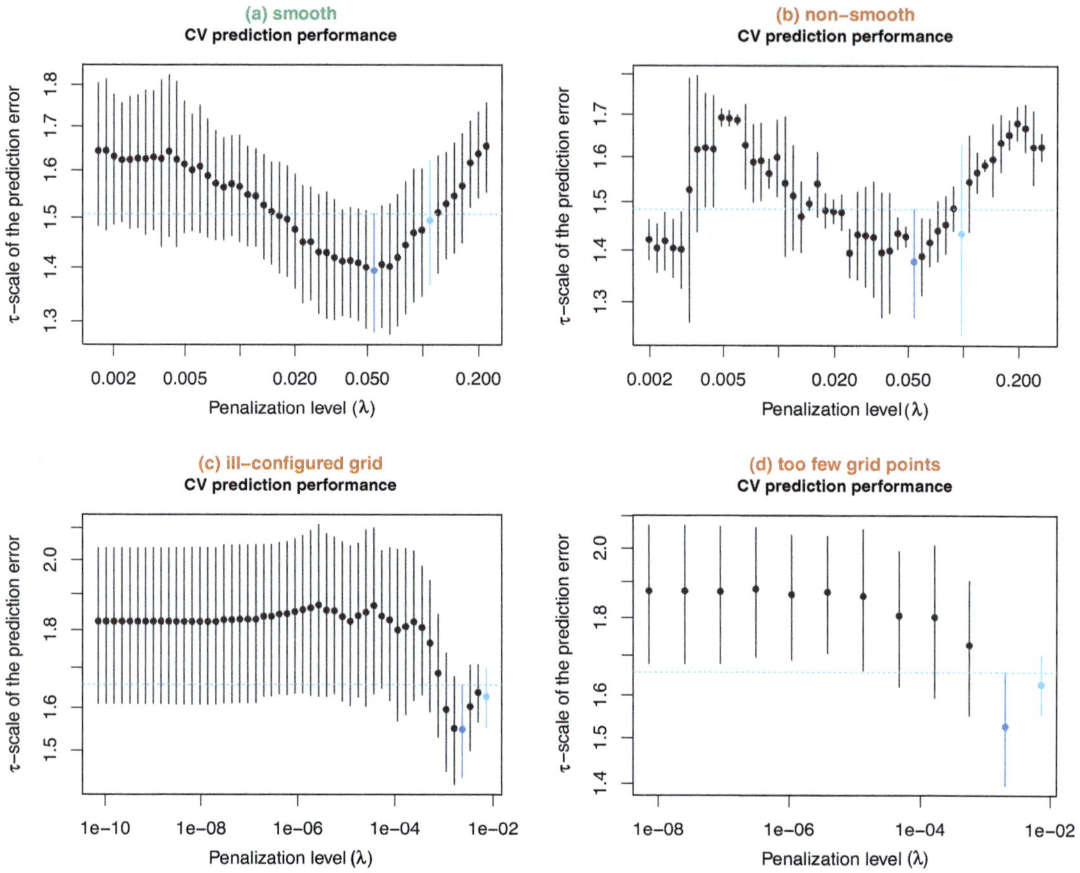

Fig. 2 Cross-validated (CV) prediction performance (vertical axis) for a grid of penalization levels (λ). By default, prediction performance is measured in terms of a robust scale estimate (τ-scale) of the prediction error. **(a)** Smooth estimate of prediction performance with enough CV replications covering the appropriate range of penalization levels. **(b)** Non-smooth estimate of prediction performance because of too few CV replications. Increase the number of CV replications with argument `cv_repl` to get better estimates. **(c)** The grid is too wide. It covers many penalization levels in a range where there is no change in the prediction performance ($\lambda < 10^{-4}$) and not enough penalization levels around the minimum. Focus the grid of penalization levels to the relevant range by increasing the argument `lambda_min_ratio`. **(d)** Too few grid points, in particular around the minimum, lead to a sub-optimal selection of the penalization level. Increase the number of penalization levels with argument `nlambda` for a more precise evaluation of the prediction performance

calibrated as in the two bottom panels of Fig. 2 (*see* **Note 14**), the model needs to be refitted with different arguments.

4. If the grid of penalization levels is well calibrated, the curve is smooth and exhibits only a single noticeable dip (*see* **Note 15**), extract coefficients, and relevant proteins with the R command:

```
summary(fit)
```

```
Adaptive PENSE fit with prediction performance
estimated by 50 replications of 5-fold
cross-validation.

8 out of 40 predictors have non-zero
coefficients:

              Estimate
(Intercept)     0.7183
ZNF883          0.5324
CHRD            0.3420
TP53TG3B        0.6336
BORCS8          0.2275
KRTAP2.3       -0.6889
ANKRD13D        1.5522
NUTM2D          0.9669
SPATA1         -1.0583
---

Hyper-parameters: lambda=0.0543, alpha=0.75, exponent=1
Estimated scale of the prediction error: 1.395
```

The output of the above command shows the eight relevant proteins in this model: `ZNF883`, `CHRD`, `TP53TG3B`, `BORCS8`, `KRTAP2.3`, `ANKRD13D`, `NUTM2D`, and `SPATA1`. The estimated coefficients of these relevant proteins are given to the right of each protein name. The output also includes the selected hyper-parameters as well as the estimated prediction accuracy (the same value as the dark blue dot shown in the top left plot of Fig. 2).

5. In applications where the goal is to have as few proteins in the predictive model as possible, it may be appropriate to use the model with prediction performance indistinguishable from the best model but using a smaller panel (*see* **Note 16**). This "second-best" model, indicated by a light blue dot in the top left panel of Fig. 2, has similar estimated prediction performance as the best model but achieves this performance with a smaller number of proteins. The coefficients and relevant proteins for the second-best model can be extracted by using the penalization level "within one standard error" of the best model, selected with argument `lambda = "se"` to the `summary()` function:

```
summary(fit, lambda = "se")
```

```
Adaptive PENSE fit with prediction performance
estimated by 50 replications of 5-fold
cross-validation.

5 out of 40 predictors have non-zero
coefficients:

              Estimate
(Intercept)     1.5536
ZNF883          0.6044
KRTAP2.3       -0.3318
ANKRD13D        0.8980
NUTM2D          0.7949
SPATA1         -0.2694
---

Hyper-parameters: lambda=0.1092, alpha=0.75, exponent=1
Estimated scale of the prediction error: 1.495
```

Using the second-best model is only sensible if it uses fewer proteins than the best model. In the example above, the second-best model substantially reduces the number of relevant proteins from 8 in the best model to only 5 proteins.

3.4 Selecting Hyper-Parameters

1. Fit predictive models for different values of the `alpha` and `exponent` arguments (*see* **Note 17**), for example,

```
set.seed(123)
fit_075_1 <- adapense_cv(protein_levels, response,
                         alpha = 0.75, exponent = 1,
                         cv_k = 5, cv_repl = 50,
                         cl = cluster)
set.seed(123)
fit_100_1 <- adapense_cv(protein_levels, response,
                         alpha = 1.00, exponent = 1,
                         cv_k = 5, cv_repl = 50,
                         cl = cluster)
set.seed(123)
fit_075_2 <- adapense_cv(protein_levels, response,
                         alpha = 0.75, exponent = 2,
                         cv_k = 5, cv_repl = 50,
                         cl = cluster)
set.seed(123)
fit_100_2 <- adapense_cv(protein_levels, response,
                         alpha = 1.00, exponent = 2,
                         cv_k = 5, cv_repl = 50,
                         cl = cluster)
```

Each call to `adapense_cv()` performs cross-validation to estimate prediction performance. Therefore, it is important to fix the seed before calling `adapense_cv()` to make the different fits comparable.

2. Check the stability of the estimated prediction performance for each of the fitted models. If necessary, increase the number of cross-validation replications or adjust the grid of penalization levels (*see* Subheading 3.3, **Step 4**).

3. Compare the prediction performance of the different models by running the following command in the R console:

```
prediction_performance(fit_075_1, fit_100_1,
                       fit_075_2, fit_100_2,
                       lambda = "min")
```

```
Prediction performance estimated by cross-validation:

       Model Estimate Std. Error Predictors alpha exp.
1 fit_100_2     1.313     0.1290          5  1.00    2
2 fit_075_2     1.315     0.1298          5  0.75    2
3 fit_075_1     1.395     0.1131          8  0.75    1
4 fit_100_1     1.399     0.1132          8  1.00    1
```

This compares the best models for each combination of hyper-parameters. To compare the second-best models, use argument `lambda = "se"` instead.

4. Based on the goals of the analysis, the user can now select the predictive model with the most appropriate balance of prediction performance (shown in column *Estimate*) and the number of proteins in the model (shown in column *Predictors*). In this example, the two models with best prediction performance both have the same number of relevant proteins and almost identical prediction performance. The comparison reveals that a value of 2 for the hyper-parameter `exponent` leads to better prediction performance for the considered data set, while the value for argument `alpha` has less of an effect. In fact, models `fit_075_2` and `fit_100_2` use the identical panel of proteins and have very similar coefficient estimates. The relevant predictors and their coefficient estimates can be printed with the `summary()` function:

```
summary(fit_100_2, lambda = "min")
```

```
Adaptive PENSE fit with prediction performance
estimated by 50 replications of 5-fold
cross-validation.

5 out of 40 predictors have non-zero
coefficients:

               Estimate
(Intercept)      1.1809
TP53TG3B         0.9609
KRTAP2.3        -1.3102
ANKRD13D         2.1592
NUTM2D           1.1739
SPATA1          -0.6123
---

Hyper-parameters: lambda=0.02981, alpha=1, exponent=2
Estimated scale of the prediction error: 1.313
```

4 Notes

1. There is no hard rule about the required number of samples, but the more samples, the better the predictive model. While 30 samples are the absolute minimum for meaningful results, the suggested number of samples increases in tandem with the number of candidate proteins. In studies with less than 100 candidate proteins, 30 samples may be sufficient. If there are more candidate proteins, more samples are necessary to get an accurate predictive model. As rough guideline, the number of candidate proteins should be less than three times the number of samples.
2. The latest version of R is recommended to ensure availability of all features and improvements in the pense R package.
3. The actual location and format of the input files are not important for the adaptive PENSE method. The only requirement is for the data finally being available in two separate R objects: one numeric matrix with the protein levels and one numeric vector (or matrix with 1 column) with the response values.
4. Proteins with a large number of missing measurements should be excluded from the analysis prior to omitting samples with missing values. The following command computes the proportion of samples with missing values for each of the candidate proteins, and prints them in decreasing order:

```
sort(colMeans(apply(protein_levels, 2, is.na)),
     decreasing = TRUE)
```

```
      IL7    ZC3H10    LANCL2    ZNF883     MEF2D
  0.09756   0.07317   0.02439   0.00000   0.00000
     SRMS      DLX6  ARHGAP12      MALL    FAM83E
  0.00000   0.00000   0.00000   0.00000   0.00000
   MPV17L      CHRD    PCDHA4       HBZ     BEND6
  0.00000   0.00000   0.00000   0.00000   0.00000
  SLC41A2      LRP8    TRIM45   TP53TG3  TP53TG3B
  0.00000   0.00000   0.00000   0.00000   0.00000
    HOOK1    BORCS8  KRTAP2.3  ANKRD13D  SEPTIN14
  0.00000   0.00000   0.00000   0.00000   0.00000
 KRTAP7.1    NUTM2D     LCE1E    SPATA1   FAM241B
  0.00000   0.00000   0.00000   0.00000   0.00000
   NSMCE1    ATXN2L     RPP40     CPXM1    ZNF727
  0.00000   0.00000   0.00000   0.00000   0.00000
   PRSS51     ZNF17    HP1BP3     IGHG4       IPP
  0.00000   0.00000   0.00000   0.00000   0.00000
```

If any of the proteins has a large proportion of missing values, it may be necessary to exclude these proteins from the analysis. The user should aim at excluding as few proteins from the analysis as possible. One recommended strategy is to exclude the protein with highest proportion of missing values and repeat (*see* Subheading 3.2, **Step 3**). If the input data still has too few samples after omitting samples with missing values, the user should exclude the protein with the next-highest proportion of missing values and continue the cycle until the number of samples is large enough.

5. It is generally advised to use one core less than available on the system. If the number of CPU cores on the computer running adaptive PENSE is not known, the function `detectCores(logical = FALSE)` is helpful. On most systems, this function reports the number of available cores. If, for example, the system has a total of 4 CPU cores, it is recommended to use only 3 of these to ensure the system has enough resources for all other necessary tasks.

6. The adaptive PENSE method by default standardizes the input data such that all protein levels are on the same scale. The scale is measured robustly to ensure it is unaffected by a few anomalous values. Scaling the data serves two important roles: (a) ensuring all proteins are treated equally when identifying proteins relevant for prediction and (b) stabilizing the internal selection of the penalization level (*see* Subheading 3.3, **Step 4**). It is therefore strongly discouraged to disable standardization of the input data, unless the user has very specific reasons. If necessary, instead of disabling standardization completely, it is suggested to set the argument `standardize = "cv_only"` in the call to `adapense_cv()`. This retains stability of the cross-validation for selecting the penalization level.

7. Computing adaptive PENSE estimates can take a long time, but the algorithm can be changed to reduce computation times for certain data sets. For data sets with more samples than proteins, the default algorithms will generally perform best. For other data sets, faster computation may be achieved by changing the algorithms in the `adapense_cv()` function call. For data sets with less than 100 samples but more than 200 proteins, it is suggested to use the DAL algorithm via

```
algorithm_opts <- mm_algorithm_options(
  en_algorithm_opts = en_dal_options())
enpy_opts <- enpy_options(
  en_algorithm_opts = en_dal_options())
set.seed(123)
fit <- adapense_cv(protein_levels, response,
                   alpha = 0.75, cv_k = 5,
                   cl = cluster, cv_repl = 50,
                   algorithm_opts = algorithm_opts,
                   enpy_opts = enpy_opts)
```

The DAL algorithm is generally recommended for data sets with substantially more proteins than samples. In cases where the number of samples is large (>200) and the number of proteins is larger than the number of samples, the ADMM algorithm is recommended. The ADMM algorithm can be selected with function `en_admm_options()`:

```
algorithm_opts <- mm_algorithm_options(
  en_algorithm_opts = en_admm_options())
enpy_opts <- enpy_options(
  en_algorithm_opts = en_admm_options())
set.seed(123)
fit <- adapense_cv(protein_levels, response,
                   alpha = 0.75, cv_k = 5,
                   cl = cluster, cv_repl = 50,
                   algorithm_opts = algorithm_opts,
                   enpy_opts = enpy_opts)
```

8. Adaptive PENSE is a robust regularized estimator for the coefficients in a linear regression model. The method considers both the fit to the response values and the overall size of the coefficients. The penalization level (λ) controls the balance between these two: a small penalization level puts more emphasis on a good fit to the observed response values, while a large penalization level gives more importance to small coefficients. If two estimates would yield the same fit to the response values, adaptive PENSE always favors the estimates with a smaller size. The fit to the observed response values is measured by the

S-loss, a robust scale of the residuals (the differences between the observed and the fitted response values). This S-loss is the key for adaptive PENSE to be robust toward arbitrary contamination. The size of the coefficients is measured by the adaptive elastic net penalty [5], which favors sparse estimates, i.e., estimates where some coefficients are 0. Therefore, the larger the penalization level, the more proteins will be omitted from the model.

9. The `alpha` hyper-parameter controls the balance between aggressiveness and stability of protein selection. With the most aggressive setting, `alpha` = `1`, fewer proteins tend to be selected for the predictive model, usually including only a single "representative" protein from any group of highly correlated proteins. For a slightly different penalization level, however, another protein from each group of highly correlated protein may be the representative protein and hence included in the model. This leads to unstable protein selection in the presence of correlated proteins. With smaller values for `alpha`, protein selection is more stable, but typically more proteins will be deemed relevant for prediction. Setting `alpha` = `0` is not recommended as it does not lead to any selection of proteins, i.e., all proteins are deemed relevant.

10. The `cv_k` argument determines the number of chunks the data set is split into for each CV run. Each chunk contains approximately n`cv_k` samples, where n is the total number of samples in the data set. The argument `cv_k` should be between 3 and 10, with larger values leading to a more stable selection of the penalization level. At the same time, `cv_k` should be chosen such that each chunk contains at least 5 samples. A single CV run may not be reliable and it is advised to perform several (>10) CV runs, specified via the `cv_repl` argument. The more cross-validation runs, the more reliable the estimated prediction performance for different penalization levels, but computation time increases with every additional run.

11. By default, adaptive PENSE can tolerate up to 25% of samples with anomalous values either in their response value or in the protein levels. If it is suspected that more than 25% of samples may have anomalous values, the breakdown point (argument `bdp` in the `adapense_cv()` call) can be increased to up to 0.5 (i.e., 50%). A higher breakdown point leads to reliable results in applications with higher proportion of contaminated samples, but this will in general also reduce the precision of the coefficient estimates and in turn prediction performance of the estimated model.

12. The "seed" determines the random numbers generated by R. After setting the seed with `set.seed()`, the generated random numbers are deterministic and hence reproducible. Adaptive PENSE estimates the prediction accuracy using

cross-validation, which is randomly shuffling the order of samples in the data set before splitting the data set into several chunks. By setting the seed prior to the call to `adapense_cv()`, the random shuffling of the data set can be reproduced.

13. If the shown curve is highly non-smooth (as in the top right of Fig. 2), the number of cross-validation replications should be increased. To increase the number of CV replications, change the argument `cv_repl` in the call to `adapense_cv()`. If the curve is still non-smooth, other parameters of the adaptive PENSE method may need to be adjusted. The `adapense_cv()` function uses several computational shortcuts for reducing computation time [1]. These shortcuts can lead to higher variability over the grid of penalization levels in some applications. In these cases, increasing the number of initial estimates with argument `nlambda_enpy` may lead to higher stability.
14. If the curve is mostly flat (as in the bottom left of Fig. 2), the grid is not focused enough on the proper range of penalization values. In these cases, differences in prediction performance are mainly observed for larger values of the penalization level, while smaller penalization levels all lead to similar predictions. The argument `lambda_min_ratio` in the call to `adapense_cv()` determines the minimum penalization level to consider, i.e., the left endpoint of the grid of penalization levels. By increasing this value, the left endpoint is moved farther right to focus the grid on larger penalization levels. The right endpoint of the grid is by default fixed at the penalization level at which no protein is selected in the model (the intercept-only model). On the other hand, if the CV curve does not properly capture the change in prediction performance in the area around the minimum (as in the bottom right of Fig. 2), more grid points may be necessary. The default number of penalization levels is 50. To increase the number of penalization levels, set the argument `nlambda` in the call to `adapense_cv()`. The more penalization levels are considered, the longer the computation will take.
15. If the point at the far right on the plot is the minimum of the curve, the robust average of the response values is the best possible prediction. In this case, a linear combination of the protein levels, as fitted by adaptive PENSE, may not be appropriate for the data at hand.
16. By default, the adaptive PENSE method considers models as indistinguishable if their prediction performance is within 1 standard error of the estimated prediction performance of the best model (often called the "one-standard-error rule" [4 p. 214]). The user can adjust the rule to an *xx*-standard-error rule (e.g., 1.5-standard-error rule) by setting a different value

for the `se_mult` argument when calling functions `plot()`, `summary()`, or `prediction_performance()`.

17. The `exponent` argument determines the aggressiveness of screening of proteins with small coefficient estimates. A large value for `exponent` typically leads to a sparser model, i.e., a model with only a small number of relevant proteins.

Acknowledgements

We thank Matías Salibián-Barrera for comments on a draft of this chapter and contributions to the method and software and Ezequiel Smucler for his help with the theoretical underpinning. GCF was supported by an NSERC Discovery grant, and the computational infrastructure was funded by the Canada Foundation for Innovation (CFI).

References

1. Cohen Freue GV, Kepplinger D, Salibián-Barrera M, Smucler E (2019) Robust elastic net estimators for variable selection and identification of proteomic biomarkers. Ann Appl Stat 13(4):2065–2090. https://doi.org/10.1214/19-AOAS1269
2. Kepplinger D (2020) Robust estimation and variable selection in high-dimensional linear regression models. PhD thesis, University of British Columbia, Vancouver, BC. https://doi.org/10.14288/1.0392915
3. Lazar C, Gatto L, Ferro M, Bruley C, Burger T (2016) Accounting for the multiple natures of missing values in label-free quantitative proteomics data sets to compare imputation strategies. J Proteome Res 15(4):1116–1125. https://doi.org/10.1021/acs.jproteome.5b00981
4. James G, Witten D, Hastie T, Tibshirani R (2013) An introduction to statistical learning. Springer texts in statistics. Springer, New York. https://doi.org/10.1007/978-1-4614-7138-7
5. Zou H, Zhang HH (2009) On the adaptive elastic-net with a diverging number of parameters. Ann Stat 37(4):1733–1751. http://dx.doi.org/10.1214/08-AOS625

Chapter 15

Multivariate Analysis with the R Package mixOmics

Zoe Welham, Sébastien Déjean, and Kim-Anh Lê Cao

Abstract

The high-dimensional nature of proteomics data presents challenges for statistical analysis and biological interpretation. Multivariate analysis, combined with insightful visualization can help to reveal the underlying patterns in complex biological data. This chapter introduces the R package `mixOmics` which focuses on data exploration and integration. We first introduce methods for single data sets: both Principal Component Analysis, which can identify the patterns of variance present in data, and sparse Partial Least Squares Discriminant Analysis, which aims to identify variables that can classify samples into known groups. We then present integrative methods with Projection to Latent Structures and further extensions for discriminant analysis. We illustrate each technique on a breast cancer multi-omics study and provide the R code and data as online supplementary material for readers interested in reproducing these analyses.

Key words mixOmics, Multivariate analysis, Dimension reduction, Feature selection, Principal Component Analysis, Projection to Latent Structures, PLS-Discriminant Analysis, Multi-block PLS-DA

1 Introduction

In recent years, high-throughput technologies have advanced new discoveries in proteomics, typically through liquid chromatography coupled to mass spectrometry (LC-MS). This has generated quantitative data sets containing information for potentially hundreds of thousands of peptides. However, it is still challenging to fully unravel information from those assays. Firstly, the high-dimensional nature of proteomic data is not conducive to easy biological interpretation, thus motivating an interest in summarizing the data, identifying those peptides that most contribute to the biology of interest, or enabling the classification of samples into known groups. Secondly, biological systems are regulated at many functional levels [1], and molecules of different types act in concert to drive biological systems, prompting a desire to understand how proteins interact with other relevant molecules, (e.g., messenger or micro RNA), which requires integration with other types of data.

Thomas Burger (ed.), *Statistical Analysis of Proteomic Data: Methods and Tools*,
Methods in Molecular Biology, vol. 2426, https://doi.org/10.1007/978-1-0716-1967-4_15,

Interpreting large data sets has been achieved through knowledge-driven approaches by including external information (e.g., Protein Data Bank) to the statistical analyses to identify proteins, perform enrichment analyses, or perform network modeling [2, 3]. These approaches often employ *univariate* methods to identify relevant proteins which are assessed independently.

While these approaches can identify proteins of interest, they often rely on, and generate, known information. An alternative is to take a data-driven approach to understand biological systems, using mathematical and statistical models that can describe the relationships between molecules. Furthermore, by adopting a multivariate analysis, we can examine multiple variables (e.g., proteins) simultaneously to uncover patterns in data sets, and understand how these variables correlate with other molecule types according to the biology of interest.

In this chapter, we introduce the application of the `mixOmics` package [4], an R toolkit dedicated to the exploration and integration of biological data sets. The package employs *multivariate projection-based* methods to manage high-dimensional data by aggregating original variables (e.g., genes, transcripts, or proteins) into artificial summary *components* to extract the main sources of variation in the data. Several analytical frameworks are proposed. *Unsupervised* analyses identify patterns in data while ignoring a priori any information about sample group membership (e.g., disease status, cancer subtype, treatment intake), to inform about how the data "naturally" forms subgroups. A popular approach is Principal Component Analysis (PCA) [5] for the unsupervised analysis of one data set. *Supervised* analyses include information about the class membership of the samples to visualize the relationship between the measured variables and the outcome, and potentially predict the class of new samples. We will illustrate Partial Least Squares (a.k.a. Projection to Latent Structures)-Discriminant Analysis (PLS-DA, [6]).

Since biological systems are regulated at many functional levels [1] and cannot be understood through the analysis of a single data type, integrative analyses focus on identifying relationships between two or more data sets (e.g., proteins and messenger RNA). Here, we will introduce the PLS regression method [7, 8] and a supervised method for multiple data set integration (DIABLO, [9]).

The mixOmics package also offers the ability to select key variables that can best explain the variance in the data, the outcome of interest, or the covariance between data sets. This allows for the generation of biomarker panels for further analysis. Protein data in particular are high dimensional and as such, comprise large amounts of inherent noise, with abundances that can vary across a wide dynamic range. The methods in `mixOmics` are designed to manage these issues as well as the highly correlated structure of the data. They are also suitable for a wide range of data types, such as transcriptomics, metabolomics, and metagenomics measured on a continuous scale (e.g., *see* [10, 11]).

We illustrate the different techniques introduced in this chapter on a subset of a breast cancer multi-omics data set from The Cancer Genome Atlas (TCGA) (*see* **Note 1**) that is available in the `mixOmics` package. The reproducible R code is provided as an online supplementary material to re-run all analyses.

2 Material

2.1 Hardware and Software Requirements

The `mixOmics` package is freely available on Bioconductor (https://bioconductor.org) and can be installed on a Linux, Mac OS, or Windows desktop machine [12]. We recommend to install the most recent version of the R software, and the integrated development environment RStudio for ease of script writing and plot visualization.

To install `mixOmics`, enter the following instructions into the R console to install the latest Bioconductor version (*see* **Note 2**):

```
if (!requireNamespace("BiocManager", quietly = TRUE))
    install.packages("BiocManager")
BiocManager::install("mixOmics")
```

A development version is also available on GitHub (*see* https://github.com/mixOmicsTeam/mixOmics for further instructions). For Mac OS Users, ensure first that you have installed the XQuartz software (*see* **Note 3**) for 3D plots.

2.2 Data Pre-processing

Data pre-processing steps are crucial prior to data analysis. Raw data frequently contains biases introduced by the experimental methods and technological platforms employed, uninformative variables that introduce noise and increased computational time, and missing values. The data may also be incorrectly formatted for analysis. These issues must be managed to optimize the statistical analysis. We describe several pre-processing steps, including normalization that is not part of the package.

2.2.1 Normalization

Normalization aims to improve the ability to compare samples by reducing the impact of inherent technical biases. As normalization is omics- and platform-specific, this process is not part of the package. However, we assume the input data are normalized using techniques that are appropriate for the type of data. Normalization has been described as a strategy that employs a variety of methods, both pre- and post-data acquisition. This can include quantifying proteins prior to sample digestion, increasing technical replicates, and taking into account random variance and batch effects to ensure that technical biases are minimized across samples [13]. Several methods have been developed for the normalization of mass spectrometry-based, label-free proteomics, including post-digestion techniques such as spiking samples with internal standards (ISTDs) [13] and using endogenous molecules such as histone abundance (being proportional to DNA) as an internal standard (e.g., [14]). Post-analysis techniques may also be used, such as linear and local regression; total, average, or median intensity; or variance stabilization normalization [15]. Methods for non-targeted metabolomics are also being developed [16].

2.2.2 Data Filtering

While `mixOmics` methods can manage large data sets, we recommend filtering out proteins that do not vary in abundance across samples, as they are unlikely to provide useful information. Filtering can be based on measures such as median absolute deviation, or the variance calculated per variable. The `mixOmics` function `nearZeroVar()` can also be used. It detects variables with very few unique values relative to the number of samples, and thus variables with a variance close to zero. These filtering methods can lessen the computational time if you expect to perform parameter tuning (*see* **Note 4**).

2.2.3 Managing Zeros and Missing Values

Protein data sets often contain a large number of zeros, which require care in interpretation. *Structural zeros*, or true zeros, reflect a true absence of the variable in the data environment, while *sampling zeros*, or false zeros, may not reflect reality due to issues such as experimental error, technological reasons, or an insufficient sample size [17].

In biological terms, missing values refer to data that are not measured. However, in practice, the definition of *missing* is often unclear. For example, a value might be missing because it did not pass the detection threshold in a mass spectrometry experiment. Therefore, the data analyst must make the choice of setting the missing value to a numerical 0 (i.e., the "absence" is biologically relevant), or "NA" (i.e., technically missing). The decision on how to assign missing values will affect the statistical analysis.

The PLS-based methods in `mixOmics` manage missing values by using the non-linear iterative partial least squares algorithm (NIPALS, [7]). Note however that when the amount of missing values is too large (e.g., > 20% of the whole data set), filtering out variables or imputing the missing values might be required (*see* http://mixomics.org/methods/missing-values/, as well as Chapters 6 and 7).

2.3 Get Started with mixOmics

Follow these steps to get started with the package (*see* **Note 5**). Before using `mixOmics`, data can be imported into R as a comma separated value (*.csv), text (*.txt), or excel (*.xls) file. When using `mixOmics`, the data set should be a normalized (*see* Subheading 2.2) matrix, where each row indicates one sample and each column indicates one variable, such that each cell has one value. Phenotype information should be stored in a separate data frame or vector. When including phenotype information, or when integrating several data sets, samples should be in the same order across data sets:

1. Launch `mixOmics` with the following R code:

```
library(mixOmics)
```

2. Set the working directory: This is the folder where the R scripts will be saved, and where the data are stored. Use either the `setwd()` function, or by specifying a specific folder where your data will be uploaded from, for example:

```
setwd("C:/path/to/my/data/")
```

Or in RStudio by navigating from Session » Set Working Directory » Choose Directory.

3. Load or read the data: The data should be stored in the working directory specified in the previous step, or the code should specify your file path. For example:

   (a) When sample names are in the first column and variable names are in the first row in a .csv file:

```
data <- read.csv("my_data.csv", row.names = 1,
                 sep = ",", header = TRUE)
```

   (b) When the data contains header information and row names of the samples in the first column in a .txt file:

```
data <- read.table("C:/path/to/my/data/my_data.txt",
                   sep = "\t", row.names = 1, header = TRUE)
```

   (c) When sample names are in the first column and variable names are in the first row in a .xls file:

```
install.packages("readxl")
library(readxl)
data <- read_excel("my_data.xls")
data <- data.frame(data, row.names = 1)
```

Data may also be imported into RStudio by navigating from Environment » Import Dataset.

4. Check the data dimensions: The data should show samples in rows and variables (e.g., proteins) in columns:

```
dim(data)
```

3 Methods

3.1 The Mechanics of mixOmics

3.1.1 What Do the Methods in `mixOmics` Aim to Achieve?

The `mixOmics` package employs multivariate techniques that aim to identify biological or technical patterns in data sets by reducing the dimensions of the data. *Dimension reduction* techniques consist of constructing artificial variables (called *components* or *variates*) from the original variables so that the variance in a data set (with PCA) or the covariance (*see* **Note 6**) between two data sets (with PLS) is maximized.

These "summary" components are constructed sequentially, with the first component describing the largest source of variance, or covariance, in the data, and each subsequent component explaining less information. Statistically, these components are defined as *orthogonal*: They each explain information that is not shared by the other components. The analyst can then choose to retain only those first few components that identify the largest sources of (co)-variance in the data set(s), hence moving from describing a data set with thousands of proteins to, for example, two or three components. By doing so, we assume that the discarded components contain noise or information that is irrelevant, or less relevant, to the biology of interest. Samples are then represented on a smaller space spanned by these components.

3.1.2 How Are These Components Calculated?

Components are *linear combinations* of the protein abundances. For example, for the first principal component (PC) in PCA, each protein is weighted so as to explain as much variance in the data as possible. Each weight corresponding to each protein is consistent across all samples. These weights are stored in *loading vectors.* Each protein abundance is multiplied by its assigned weight, then all weighted proteins are summed to produce a PC score for each sample.

As an example of a linear combination, imagine we have three protein abundances for "sample 1" (i.e., 5, 162, 60). Each protein abundance is multiplied by a corresponding loading weight in the loading vector (i.e., 0.06, 0.89, 0.46). The weighted amounts are then summed to obtain a final score: $(0.06 \times 5) + (0.89 \times 162) + (0.48 \times 60) = 172.08$. Thus sample 1 has a score of 172.08 on this component.

The second PC attempts to explain some of the remaining variance present in the data that was not explained by the first component, and the weights assigned to each protein (across all samples) will differ from those assigned for the first component. The weighted scores are summed to produce the second component score. This process is repeated for the remaining components.

Thus, each sample can be described by a score on each component. If we consider the first two components, we can *project* each sample into the coordinate space spanned by those components to visualize how the samples may cluster according to the data set variance.

For the PLS methods, components are calculated in a similar manner, but the components are defined to maximize the covariance between the two data sets. In PLS-DA, the components are calculated to maximize the discrimination between known groups of samples.

3.1.3 How Are the Loading Vectors Calculated?

We use *matrix decomposition* techniques to solve the set of linear equations required to obtain the loading weights for each component. Matrix decomposition techniques can divide a matrix into constituent parts to facilitate complex matrix operations, often using Singular Value Decomposition (SVD, [18]).

3.1.4 How Do We Identify Important Variables?

For each component, a large loading value (weight) indicates a variable that is important to define the component, while a loading value close to zero indicates a less important variable. Variables with weights of the same sign and with a large absolute value are likely to be highly positively correlated, while variables with different signs but with a large absolute value are likely to be negatively correlated.

To identify important proteins, we employ penalties such as lasso (Least Absolute Shrinkage and Selection Operator [19]) that shrinks many of the proteins' weights in the loading vectors to zero while other weights assigned to important proteins are increased. These penalized loading vectors are called *sparse*, as they contain many zeros corresponding to irrelevant variables regarding the aim of the method. Thus, the components are calculated based on a small number of important variables, and in addition, we can also identify a molecular signature, i.e., variables with non-zero coefficients.

3.2 Interpreting Graphical Outputs

The `mixOmics` package emphasizes the use of sample and variable plots, not only to understand the relationships between omics data but also the correlations between the different biological entities. These plots are especially relevant when data sets contain information from different biological sources.

3.2.1 Sample Plots

To visualize how samples cluster according to the main sources of variation or covariance in the data, the samples can be projected into the space defined by the retained components, in a simple scatterplot of two or three components. These plots allow us to visualize any similarities between samples, or differences between, for example, experimental and control samples (*see* **Note 7**). Because the components are artificial variables, the biological meaning of these components must be assessed jointly by the data analyst and biologist. Examples of sample plots are provided with each method described in this chapter.

3.2.2 Variable Plots

The mixOmics package also provides several graphical outputs to visualize the relationships between variables. These include correlation circle plots, loading plots, and also heatmaps and network representations that are based on similarity matrices.

3.2.3 Correlation Circle Plots

Correlation circle plots visualize the *contribution of each variable* in defining each component, and the *correlation between variables* in PCA. They can also represent variables of two different types when using integrative approaches (e.g., proteins and metabolites in PLS).

Protein plot coordinates are obtained by calculating the correlation between each original protein and their associated component, as shown in Fig. 1. Note that the variables need to be centered and scaled in the method to fully interpret these plots [20].

When performing data integration (e.g., with PLS or multi-block PLS), variables of different types can be overlaid on the correlation circle plots to visualize cross-correlation.

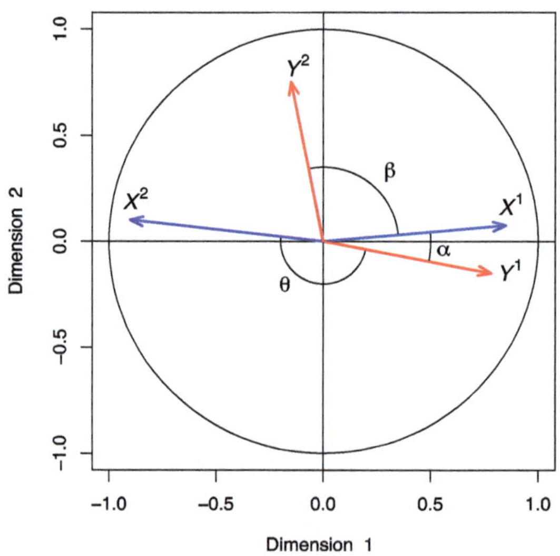

Fig. 1 Correlation circle plot interpretation. Assume the data are centered and scaled (arguments provided in the method function). The coordinate of each protein (indicated by the tip of the arrow) is defined as the correlation between the original protein abundance and a given component (here component/dimension 1 on the *x*-axis, and component/dimension 2 on the *y*-axis). Each coordinate value indicates how influential the protein is to define a given component, as informed by the length of the arrow (close to the large circle of radius 1). Correlations between two variables can be assessed by looking at the cosine angle value between two variables. If the angle is sharp (e.g., between X^1 and Y^1), the correlation is strong and positive, if the angle is obtuse (X^1 and X^2) the correlation is strong and negative, and if the angle is right, the correlation is null. (Y^1 and Y^2). In practice, in the package we do not show the whole arrows but the tips of the arrows to declutter the graphic

3.2.4 Loading Plots

A bar plot represents the loading weight of each variable ranked from the most important (largest weight, bottom of the plot) to the least important (smallest weight, top of the plot). Positive or negative signs often reflect how specific groups of samples are separated from other groups of samples when projected on a given component. Bars can be colored to indicate whether a particular variable is over/under abundant in known groups of samples in supervised analyses.

4 Case Studies

As an illustrative example for these methods, we consider the `breast.TCGA` study available directly in `mixOmics`, which includes a subset of the full data set from The Cancer Genome Atlas [21]. It includes 150 breast cancer samples comprising three subtypes: 45 Basal, 30 Her2, and 75 Luminal A. The omics data sets are divided into training and test sets, with the training sets including the expression levels of 520 mRNAs, 184 miRNAs, and the abundances of 142 proteins. This chapter will only use the training set but further tutorials are available on www.mixOmics.org to illustrate prediction on the test set.

4.1 Unsupervised Exploration of One Data Set: PCA and sPCA

PCA is used to observe the major trends or patterns in the data and discover whether the samples cluster according to biological conditions of interest. Sparse PCA is useful to identify which proteins contribute the most to explaining the variance in the data (*see* **Note 8**). We will analyze the proteomics data from the `breast.TCGA` training set. The following R code can be adapted for your own data set (*see* Subheading 2.3):

1. Load the data from the package:

```
data(breast.TCGA)
```

2. Specify ***X*** as the data set. We print out a summary of the dimensions of the data set to check the input data:

```
X <- breast.TCGA$data.train$protein
dim(X)
```

   The code `dim(X)` indicates that there are 150 samples and 142 proteins.

3. Run the full PCA on the protein data (X):

```
protein.pca <- pca(X, scale = TRUE, ncomp = 10)
```

Although we use `ncomp = 10` here for explanatory purposes, the default values for `pca` include (type `?pca` in the R console to see the help file):

(a) `ncomp = 2`: The first two principal components are calculated and used for graphical outputs,

(b) `center = TRUE`: Data are centered (each variable has a zero mean),

(c) `scale = FALSE`: Data are not scaled. If scale = TRUE is employed, each variable has a variance of 1.

4. Examine the scree (bar) plot of the proportion of explained variance relative to the total amount of variance in the data for each PC, to choose how many components to retain (*see* Fig. 2). The rule of thumb is not so much to set a hard threshold based on the cumulative proportion of explained variance (as this is data-dependent), but to observe when a drop, or elbow, appears on the scree plot. The elbow indicates that the remaining variance is spread over many PCs and is not relevant in obtaining a low dimensional "snapshot" of the data.

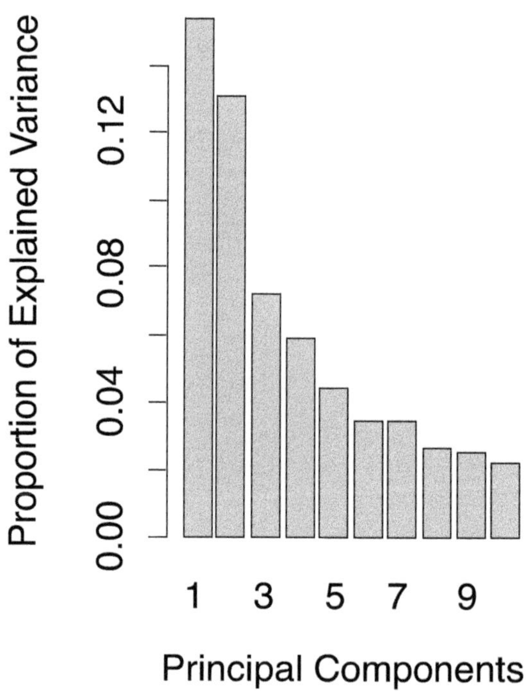

Fig. 2 PCA scree plot applied on the `breast.TCGA` protein data set. This plot shows the proportion of explained variance relative to the total amount of variance in the data for each PC. The "elbow" appears after the first two components, suggesting that these two may be sufficient to summarize the main sources of variation in the data

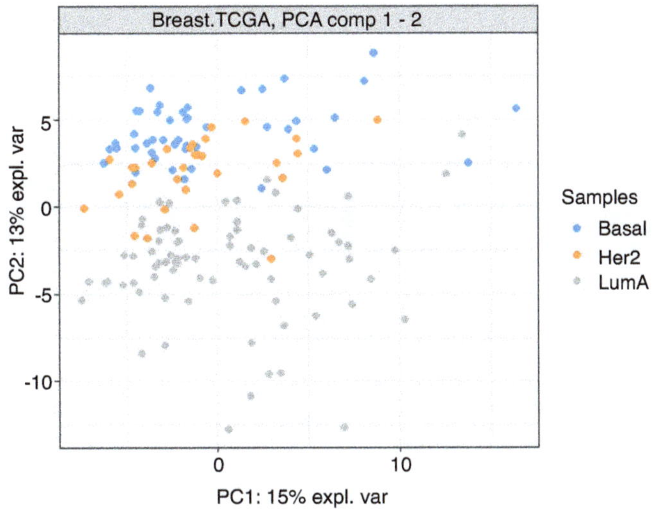

Fig. 3 PCA sample plot applied on the `breast.TCGA` proteomics data set. Samples are projected onto the first two principal components and colored after the analysis according to their known cancer subtype to assist in interpretation

Choosing the number of components to retain can also rely on our ability to interpret the components:

```
pca.scree <- tune.pca(X, ncomp = 10, scale = TRUE)
```

5. Plot the samples on the first two components (*see* Fig. 3). Although this method is unsupervised and does not take the subtype of breast cancer into account, we can color the samples according to their subtype (argument `group = ()`) to aid in interpretation (although *see* **Note 7** for a word of caution). To visualize sample names, set `ind.names = TRUE`:

```
plotIndiv(protein.pca, ind.names = FALSE,
    group = breast.TCGA$data.train$subtype, pch = 20,
    legend = TRUE, legend.title = "Samples",
    title = "Breast.TCGA, PCA comp 1 - 2",
    size.title = rel(0.8))
```

The sample plot in Fig. 3 suggests that the first two principal components of the PCA can distinguish the Luminal A breast cancer subtype (gray) from the Her2 (orange) and basal (blue) subtypes.

6. Generate a correlation circle plot on the first two components to identify the proteins that contribute the most toward defining each component, and thus explaining the largest sources of variation in the data (*see* Fig. 4), as well as their correlation. For ease of visualization, we can specify a cutoff (argument `cutoff = 0.5`) to show only those proteins that correlate with each

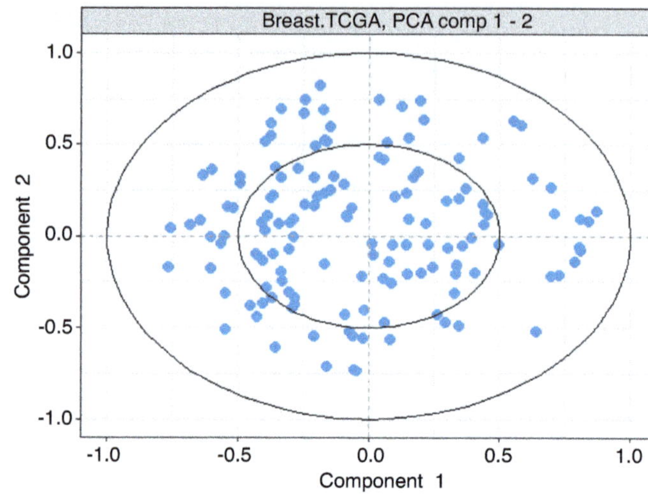

Fig. 4 PCA correlation circle plot applied on the `breast.TCGA` proteomics data set. The plot enables us to identify proteins that are important in defining the components and the variance in the data, and their correlation (for simplicity, their ID names are not shown here)

component at greater than 0.5. The alternative is to use a sparse PCA:

```
plotVar(protein.pca, var.names = FALSE, pch = 20,
        title = "Breast.TCGA, PCA comp 1 - 2")
```

The proteins closest to the outer circle along the *x*- (Component 1) and *y*- (Component 2) axes are highly correlated with the first and second components respectively and contribute the most toward explaining the primary sources of variation in the data, while those in the inner circle do not contribute much explanatory power. Clusters of proteins indicate that they are strongly correlated.

7. Run a sparse PCA. In the `spca()` function we specify the number of proteins `keepX()` to select on each component. Here, we arbitrarily choose to select 20 proteins for the first component, and 15 for the second component. You can choose more proteins for further exploratory analyses, or fewer proteins for biomarker discovery. A tuning function is also available to decide on the optimal number of variables to select on each component (*see* **Note 4**):

```
protein.spca <- spca(X, keepX = c(20, 15), ncomp = 2)
```

The default values of `spca()` are (*see* results from `?spca()`):

(a) `center` = `TRUE`: Data are centered (each variable has a zero mean),

(b) `scale` = `TRUE`: Data are scaled (each variable has a variance of 1).

8. Plot the samples projected onto the space spanned by the first two sparse principal components (*see* Fig. 5):

```
plotIndiv(protein.spca, ind.names = FALSE,
          group = breast.TCGA$data.train$subtype,
          legend = TRUE, legend.title = "Samples", pch = 20,
          title = "Breast.TCGA, sPCA comp 1 - 2",
          size.title = rel(0.8))
```

The sample plot in Fig. 5 shows the Luminal A breast cancer subtype (gray) separating from the Her2 (orange) and Basal (blue) subtypes, suggesting that there are differences in protein expression between Luminal A and the other breast cancer subtypes.

9. Generate a correlation circle plot to identify the proteins that contribute the most toward defining each component, and thus explaining the largest sources of variation in the data, as well as their correlation (*see* Fig. 6):

```
plotVar(protein.spca, cex = 4,
        title = "Breast.TCGA, sPCA comp 1 - 2")
```

From the correlation circle plot (*see* Fig. 6), we observe proteins such as Caveolin-1, FOXO3a, and EGFR contributing

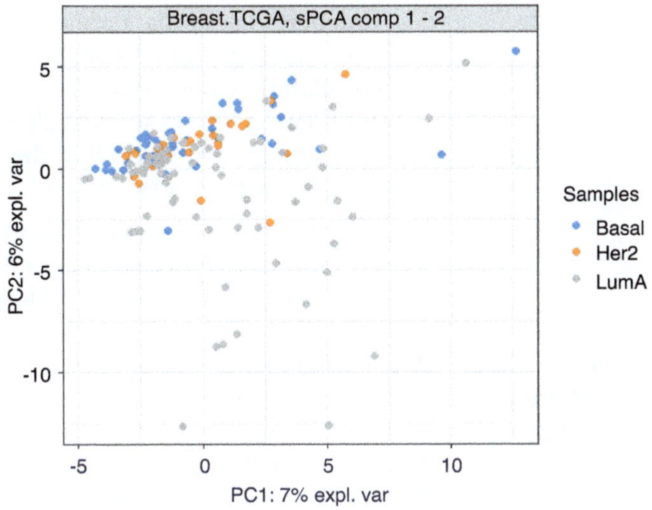

Fig. 5 sPCA sample plot applied on the `breast.TCGA` protein data set representing the samples on components 1 and 2. The sparse PCs are calculated based on the top 20 (15) most influential proteins for component 1 (2)

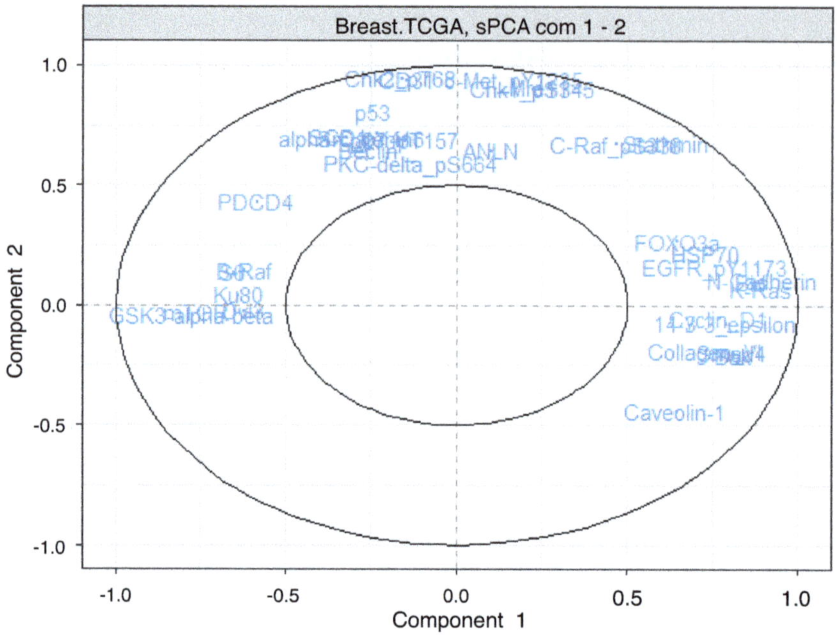

Fig. 6 sPCA variable plot applied on the `breast.TCGA` protein data set representing the proteins on components 1 and 2

positively to component 1 (*x*-axis), and GSK3 and B-Raf contributing negatively. Meanwhile, proteins such as p53, c-Met, and ANLN contribute positively to component 2 (*y*-axis).

10. Both the sample and variable plots can be examined together to interpret the results. In this example, the main sources of variability in the data are due to Luminal A subtype samples which are segregated along component 1. Overlaying with the correlation circle plot, the samples are likely to have high abundances of proteins such as Caveolin-1 and FOXO3a—such a statement should be checked with additional inspection or visualization of the data.

11. The `mixOmics` package offers tuning functions to statistically optimize the choice of the number of parameters (components and variables) to retain (*see* **Note 4**).

4.2 Supervised Exploration of One Data Set—PLS-DA and sPLS-DA

PLS-Discriminant Analysis (PLS-DA) aims to discriminate samples based on their known outcome category. Similar to PCA, PLS-DA reduces high-dimensional data to a smaller number of "summary" variables (components). The sparse variant (sPLS-DA, [22]) enables the selection of the most predictive or discriminating features in the data to classify the samples.

We consider the same proteomics data set as in PCA, but we add to the analysis information about the cancer subtypes. We apply sparse PLS-DA to identify the proteins that best discriminate the three cancer subtypes. A classical PLS-DA can also be considered if the aim is mainly to separate samples according to the outcome without performing variable selection. The following R code can be adapted for your own data set (*see* Subheading 2.3):

1. Specify $\boldsymbol{X}$ as the data set. The outcome $\boldsymbol{Y}$ is set as a factor that indicates the class of each sample. We print out a summary of the number of samples per group, the dimensions of the data set, and the number of samples in $\boldsymbol{Y}$ to check all input data are correct:

```
X <- breast.TCGA$data.train$protein
Y <- breast.TCGA$data.train$subtype
length(Y); summary(Y)
dim(X)
```

The code `dim(X)` and `length(Y)` indicate that there are 150 samples and 142 proteins.

2. Run the sPLS-DA. We specify the selection of 30 proteins on the first component and 20 on the second component of the PLS-DA (arbitrary choice):

```
protein.splsda <- splsda(X, Y, keepX = c(30,20))
```

This code will employ the following default values:

(a) `ncomp` = 2: The first two PLS components are calculated and are used for graphical outputs,

(c) `scale` = `TRUE`: Data are scaled (default parameter advised in PLS-DA).

3. Plot the samples on the first two components to visualize how they cluster. We include `ellipse` = `TRUE` to further visualize the sample grouping according to the outcome Y (*see* Fig. 7):

```
plotIndiv(protein.splsda,ellipse = TRUE, var.names=FALSE,
          pch=20, legend = TRUE, legend.title = "Samples",
          title = "Breast.TCGA, sPLS-DA comp 1 - 2",
          size.title = rel(0.8))
```

In the sample plot, the first component (across the x-axis) appears to separate the Basal (blue) and Luminal (gray) breast cancer subtypes, while the second component (across the y-axis) appears to separate out the Her2 (orange) subtype. Her2 reveals a higher variability than the other two cancer subtypes.

4. Plot the proteins identified by the sPLS-DA onto the first two components (*see* Fig. 8):

```
plotVar(protein.splsda, cex=4,
        title = "Breast.TCGA, sPLS-DA comp 1 - 2")
```

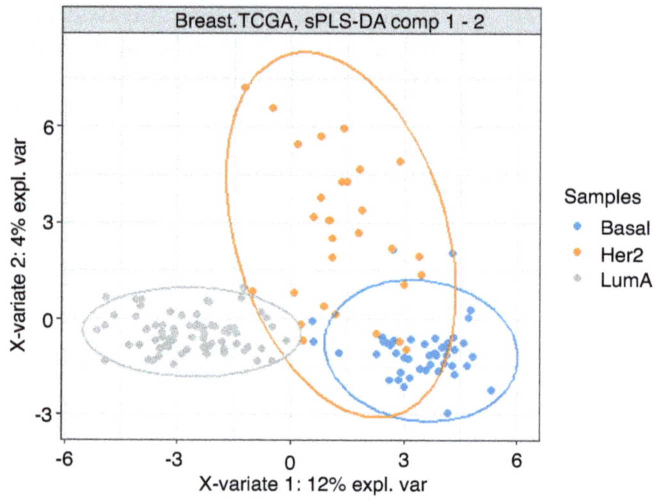

Fig. 7 sPLS-DA sample plot applied on the `breast.TCGA` protein data set, representing the samples on components 1 and 2

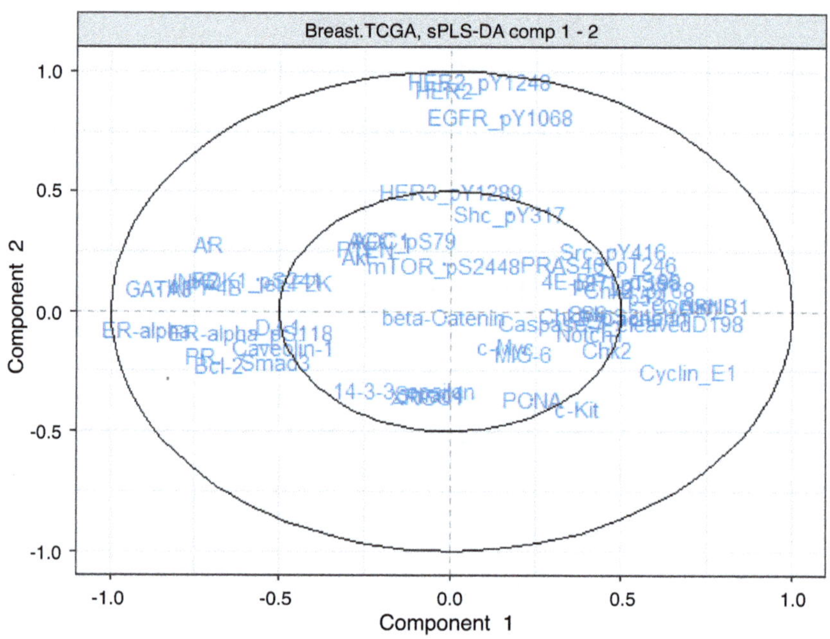

Fig. 8 sPLS-DA variable plot applied on the `breast.TCGA` protein data set. The contribution of selected proteins and their correlation structure can be visualized

In the correlation circle plot, proteins such as Cyclin-E1 and Cyclin-B1 contribute positively to component 1, while Smad3, Bcl-2, and ER-alpha also contribute toward component 1, but negatively. Proteins such as HER2, HER2_pY1248, and EGFR_1068 all contribute to component 2, and are also positively correlated to each other.

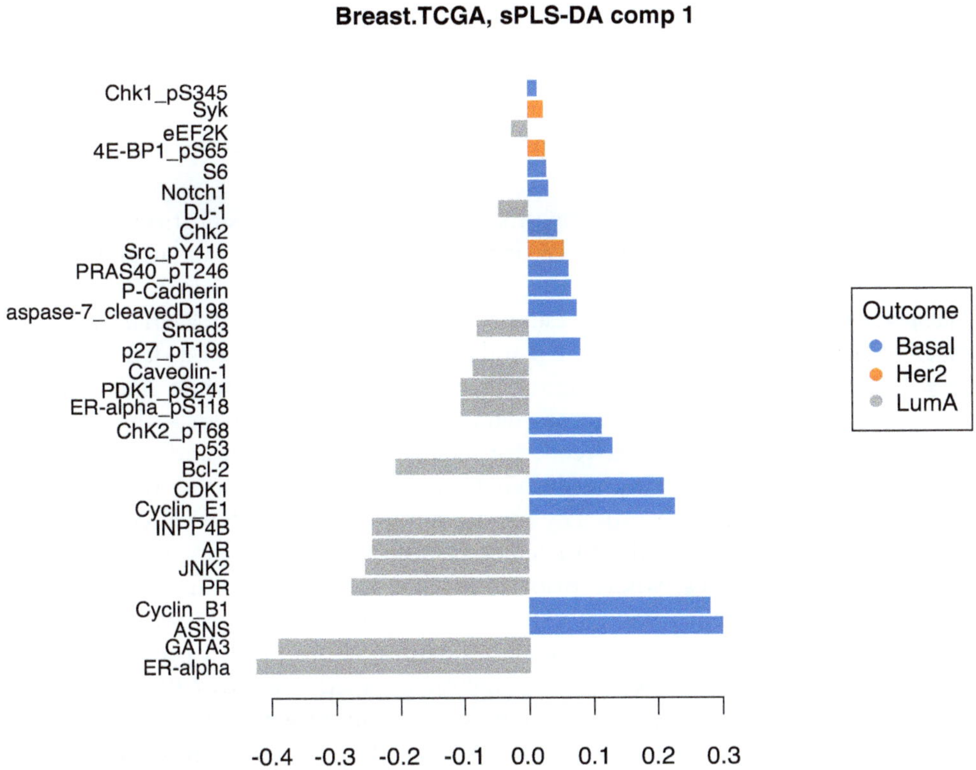

Fig. 9 sPLS-DA loading plot applied on the `breast.TCGA` protein data for component 1. Colors indicate the class for which a particular protein is maximally expressed according to the median

5. Use the `plotLoadings()` function to examine the loading weight of each variable selected by the sparse PLS-DA, ranked from the most important (largest weight in absolute value, bottom of the plot) to the least important (smallest weight in absolute value, top of the plot). Here the color represents which sample group has maximum median abundance for each protein. (*see* Fig. 9):

```
plotLoadings(protein.splsda, comp = 1, method = 'median',
             contrib = 'max', size.title = rel(0.7),
             title = "Breast.TCGA, sPLS-DA comp 1")
```

6. The weights for each protein selected on component 1 (`comp = 1`) can also be examined using the following code:

```
selectVar(protein.splsda, comp=1)
```

The loading plot and selected variable output suggest that, for example, ER-alpha and GATA3 are influential proteins for the Luminal A breast cancer subtype, while ASNS and Cyclin_B1 are influential proteins for the Basal subtype.

7. Both sample and variable plots can be examined together to interpret the results. The results suggest that proteins such as Cyclin-E1 and Cyclin-B1 are associated with the Luminal A subtype. Meanwhile, proteins such as HER2, HER2_pY1248, and EGFR_pY1068 are associated with the Her2 breast cancer subtype.
8. It is also possible to examine the classification and prediction performance of the model using cross-validation (*see* **Note 4**).

4.3 Unsupervised Exploration of Two Data Sets—PLS and sPLS

Projection to Latent Structures, also called Partial Least Squares regression [7, 8], reduces the dimensions of a data set into summary components (here called *variates*) while maximizing their covariance across data sets. Thus PLS can describe the common pattern between, for example, proteomics and metabolomics by identifying underlying factors in both data sets that best describe this pattern. The most common use of PLS is called *regression* PLS to model and predict response variables from the predictor variables, where prior biological knowledge indicates which type of omics data is expected to explain the other type (e.g., we may expect messenger RNA expression to predict protein abundance). The other type of analysis is called *canonical* PLS when the relationship between data types is unknown, or to be explored.

Here we illustrate sparse PLS to examine how messenger RNA expression may predict protein abundance for the breast cancer samples. We will analyze the transcriptomics and proteomics data from the `breast.TCGA` training set. The following R code can be adapted for your own data set (*see* Subheading 2.3):

1. As we expect messenger RNA to predict protein expression, specify the messenger RNA data set as X and the protein data set as the outcome data set Y. Print out a summary of the dimensions of the data sets to check all input data:

```
X <- breast.TCGA$data.train$mrna
Y <- breast.TCGA$data.train$protein
dim(X)
dim(Y)
```

The code `dim(X)` and `dim(Y)` indicate that there are 150 samples, 200 messenger RNA and 142 proteins.

2. Run the sPLS analysis. In this analysis, we have arbitrarily chosen 25 mRNA and 20 proteins on the first variate and 15 mRNA and 10 proteins on the second variate, as indicated in the `keepX` and `keepY` arguments:

```
protein.spls <- spls(X,Y, keepX = c(25, 15),
                    keepY = c(20,10))
```

This code employs the following default values:

(a) ncomp = 2: The first two PLS components are calculated and are used for graphical outputs,

(b) scale = TRUE: Data are scaled (variance = 1, strongly advised for a PLS integrative model),

(c) mode = "regression": By default a PLS regression mode should be used.

3. Plot the samples on the first two components (*see* Fig. 10). Although this is an unsupervised technique, where no information about sample grouping is used in the method, we can color the samples by their breast cancer subtype for biological interpretation:

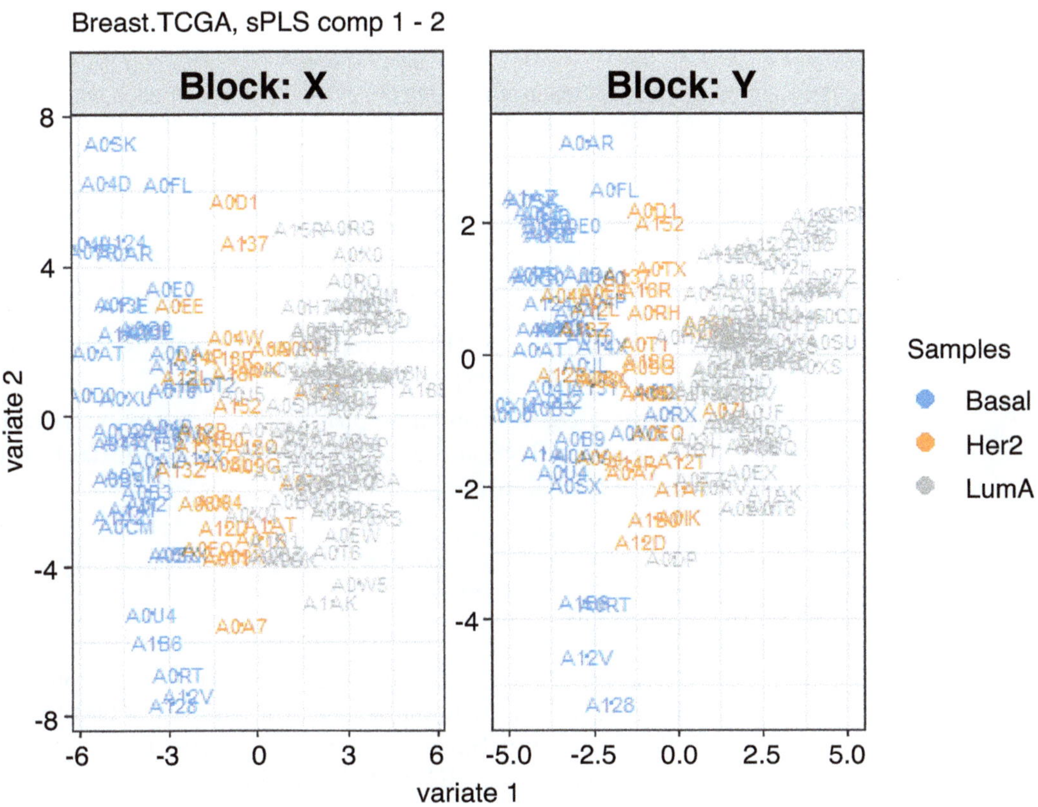

Fig. 10 sPLS sample plot applied on the `breast.TCGA` data set, representing the proteins and transcripts on variates 1 and 2, calculated for both the mRNA (block: ***X***) and protein (block:***Y***) data sets

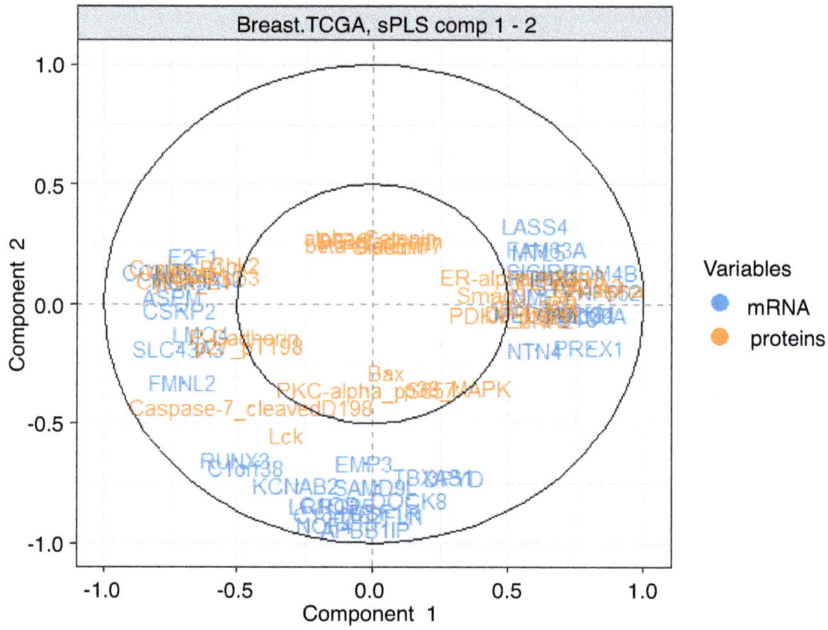

Fig. 11 sPLS variable plot applied on the `breast.TCGA` data. The plot highlights correlations *within* selected proteins and selected mRNA, and correlations *between* proteins and mRNA on each dimension of the sPLS

```
plotIndiv(protein.spls, legend = TRUE,
          group= breast.TCGA$data.train$subtype,
          legend.title = "Samples", size.title = rel(1),
          title = "Breast.TCGA, sPLS comp 1 - 2")
```

Variate 1 in the sPLS sample plot appears to distinguish between cancer subtypes for both data types.

4. Generate a correlation circle plot for the first two variates for the selected mRNA and proteins (*see* Fig. 11) to visualize their correlation structure:

```
plotVar(protein.spls, cex=c(4,4),
        legend =  c("mRNA", "proteins"),
        legend.title = "Variables",
        title = "Breast.TCGA, sPLS comp 1 - 2")
```

The correlation circle plot suggests that, for example, proteins GATA3 and ER_alpha and the mRNA KDM4B and ZNF552 are contributing positively to variate 1, and proteins cyclin_B1 and cyclin_E1 and the mRNA CCNA2 and ASPM are contributing negatively to variate 1.

Moreover, the proteins ER_alpha and GATA3 are positively correlated with mRNA KDM4B and and ZNF552, while the proteins cyclin_B1 and cyclin_E1 are positively correlated with mRNA CCNA2 and ASPM.

5. The selected proteins and mRNA on variate 1 (comp = 1) can be extracted with the following code:

```
selectVar(spls.result, comp=1)
```

6. Both sample and variable plots can be examined together to interpret the results. For example, the results suggest that proteins such as GATA3 and ER_alpha correlate with messenger RNA KDM4B and ZNF552, and proteins Cyclin_B1 and cyclin_E1 correlate with messenger RNA CCNA2 and ASPM to distinguish between breast cancer subtypes.
7. Other types of plots can be examined in the package [20], such as relevance networks (?network) and clustered image maps (?cim).
8. The mixOmics package offers tuning functions to statistically optimize the choice of the number of parameters (components and variables) to retain (*see* **Note 4**).

4.4 Supervised Exploration of More Than Two Data Sets: Multi-Block sPLS-DA

Integrative methods for more than two data sets are also available in mixOmics. The methods use variants of the PLS method, called regularized and sparse generalized canonical correlation analysis (RGCCA and SGCCA from [23] and [24]) that extend the methods PLS and PLS-DA. In this chapter we consider the multi-block sparse PLS-DA extension called **D**ata **I**ntegration **A**nalysis for **Bio**marker discovery using **L**atent variable approaches for **O**mics (DIABLO, [9]), which can be employed when the aim is to understand the relationships between omics and to identify a molecular signature that explains the outcome of interest.

In DIABLO we aim to maximize the sum of covariances or the weighted sum of covariances between pairs of components associated with two data sets. Thus we need to create a design matrix to specify which pairs of data sets we are especially interested in maximizing the covariance. This can be based on prior knowledge ("I expect mRNA and miRNA to be highly correlated") or data-driven (e.g., based on a prior analysis, such as pairwise PLS shown above that examines the agreement between data sets. Values in the design matrix may range between 0 (no relationship is sought) to 1 (agreement to maximize). Since DIABLO is supervised, we are interested in discriminating the sample groups irrespective of the design (*see* **Note 9**).

We will analyze the transcriptomics, epigenomics, and proteomics data from the breast.TCGA training data set to illustrate the integration of omics data for breast cancer subtype discrimination. The following R code can be adapted for your own data sets (*see* Subheading 2.3):

1. Specify $\boldsymbol{X}$ as a list of omics data matrices, and the outcome $\boldsymbol{Y}$ as a factor indicating the class membership of each sample. Each data frame in $\boldsymbol{X}$ should be named as we will match these names with the `keepX` parameter for the sparse method. Print out a summary of the dimensions of the data sets to check all input data:

```
data(breast.TCGA)
# extract training data,  name each data frame
# store as a list
X <- list(mRNA = breast.TCGA$data.train$mrna,
    miRNA = breast.TCGA$data.train$mirna,
    protein = breast.TCGA$data.train$protein)

# outcome
Y <- breast.TCGA$data.train$subtype

summary(Y)
```

Similar to previous methods, the method `block.splsda` uses the following default values:

(a) `ncomp` = 2: The first two sets of components are calculated and are used for graphical outputs,

(b) `design`: By default a full design is implemented,

(c) `scale` = `TRUE`: Data are scaled (variance = 1, strongly advised here for data integration).

2. Set the design matrix. For this analysis, we chose a weight of 1 between all data sets, meaning that we want to maximize the correlation between data sets:

```
design <- matrix(1, ncol = length(X), nrow = length(X),
                 dimnames = list(names(X), names(X)))
diag(design) <- 0
design
```

3. Set the number of components and number of variables to retain for each block. Although as previously mentioned, `mixOmics` provides methods to optimize these parameters, for simplicity here we choose these values arbitrarily, retaining 2 components, and a selection of 12 mRNA, 8 miRNA, and 12 proteins for the first component, and 8 mRNA, 6 miRNA, and 10 proteins for the second component. The method will select those variables that best maximize the covariances between data sets. The `keepX` values should be set as a list, with names matching the data set names:

```
ncomp <- 2
list.keepX <- list(mRNA = c(12, 8), miRNA = c(8,6),


                protein = c(12, 10))
```

4. We then run the DIABLO method:

```
tcga.diablo <- block.splsda(X, Y, ncomp = ncomp,
                            keepX = list.keepX,
                            design = design)
```

5. The variables selected by the method can be extracted with the function `selectVar()`, for example, in the protein block:

```
selectVar(tcga.diablo, block = 'protein', comp = 1)
```

6. Once the model is run, we can then use plots to visualize how the samples group or separate, and which variables are correlated and associated with cancer subtypes. Firstly, we use the diagnostic plot `plotDiablo()` (*see* Fig. 12) to assess whether the correlations between components from each data set were maximized as we requested in the design matrix. We specify the dimension to be assessed with the `ncomp` argument:

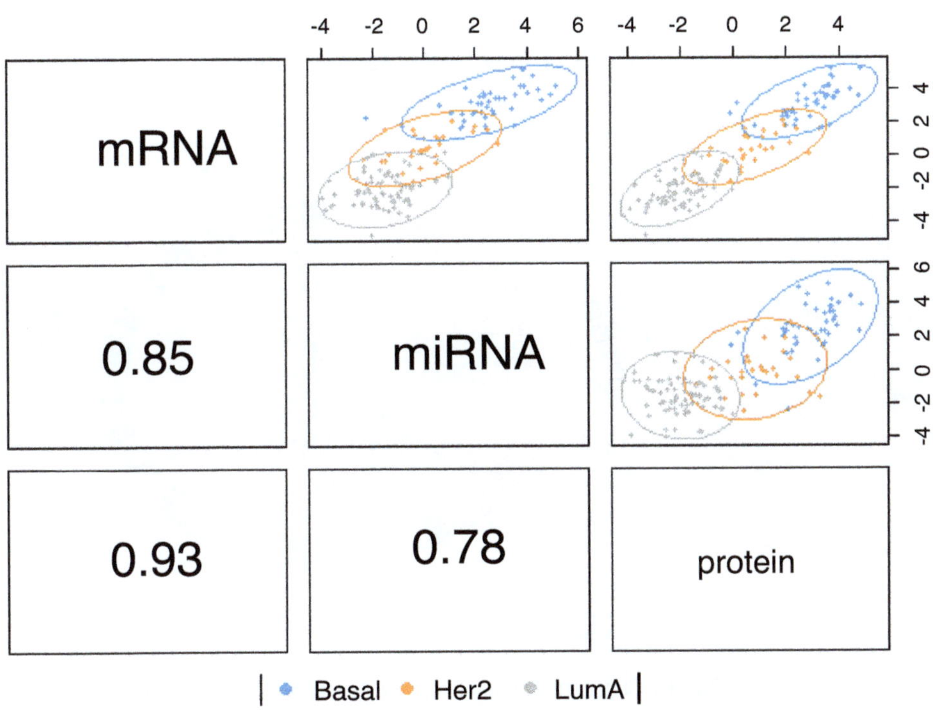

Fig. 12 Diagnostic plot applied on the `breast.TCGA` study for the first component with multi-block sPLS-DA. Samples are represented across each data set (mRNA, miRNA, and protein) and colored by breast cancer subtype (Basal, Her2, and LumA). 95% confidence ellipse plots are represented. The bottom left numbers indicate the correlations between the first components from each data set

```
plotDiablo(diablo.tcga, ncomp = 1)
```

In the diagnostic plot, the first components associated with mRNA expression and protein abundance are highly correlated (correlation of 0.93), indicating that the method can extract a high level of agreement between these data sets. In addition, we can visualize some separation between the sample groups. Had we observed lower correlation values, we could revise our design matrix and start a new analysis.

7. For variable visualization, in addition to the correlation circle plots and loading plots described in the previous case studies, circos plots can be used to represent the correlations between variables of different types, represented on the side quadrants. Both within and between connections between blocks can be displayed, and expression levels of each variable according to each class (argument `line = TRUE`). A `cutoff` argument can be included to visualize only selected variables above a specified correlation threshold in the multi-omics signature (*see* Fig. 13). Here the cutoff value is set to 0.7 in absolute value:

```
circosPlot(diablo.tcga, cutoff = 0.7, size.labels = 1.5,
        title = 'Breast.TCGA, DIABLO comp 1 - 2')
```

On the circos plot we observe that correlations > 0.7 are between a few mRNAs and some proteins, whereas the

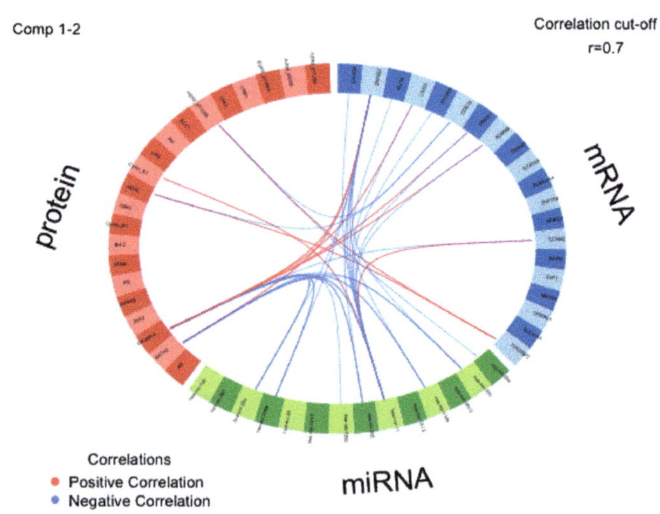

Fig. 13 Circos plot applied on the `breast.TCGA` study. Only the selected variables with a correlation greater than 0.7 (names indicated on the side quadrants) are shown. The internal connecting lines show the positive (negative) correlations in red (blue). This plot enables us to visualize cross-correlations between data types, and the nature of these correlations (positive or negative)

majority of strong (negative) correlations are observed between miRNA and mRNA or proteins.

8. Relevance networks (`?network`) and clustered image maps (`?cim`) are also available in `mixOmics` as complimentary methods of variable visualization.
9. It is also possible to examine the classification and prediction performance of the model using cross-validation (*see* **Note 4**).

5 Notes

1. http://cancergenome.nih.gov/.
2. https://bioconductor.org/packages/release/bioc/html/mixOmics.html.
3. https://www.xquartz.org/.
4. Although this chapter arbitrarily chooses the number of components and variables to retain for each analysis, the `mixOmics` package has options for tuning both these parameters based on a statistically optimal model which yields the lowest possible error, which can be useful when the biological aim is to obtain a molecular signature. More information can be found at http://mixomics.org/.
5. It is strongly recommended to save all R command lines into a script (in RStudio: File › New File › Script), rather than typing code into the console. Using a script will allow modification and re-run of the code at a later date. The best reproducible coding practice is to use `Rmarkdown` or `R Notebook` (in RStudio: File › New File › R Markdown › R Notebook4) as a transparent and reproducible way of analyzing data [25].
6. Covariance and correlation both measure the linear relationship between two numerical variables. Covariance indicates how these variables change together, or, the direction of the relationship: A positive relationship is indicated by an increase (or decrease) in the measurement of *both* variables. If one variable increases and the other variable decreases, the association is negative. If one variable increases but the other does not change, there is no relationship between the variables. Because variables may be measured on different scales, the covariance changes when the scale changes, thus covariance does not indicate the strength of the relationship. Correlation *standardizes* the scales of the variables, thus allowing comparison between two different variables, and indicating the strength or magnitude of the relationship. A correlation of −1 indicates

a strong negative relationship, 0 no relationship, and +1 a very strong relationship.

7. However, plots can be misleading. In unsupervised methods (e.g., PCA, PLS) the user can choose to add colors and symbols to the samples according to a phenotype of interest. This may result in mis- or over-interpretation about sample clustering. Sometimes it is best to examine a plot with a single color and symbol first, to critically assess clusters of samples.
8. PCA may not be the ideal choice for analysis when: (*i*) There are too many missing values. This can occur in proteomics data, and can be identified on a scree plot, where the amount of explained variance does not decrease with the components. (*ii*) When too many noisy variables contribute to the variance. (*iii*) When samples are not independent (e.g., time course data, repeated measures) and the subject variation is greater than the time/repeated treatment variation. (*iv*) When variance is not an appropriate measure to explain information in your data. In this case, Independent Components Analysis (ICA, [26]) may be preferred. See [27] for an example of ICA performed on MALDI-TOF data.
9. For more information, see [9], the mixOmics bookdown manual https://mixomicsteam.github.io/Bookdown/ and tutorials on www.mixOmics.org.

Acknowledgements

KALC was supported in part by the National Health and Medical Research Council (NHMRC) Career Development fellowship (GNT1159458).

References

1. Zhang W, Li F, Nie L (2010) Integrating multiple 'omics' analysis for microbial biology: application and methodologies. Microbiology 156(2):287–301. https://doi.org/10.1099/mic.0.034793-0
2. Joyce AR, Palsson BØ (2006) The model organism as a system: integrating 'omics' data sets. Nat Rev Mol Cell Biol 7(3):198–210. https://doi.org/10.1038/nrm1857
3. Gomez-Cabrero D, Abugessaisa I, Maier D, Teschendorff A, Merkenschlager M, Gisel A, Ballestar E, Bongcam-Rudloff E, Conesa A, Tegnér J (2014) Data integration in the era of omics: current and future challenges. https://doi.org/10.1186/1752-0509-8-S2-I1
4. Rohart F, Gautier B, Singh A, Lê Cao KA (2017) mixOmics: an R package for 'omics feature selection and multiple data integration. PLoS Comput Biol 13(11):e1005752. https://doi.org/10.1371/journal.pcbi.1005752
5. Jolliffe I (2005) Principal component analysis. Wiley Online Library
6. Barker M, Rayens W (2003) Partial least squares for discrimination. J Chemom 17(3):166–173. https://doi.org/10.1002/cem.785
7. Wold H (1966) Estimation of principal components and related models by iterative least squares. Academic Press, New York
8. Wold S, Sjöström M, Eriksson L (2001) PLS-regression: a basic tool of chemometrics. Chemom Intell Lab Syst 58(2):109–130. https://doi.org/10.1016/S0169-7439(01)00155-1

9. Singh A, Shannon CP, Gautier B, Rohart F, Vacher M, Tebbutt SJ, Lê Cao KA (2019) Diablo: an integrative approach for identifying key molecular drivers from multi-omics assays. Bioinformatics 35(17):3055–3062. https://doi.org/10.1093/bioinformatics/bty1054
10. Shah AK, Lê Cao KA, Choi E, Chen D, Gautier B, Nancarrow D, Whiteman DC, Saunders NA, Barbour AP, Joshi V et al. (2015) Serum glycoprotein biomarker discovery and qualification pipeline reveals novel diagnostic biomarker candidates for esophageal adenocarcinoma. Mol Cell Proteomics 14(11): 3023–3039. https://doi.org/10.1074/mcp.M115.050922
11. Gavin PG, Mullaney JA, Loo D, Lê Cao KA, Gottlieb PA, Hill MM, Zipris D, Hamilton-Williams EE (2018) Intestinal metaproteomics reveals host-microbiota interactions in subjects at risk for type 1 diabetes. Diabetes Care 41(10):2178–2186. https://doi.org/10.2337/dc18-0777
12. Rohart F, Gautier B, Singh A, Lê Cao K-A (2017) mixOmics: an R package for 'omics feature selection and multiple data integration. PLoS Comput Biol 13(11):e1005752. http://www.mixOmics.org, https://doi.org/10.1371/journal.pcbi.1005752
13. O'Rourke MB, Town SE, Dalla PV, Bicknell F, Koh Belic N, Violi JP, Steele JR, Padula MP (2019) What is normalization? The strategies employed in top-down and bottom-up proteome analysis workflows. Proteomes 7(3):29. https://doi.org/10.3390/proteomes7030029
14. Wiśniewski JR, Hein MY, Cox J, Mann M (2014) A "proteomic ruler" for protein copy number and concentration estimation without spike-in standards. Mol Cell Proteomics 13(12):3497–3506. https://doi.org/10.1074/mcp.M113.037309
15. Välikangas T, Suomi T, Elo LL (2018) A systematic evaluation of normalization methods in quantitative label-free proteomics. Brief Bioinform 19(1):1–11. https://doi.org/10.1093/bib/bbw095
16. Mizuno H, Ueda K, Kobayashi Y, Tsuyama N, Todoroki K, Min JZ, Toyo'oka T (2017) The great importance of normalization of LC–MS data for highly-accurate non-targeted metabolomics. Biomed Chromatogr 31(1):e3864. https://doi.org/10.1002/bmc.3864
17. Blasco-Moreno A, Pérez-Casany M, Puig P, Morante M, Castells E (2019) What does a zero mean? Understanding false, random and structural zeros in ecology. Methods Ecol Evol 10(7):949–959. https://doi.org/10.1111/2041-210X.13185
18. Alter O, Brown PO, Botstein D (2000) Singular value decomposition for genome-wide expression data processing and modeling. Proc Natl Acad Sci 97(18):10101–10106. https://doi.org/10.1073/pnas.97.18.10101
19. Tibshirani R (1996) Regression shrinkage and selection via the lasso. J R Stat Soc Ser B Methodol 267–288. https://doi.org/10.1111/j.2517-6161.1996.tb02080.x
20. González I, Lê Cao KA, Davis MJ, Déjean S et al. (2012) Visualising associations between paired 'omics' data sets. BioData Min 5(1):19. https://doi.org/10.1186/1756-0381-5-19
21. Cancer Genome Atlas Network et al. (2012) Comprehensive molecular portraits of human breast tumours. Nature 490(7418):61–70. https://doi.org/10.1038/nature11412
22. Lê Cao KA, Boitard S, Besse P (2011) Sparse PLS discriminant analysis: biologically relevant feature selection and graphical displays for multiclass problems. BMC Bioinform 12(1):253. https://doi.org/10.1186/1471-2105-12-253
23. Tenenhaus A, Tenenhaus M (2011) Regularized generalized canonical correlation analysis. Psychometrika 76(2):257–284. https://doi.org/10.1007/s11336-011-9206-8
24. Tenenhaus A, Philippe C, Guillemot V, Lê Cao KA, Grill J, Frouin V (2014) Variable selection for generalized canonical correlation analysis. Biostatistics p kxu001. https://doi.org/10.1093/biostatistics/kxu001
25. Baumer B, Udwin D (2015) R markdown. Wiley Interdiscip Rev Comput Stat 7(3): 167–177. https://escholarship.org/uc/item/90b2f5xh
26. Comon P (1994) Independent component analysis, a new concept? Signal Process 36(3): 287–314. https://doi.org/10.1016/0165-1684(94)90029-9
27. Mantini D, Petrucci F, Del Boccio P, Pieragostino D, Di Nicola M, Lugaresi A, Federici G, Sacchetta P, Di Ilio C, Urbani A (2008) Independent component analysis for the extraction of reliable protein signal profiles from MALDI-TOF mass spectra. Bioinformatics 24(1):63–70. https://doi.org/10.1093/bioinformatics/btm533

Chapter 16

Integrating Multiple Quantitative Proteomic Analyses Using MetaMSD

So Young Ryu, Miriam P. Yun, and Sujung Kim

Abstract

MetaMSD is a proteomic software that integrates multiple quantitative mass spectrometry data analysis results using statistical summary combination approaches. By utilizing this software, scientists can combine results from their pilot and main studies to maximize their biomarker discovery while effectively controlling false discovery rates. It also works for combining proteomic datasets generated by different labeling techniques and/or different types of mass spectrometry instruments. With these advantages, MetaMSD enables biological researchers to explore various proteomic datasets in public repositories to discover new biomarkers and generate interesting hypotheses for future studies. In this protocol, we provide a step-by-step procedure on how to install and perform a meta-analysis for quantitative proteomics using MetaMSD.

Key words Proteomic software, Quantitative proteomics, Mass spectrometry data analysis, Meta-analysis, Integrating multiple differential analyses

1 Introduction

Mass spectrometry is a powerful tool that identifies and quantifies hundreds to thousands of proteins in complex biological samples. For a few decades, various mass spectrometry techniques (e.g., stable isotope-labeling methods and label-free methods) and bioinformatic tools have been developed to accurately measure protein expression changes between normal and disease groups [1–8]. These quantitative mass spectrometry techniques have been used for the discovery of protein biomarkers associated with cancer, diabetes, cardiovascular diseases, and various biological investigations [9]. However, biomedical researchers, especially those in the laboratories with limited resources, encounter several challenges in their biomarker studies due to the small number of clinical samples and limited technical capacities. Recognizing these challenges, our computational group recently developed MetaMSD (Meta-analysis for Mass Spectrometry Data), which integrates multiple

Thomas Burger (ed.), *Statistical Analysis of Proteomic Data: Methods and Tools*,
Methods in Molecular Biology, vol. 2426, https://doi.org/10.1007/978-1-0716-1967-4_16,

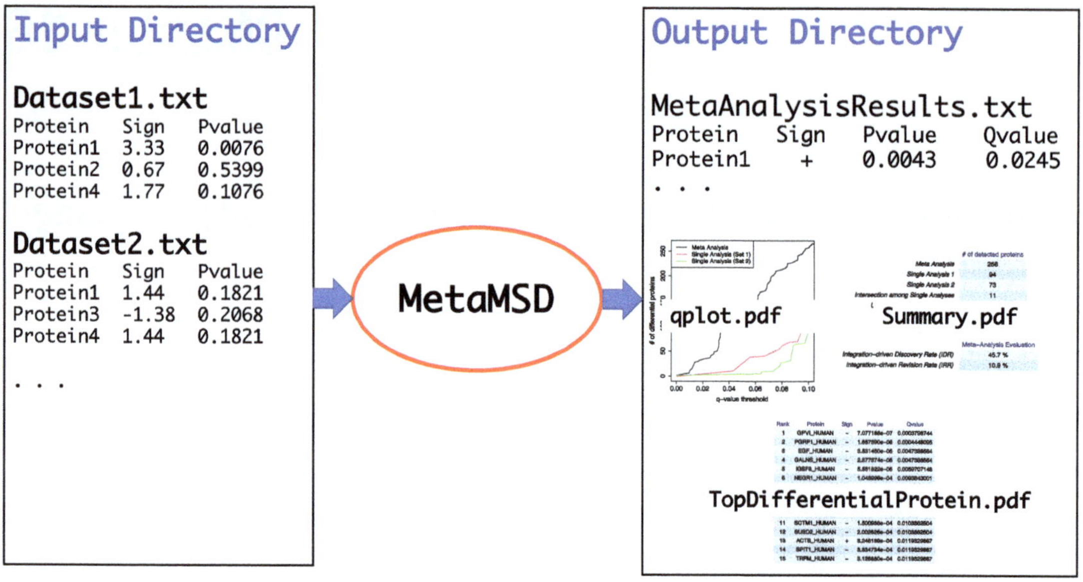

Fig. 1 MetaMSD inputs and outputs

quantitative mass spectrometry datasets [10]. Using this tool, researchers can fully utilize multiple small datasets that may be generated at different study stages (e.g., pilot and main studies).

MetaMSD integrates multiple quantitative proteomic datasets using one of the two meta-analysis techniques—either Stouffer's test (*see* **Note 1**) or Pearson's test (*see* **Note 2**). As shown in Fig. 1, MetaMSD takes multiple protein significance analysis results as input data and outputs meta-analysis results. It can also combine multiple datasets generated by various labeling techniques/instruments while controlling false discovery rates [11]. Here, we provide a detailed protocol on how to perform a meta-analysis with MetaMSD.

2 Material

2.1 Hardware Requirements

MetaMSD can be installed on a desktop computer or a laptop with one of the following operating systems: Windows, Mac OS X, or Ubuntu Linux. A minimum of 8GM RAM is recommended. The computer should have enough free disk space for raw and processed data. MetaMSD can also be installed on server, but this present protocol focuses on a desktop computer and a laptop.

2.2 Software Installation

Users need to have read and write permissions for the directory where MetaMSD will be located:

1. Install a recent version of R software: R software can be downloaded from CRAN at https://www.r-project.org/.

Downloading and installing pre-compiled binary distributions of the base system and contributed packages are recommended for beginners. MetaMSD.Rscript was tested using R v4.0.3.

2. Install a recent version of RStudio: RStudio can be downloaded from https://rstudio.com/products/rstudio/download/.
3. Create the directory where users want to locate MetaMSD. Rscript and name the directory as MetaMSD.
4. Download MetaMSD.Rscript in the MetaMSD directory: MetaMSD.Rscript is freely available at https://github.com/soyoungryu/MetaMSD/tree/master/MetaMSD.
5. Change the mode of the script file (MetaMSD.Rscript) to enable execute permission:
   (a) For Windows users, open a command prompt window (e.g., type **cmd** after selecting the Start button and choose **Command Prompt** from the list). For Mac or Linux users, open the terminal window.
   (b) Go to the directory where MetaMSD.Rscript is located. For example, if MetaMSD.Rscript is in `C:/Users/Lisa/Documents/MetaMSD`, type the following command in the command prompt or the terminal:

```
cd C:/Users/Lisa/Documents/MetaMSD
```

   (c) Then, type the following command in the command prompt or the terminal:

```
chmod a+x MetaMSD.Rscript
```

6. Type the following script in the command prompt or the terminal to check whether MetaMSD.Rscript is properly installed. This will display parameter options for MetaMSD:

```
./MetaMSD.Rscript -h
```

When the script is run for the first time, it will automatically check for the presence of its required packages and attempt to install them if they are not already installed. This can take a few minutes. If MetaMSD.Rscript is properly installed, then the following help output will be displayed as shown below (*see* **Note 3**):

```
-m CHARACTER, --metaanalysis=CHARACTER
Specify a meta-analysis test.
Either Stouffer or Pearson. [default = Stouffer]
```

```
-c NUMBER, --cutoff=NUMBER
Specify a q-value cut off. [default = 0.05]
-t NUMBER, --top=NUMBER
Specify the number of proteins in Top-N differential
protein list. [default = 15]
-i CHARACTER, --input=CHARACTER
Specify the input folder name. [default = input]
-o CHARACTER, --output=CHARACTER
Specify the output folder name. [default = output]
-h, --help
Show this help message and exit
```

2.3 Software Installation Trouble Shooting

If the script does not generate the help output, users need to manually install the following packages: **ROCR**, **gridExtra**, **gtable**, **optparse**, and **qvalue**. There are several ways to install packages. Users can choose one of the options shown below to install necessary R packages:

1. Open RStudio by clicking an RStudio icon and type the following command in the R console:

```
install.packages("BiocManager")
install.packages("ROCR")
BiocManager::install(c("gridExtra","gtable",
                       "optparse","qvalue"))
```

2. Open RStudio by clicking an RStudio icon and choose **Tools** on the top menu bar and click **Install Packages...**. Type the necessary packages separated with a space or a comma and click the **Install** button.

After installing the necessary R packages, please enter the following command in the command prompt or the terminal to check whether MetaMSD is properly installed:

```
./MetaMSD.Rscript -h
```

2.4 Data Preparation

Multiple datasets may be generated by various instruments/labeling techniques. However, we assume that experiments in one dataset are generated by the same instrument using the same labeling or label-free technique. Prior to preparing for MetaMSD input data files, users need to quantify peptides/proteins using a proteomic software such as Skyline [12], MaxQuant [13, 14], Proteome Discover [15], MetaMorpheus (*see* **Chapter 3** and [16]), Proline (*see* **Chapter 4** and [17]), or Triqler (*see* **Chapter 5** and [18]) and perform protein significance analyses (e.g., t-test and linear mixed model). Users may use a TTEST function in Microsoft Excel after proper transformation/normalization or use specific software protein significance analysis: MSstats [19–21], Triqler, seaMass (*see* **Chapter 8**), Prostar (*see* **Chapter 9** and [22]), or msmsTest (*see* **Chapter 10**).

2.5 Data Description

1. Input data files are quantitative proteomic results that users want to integrate (*see* **Note 4**). Users need to organize protein significant analysis results (*see* Subheading 2.4) and create MetaMSD input files. Microsoft Excel can be used for creating input files (*see* **Note 5**).
2. The input files are space delimited text files (ending with .txt extension). One text file contains quantitative proteomic results from one study. Each text file contains three columns. The three column names are Protein, Sign, and Pvalue. Users can list either test statistics or signs of test statistics for the Sign column (*see* **Note 6**).
3. The example files are shown below (*see* **Note 7**). In this example, we name input files as Dataset1.txt and Dataset2.txt. However, other variations of names can be used. Dataset1.txt contains

```
Protein     Sign      Pvalue
Protein1    6.7200    0.0006
Protein2    -1.6714   0.1388
Protein4    0.6949    0.6092
```

or

```
Protein     Sign      Pvalue
Protein1     1        0.0006
Protein2    -1        0.1388
Protein4     1        0.6092
```

Dataset2.txt contains

```
Protein     Sign      Pvalue
Protein1    4.3934    0.0013
Protein3    1.9609    0.0783
Protein4    0.4343    0.6865
```

or

```
Protein     Sign      Pvalue
Protein1     1        0.0013
Protein3     1        0.0783
Protein4     1        0.6865
```

4. Dataset2.txt does not have to contain all the proteins in Dataset1.txt and vice versa. However, some overlap of quantified proteins between Dataset1.txt and Dataset2.txt is expected since they are from the same type of study (*see* **Note 8**). Dataset1.txt and Dataset2.txt do not have to be same the number of rows (*see* **Note 9**).

3 Methods

3.1 Integrative Quantitative Proteomics Data Analysis

This section describes how to run MetaMSD in a default mode. By default, MetaMSD performs Stouffer's test using datasets in `input` and writes result files in `output` with a *q*-value threshold of 5%. In this mode, the top 15 differential proteins are displayed for the purpose of visualization:

1. Create input/output directories in the MetaMSD directory:
   (a) Create an input directory in the directory where MetaMSD.Rscript is located. Please name the input directory as `input`.
   (b) Place input files (*see* Subheading 2.5 for its description) in the input directory.
   (c) Create an output directory in the directory where MetaMSD.Rscript is located. Please name the output directory as `output`.
   (d) If there are old results in `output`, then delete such files.
2. For Windows users, open the command prompt window (e.g., type **cmd** after selecting the Start button and choose **Command Prompt** from the list). For Mac or Linux users, open the terminal window.
3. Go to the directory where MetaMSD.Rscript is located. For example, if MetaMSD.Rscript is in `C:/Users/Lisa/Documents/MetaMSD`, type the following command in the command prompt or the terminal:

```
cd C:/Users/Lisa/Documents/MetaMSD
```

4. To run MetaMSD in a default mode, please type the following command:

```
./MetaMSD.Rscript
```

Running MetaMSD in a default mode with two input files (Dataset1.txt and Dataset2.txt) displays the following in the command prompt or the terminal window:

```
MetaMSD (Meta-analysis for Mass Spectrometry Data)
```

```
Loading required package: ROCR
Loading required package: qvalue
Loading required package: gridExtra
Loading required package: gtable

Found 2 in the input directory. Processing...

input/Dataset1.txt
Creating initial data frame ...
input/Dataset2.txt
Merging data frame ...

Merged data.frame contains the following columns.
Protein, Sign.1, Pvalue.1, Sign.2, Pvalue.2
```

In this example, MetaMSD reads Dataset1.txt prior to Dataset2.txt considering Dataset1.txt as the first dataset and Dataset2.txt as the second dataset.

3.2 MetaMSD Output Files

MetaMSD generates seven output files (three pdf files and four text files) as shown in Table 1.

1. **Meta-analysis Results**: MetaAnalysisResults.txt is a primary output file, which contains single analysis and meta-analysis results with confidence scores such as *p*-values and *q*-values. Specifically, it contains protein names, separate analysis results for datasets 1 and 2 (sign, *p*-value, and *q*-value), and meta-analysis results (sign, *p*-value, and *q*-value). The *q*-value cutoff and the number of proteins in Top-N differential protein list are not applied to this primary output file.
2. **Top-N Differential Protein List**: TopDifferentialProteins.pdf and TopDifferentialProtein.txt contain the Top-N differential protein list detected by a meta-analysis (*see* Fig. 2). In these files, there are the Top-N differential proteins ranked by their *q*-values, directionality of hypotheses (sign), *p*-values, and *q*-values for significant protein analyses. The directionality of

Table 1
MetaMSD output files

Visualization (.pdf)	Results (.txt)
TopDifferentialProteins.pdf	TopDifferentialProteins.txt
q-plot.pdf	MetaAnalsyisResults.txt
Summary.pdf	Summary.txt
	Diagnosis.txt

Rank	Protein	Sign	Pvalue	Qvalue
1	GPVI_HUMAN	–	7.077188e–07	0.0003798744
2	PGRP1_HUMAN	–	1.657390e–06	0.0004448095
3	EGF_HUMAN	–	3.531460e–05	0.0047388564
4	GALNS_HUMAN	–	2.877374e–05	0.0047388564
5	IGSF8_HUMAN	–	5.561822e–05	0.0059707148
6	NEGR1_HUMAN	–	1.048996e–04	0.0093843001
7	CLM9_HUMAN	–	2.321992e–04	0.0103862504
8	ICOSL_HUMAN	–	1.872730e–04	0.0103862504
9	K2C7_HUMAN	+	1.972104e–04	0.0103862504
10	PVRL4_HUMAN	–	2.221740e–04	0.0103862504
11	SCTM1_HUMAN	–	1.500986e–04	0.0103862504
12	SUSD2_HUMAN	–	2.002626e–04	0.0103862504
13	ACTB_HUMAN	+	3.246189e–04	0.0119329867
14	SPIT1_HUMAN	–	3.334734e–04	0.0119329867
15	TRFM_HUMAN	–	3.126880e–04	0.0119329867
16	CADM4_HUMAN	–	3.889783e–04	0.0122816233
17	KLK1_HUMAN	–	3.823149e–04	0.0122816233
18	2B13_HUMAN	+	5.939719e–04	0.0135106047
19	CADM2_HUMAN	–	5.813909e–04	0.0135106047
20	CATZ_HUMAN	+	6.040971e–04	0.0135106047

Fig. 2 Top 20 differential proteins list

hypothesis is "plus" when a protein is more abundant in Group 1 compared to Group2, and "minus" otherwise.

3. **Summary**: Summary.pdf contains the number of differential proteins detected by meta-analysis and single analyses as well as a meta-analysis evaluation at a given q-value threshold (*see* Fig. 3). Thus, results change as the q-value cutoff changes. The information in Summary.pdf is also found in Summary.txt (*see* Fig. 3 and also **Note 10**) and Diagnosis.txt (*see* **Note 11**).
4. **Q-plot**: The qplot.pdf contains the q-plot (*see* Fig. 4), which displays the number of differential proteins in respect to various q-value thresholds. In this figure, q-value threshold is displayed between 0 and 0.10. However, users can change its range by changing the q-value cutoff parameter option.

3.3 Meta-analysis Parameter Option

MetaMSD allows users to choose either Stouffer's test or Pearson's test for their meta-analysis using the following parameter option:

```
-m CHARACTER, --metaanalysis=CHARACTER
Specify a meta-analysis test.
Either Stouffer or Pearson. [default = Stouffer]
```

	# of detected proteins
Meta Analysis	256
Single Analysis 1	94
Single Analysis 2	73
Intersection among Single Analyses	11
Union among Single Analyses	156

	Meta–Analysis Evaluation
Integration–driven Discovery Rate (IDR)	45.7 %
Integration–driven Revision Rate (IRR)	10.9 %

Fig. 3 The number of differential proteins and meta-analysis diagnosis at the *q*-value threshold of 10%

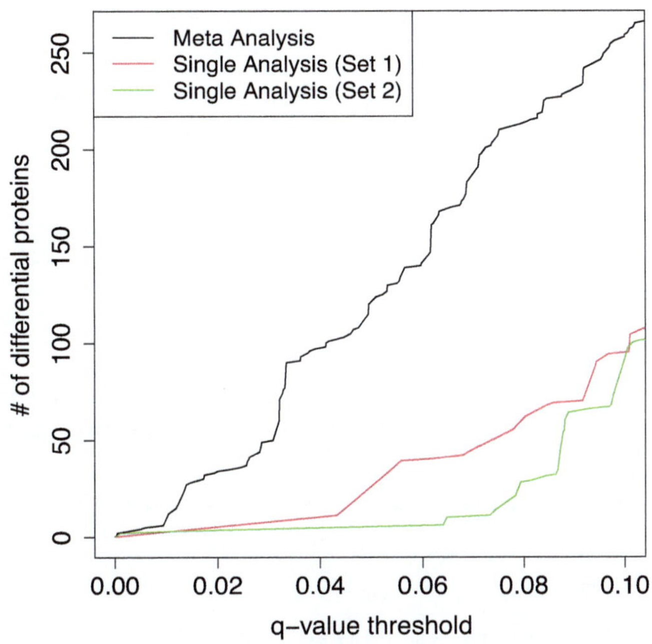

Fig. 4 The q-plot

By default, MetaMSD uses Stouffer's test. However, users can choose a meta-analysis approach by specifying -m or –metanalaysis. For example, if users want to perform Pearson's test, they can type the following in the command prompt or in the terminal window:

```
./MetaMSD.Rscript --metaanalysis = Pearson
```

or

```
./MetaMSD.Rscript -m Pearson
```

Users can also specify Stouffer's test as follows:

```
./MetaMSD.Rscript --metaanalysis = Stouffer
```

or

```
./MetaMSD.Rscript -m Stouffer
```

3.4 Q-Value Threshold Option

Users can choose a *q*-value threshold. Changing this parameter option will change the following outputs: qplot.pdf, Summary.pdf, Summary.txt, and Diagnosis.txt. Users can change the *q*-value threshold using the following parameter option:

```
-c NUMBER, --cutoff=NUMBER
Specify a q-value cut off. [default = 0.05]
```

By default, the *q*-value threshold is 5%. The following command will change the *q*-value cutoff from 5% to 10%:

```
./MetaMSD.Rscript --cutoff=0.10
```

or

```
./MetaMSD.Rscript -c 0.10
```

If users want to use Pearson's test with a *q*-value threshold of 0.10, then users can type two parameters options separated by a space:

```
./MetaMSD.Rscript --metaanalysis = Pearson  --cutoff=0.10
```

or

```
./MetaMSD.Rscript -m Pearson -c 0.10
```

3.5 Top N Differential Protein List Option

Users may change the number of proteins in TopDifferentialProteins.pdf and TopDifferentialProteins.txt using the parameter option:

```
-t NUMBER, --top=NUMBER
Specify the number of proteins in Top-N differential protein
list. [default = 15]
```

By default, MetaMSD displays 15 top proteins (ranked by their q-values). The following command changes the number of the top proteins from 15 to 10:

```
./MetaMSD.Rscript --top=10
```

or

```
./MetaMSD.Rscript -t 10
```

3.6 Input/Output Folder Name Option

Users can specify names of input/output folders using the following parameter option:

```
-i CHARACTER, --input=CHARACTER
Specify the input folder name. [default = input]
-o CHARACTER, --output=CHARACTER
Specify the output folder name. [default = output]
```

If the input directory name is 🗀 `InputKidneyTransplant` and the output directory name is 🗀 `OutputKidneyTransplant`, then use the following commands:

```
./MetaMSD.Rscript -i InputKidneyTransplant
                   -o OutputKidneyTransplant
```

or

```
./MetaMSD.Rscript --input=InputKidneyTransplant
                  --output=OutputKidneyTransplant
```

3.7 Help Message Option

The command below will generate all the parameter options that MetaMSD has.

```
./MetaMSD.Rscript -h
```

4 Notes

1. Stouffer's test option in MetaMSD utilizes inverse normal transformation of p-values from multiple datasets to properly generate a p-value for the integrative analysis [23]. The test statistic is computed by

$$Z = \sum_{i=1}^{K} Z_i/\sqrt{K}, \quad (1)$$

where Z_i is obtained from a *p*-value from the *i*th dataset for a protein of interest and *K* is the number of datasets. Then, the final *p*-value is calculated by taking a side with a smaller one-sided *p*-value and applying a Bonferroni correction [24].

2. Pearson's test is another option for integrating multiple quantitative proteomic results. It considers the directionality of the hypotheses (e.g., more abundant in Group 1 vs. Group2 or vice versa) by taking a minimum *p*-value of left- and right-sided Fisher's tests [23]. It corrects multiple testing errors using a Bonferroni correction. The *p*-value is computed as

$$min\left(1, 2Pr\left(\chi^2_{2K} \geq Q^T\right)\right), \quad (2)$$

where *K* is the number of datasets, p_i^L is the left-sided *p*-value for the ith dataset, and

$$Q^T = max\left(-2\sum_{i=1}^{K} log\left(p_i^L\right), -2\sum_{i=1}^{K} log\left(1 - p_i^L\right)\right). \quad (3)$$

3. If the users have problems with the script running for the first time and cannot generate the following help outputs, please refer to Software Installation Trouble Shooting (*see* Subheading 2.3).
4. There is no limit to how many datasets can be combined.
5. Advanced users may use their own scripts to speed up this process.
6. When no test statistic is available, users can provide a sign (1 or −1) that represents a directionality of hypothesis. For example, if a protein abundance is larger in Group 1 than 2, then the sign is positive; thus, users need to enter 1 for the sign entry. If a protein abundance of interest is smaller in Group 1 than 2, then the sign is negative, and users need to enter -1 for the sign entry.
7. Example datasets are freely available at https://github.com/soyoungryu/MetaMSD/blob/master/Data/BreastCancer/MetaMSDinput/Gemez_result.txt, https://github.com/soyoungryu/MetaMSD/blob/master/Data/BreastCancer/MetaMSDinput/TCGA_result.txt as well as https://github.com/soyoungryu/MetaMSD/blob/master/Data/KidneyTransplant/MetaMSD_input/LabelFree.txt, https://github.com/soyoungryu/MetaMSD/blob/master/Data/KidneyTransplant/MetaMSD_input/iTRAQ.txt for Breast Cancer Data and Kidney Transplant Data, respectively.

8. It is expected that datasets are from the same type of studies and are analyzed using the same protein database. A same protein in two datasets is expected to have exactly the same name (first column). Otherwise, they will be considered as different proteins.
9. The datasets can have different numbers of quantified proteins.
10. In this example, 256 proteins were detected using a meta-analysis given a user-defined *q*-value cutoff. When analyzing only one dataset, 94 and 73 differential proteins were detected for the first and second datasets. The intersection and union of single analyses are also displayed.
11. Users can evaluate meta-analysis results using IDR (Integration-driven Discovery Rate) and IRR (Integration-driven Revision Rate). IDR is a percentage of differential proteins detected only by a meta-analysis, but not by any single dataset analyses. IRR is a percentage of differential proteins detected in at least one single dataset analysis, but not by the meta-analysis.

Acknowledgements

MetaMSD software development was supported by grants from the National Institute of General Medical Sciences (GM103440 and 1R15GM126562-01).

References

1. Gygi SP, Rist B, Gerber SA, Turecek F, Gelb MH, Aebersold R (1999) Quantitative analysis of complex protein mixtures using isotope-coded affinity tags. Nat Biotechnol 17(10): 994–999. https://doi.org/10.1038/13690
2. Hardt M, Witkowska HE, Webb S, Thomas LR, Dixon SE, Hall SC, Fisher SJ (2005) Assessing the effects of diurnal variation on the composition of human parotid saliva: quantitative analysis of native peptides using iTRAQ reagents. Anal Chem 77(15):4947–4954. https://doi.org/10.1021/ac050161r
3. Ong SE, Blagoev B, Kratchmarova I, Kristensen DB, Steen H, Pandey A, Mann M (2002) Stable isotope labeling by amino acids in cell culture, SILAC, as a simple and accurate approach to expression proteomics. Mol Cell Probes 1(5):376–386. https://doi.org/10.1074/mcp.M200025-MCP200
4. Radulovic D, Jelveh S, Ryu S, Hamilton TG, Foss E, Mao Y, Emili A (2004) Informatics platform for global proteomic profiling and biomarker discovery using liquid chromatography-tandem mass spectrometry. Mol Cell Proteomics 3(10):984–997. https://doi.org/10.1074/mcp.M400061-MCP200
5. Ryu S, Gallis B, Goo YA, Shaffer SA, Radulovic D, Goodlett DR (2008) Comparison of a label-free quantitative proteomic method based on peptide ion current area to the isotope coded affinity tag method. Cancer Inf 6:CIN–S385. https://doi.org/10.4137/CIN.S385
6. Li XJ, Zhang H, Ranish JA, Aebersold R (2003) Automated statistical analysis of protein abundance ratios from data generated by stable-isotope dilution and tandem mass spectrometry. Anal Chem 75(23):6648–6657. https://doi.org/10.1021/ac034633i
7. Ryu SY, Qian WJ, Camp DG, Smith RD, Tompkins RG, Davis RW, Xiao W (2014) Detecting differential protein expression in large-scale population proteomics. Bioinformatics 30(19):2741–2746. https://doi.org/10.1093/bioinformatics/btu341

8. Ryu SY (2014) Bioinformatics tools to identify and quantify proteins using mass spectrometry data. Adv Protein Chem Struct Biol 94:1–17. https://doi.org/10.1016/B978-0-12-800168-4.00001-9
9. Washburn MP, Ulaszek RR, Yates JR (2003) Reproducibility of quantitative proteomic analyses of complex biological mixtures by multidimensional protein identification technology. Anal Chem 75(19):5054–5061. https://doi.org/10.1021/ac034120b
10. Ryu SY, Wendt GA (2019) MetaMSD: meta analysis for mass spectrometry data. PeerJ 7: e6699. https://doi.org/10.7717/peerj.6699
11. Storey JD (2002) A direct approach to false discovery rates. J R Stat Soc B 64(3): 479–498. https://doi.org/10.1111/1467-9868.00346
12. Pino LK, Searle BC, Bollinger JG, Nunn B, MacLean B, MacCoss MJ (2020) The skyline ecosystem: informatics for quantitative mass spectrometry proteomics. Mass Spectr Rev 39(3):229–244. https://doi.org/10.1002/mas.21540
13. Cox J, Hein MY, Luber CA, Paron I, Nagaraj N, Mann M (2014) Accurate proteome-wide label-free quantification by delayed normalization and maximal peptide ratio extraction, termed MaxLFQ. Mol. Cell. Probes 13(9):2513–2526. https://doi.org/10.1074/mcp.M113.031591
14. Schaab C, Geiger T, Stoehr G, Cox J, Mann M (2012) Analysis of high accuracy, quantitative proteomics data in the MaxQB database. Mol Cell Proteomics 11(3):M111–014068. https://doi.org/10.1074/mcp.M111.014068
15. Lemeer S, Hahne H, Pachl F, Kuster B (2012) Software tools for MS-based quantitative proteomics: a brief overview. Quantitative methods in proteomics. Springer, Berlin, pp 489–499. https://doi.org/10.1007/978-1-61779-885-6_29
16. Solntsev SK, Shortreed MR, Frey BL, Smith LM (2018) Enhanced global post-translational modification discovery with MetaMorpheus. J Proteome Res 17(5):1844–1851. https://doi.org/10.1021/acs.jproteome.7b00873
17. Bouyssié D, Hesse AM, Mouton-Barbosa E, Rompais M, Macron C, Carapito C, Gonzalez de Peredo A, Couté Y, Dupierris V, Burel A, Menetrey JP, Kalaitzakis A, Poisat J, Romdhani A, Burlet-Schiltz O, Cianférani S, Garin J, Bruley C (2020) Proline: an efficient and user-friendly software suite for large-scale proteomics. Bioinformatics 36(10): 3148–3155. https://doi.org/10.1093/bioinformatics/btaa118
18. The M, Käll L (2019) Integrated identification and quantification error probabilities for shotgun proteomics. Mol Cell Proteomics 18(3): 561–570
19. Choi M, Chang CY, Clough T, Broudy D, Killeen T, MacLean B, Vitek O (2014) MSstats: an R package for statistical analysis of quantitative mass spectrometry-based proteomic experiments. Bioinformatics 30(17): 2524–2526. https://doi.org/10.1093/bioinformatics/btu305
20. Chang CY, Picotti P, Hüttenhain R, Heinzelmann-Schwarz V, Jovanovic M, Aebersold R, Vitek O (2012) Protein significance analysis in selected reaction monitoring (SRM) measurements. Mol Cell Proteomics 11(4):M111–014662. https://doi.org/10.1074/mcp.M111.014662
21. Clough T, Thaminy S, Ragg S, Aebersold R, Vitek O (2012) Statistical protein quantification and significance analysis in label-free LC-MS experiments with complex designs. BMC Bioinf 13(16):1–17. https://doi.org/10.1186/1471-2105-13-S16-S6
22. Wieczorek S, Combes F, Lazar C, Giai Gianetto Q, Gatto L, Dorffer A, Hesse AM, Coute Y, Ferro M, Bruley C, Burger T (2017) DAPAR & ProStaR: software to perform statistical analyses in quantitative discovery proteomics. Bioinformatics 33(1):135–136. https://doi.org/10.1093/bioinformatics/btw580
23. Heard NA, Rubin-Delanchy P (2018) Choosing between methods of combining *p*-values. Biometrika 105(1):239–246. https://doi.org/10.1093/biomet/asx076
24. Sedgwick P (2012) Multiple significance tests: the Bonferroni correction. BMJ 344. https://doi.org/10.1136/bmj.e509

Chapter 17

Application of WGCNA and PloGO2 in the Analysis of Complex Proteomic Data

Jemma X. Wu, Dana Pascovici, Yunqi Wu, Adam K. Walker, and Mehdi Mirzaei

Abstract

In this protocol we describe our workflow for analyzing complex, multi-condition quantitative proteomic experiments, with the aim to extract biological insights. The tool we use is an R package, PloGO2, contributed to Bioconductor, which we can optionally precede by running correlation network analysis with WGCNA. We describe the data required and the steps we take, including detailed code examples and outputs explanation. The package was designed to generate gene ontology or pathway summaries for many data subsets at the same time, visualize protein abundance summaries for each biological category examined, help determine enriched protein subsets by comparing them all to a reference set, and suggest key highly correlated hub proteins, if the optional network analysis is employed.

Key words Proteomics, Pathway, Gene ontology, Functional enrichment analysis, WGCNA, Statistical R package

1 Introduction

Large quantitative proteomic experiments generate many subsets of proteins of interest, for instance, up and down regulated proteins from multiple pairwise comparisons, or groups of related proteins resulting from clustering algorithms, all of which have to be annotated functionally and compared in order to extract biological insights. The PloGO2 package performs such pathway and functional annotation of multiple protein subsets potentially including their quantitative data, and, as an optional workflow, could be coupled with the popular network correlation analysis package WGCNA [1, 2]. PloGO2 can be used to simply visualize the experimental subsets in biological context, or to indicate functionally interesting subsets of proteins for further analysis.

Both GO (Gene Ontology) [3] and KEGG pathway [4] data summaries are routinely used in proteomics experiments to provide

Thomas Burger (ed.), *Statistical Analysis of Proteomic Data: Methods and Tools*,
Methods in Molecular Biology, vol. 2426, https://doi.org/10.1007/978-1-0716-1967-4_17,

a first-glance insight into the function, process, location, interaction, and relationships between collections of genes or proteins [5–7]. We assume that most readers are familiar with both, but remind briefly that the GO data is organized as a directed acyclic graph starting from one parent node, which means that particular ontology categories can have multiple parents as well as multiple children, with thousands of different nodes available. The KEGG pathway data, on the other hand, is simply organized into broad categories such as metabolism or generic information processing.

Various user-friendly tools can be used for the analysis of a small number of protein subsets [8–11], but very few were aimed specifically at more complex experiments, and none that we are aware of take into account and summarize the abundance as well as the annotation.

The PloGO2 package is an upgrade of the PloGO package [12], and was designed for the following tasks:

1. Generate GO or KEGG summaries in text or excel format for a large number of data subsets
2. Allow for the selection of particular categories of interest for an experiment (e.g., stress response) across many annotation files, instead of working with a particular level of the GO hierarchy
3. Summarize and visualize additive protein abundance information, such as percentages of total abundance, or log fold changes, for all data sets and all categories of interest, if such data is available
4. Determine enriched subsets by comparing all sets to one of the selected sets that can serve as a background, typically all proteins identified in the experiment.

In addition, we show that our WGCNA analysis tailored for proteomics sets can be readily input into PloGO2 and further characterized functionally, in which case both enriched subsets of proteins and highly connected hub proteins within them can be quickly identified.

This workflow has been recently validated in a scientific publication [13]; the current protocol details step by step instructions, and expands on the existing vignettes in the R package [14].

2 Material

2.1 Data Type

The inputs typically include a tabbed Excel sheet containing all the data subsets of interest as a separate tab, an optional quantitative data file containing the processed data matrix for the experiment and, for the KEGG analysis, a pathway database file containing the KEGG annotation. Several full examples as well as all the code needed are included in the example data subfolders of our GitHub repository [15]: https://github.com/APAFbioinformatics/PloGO2_R_Package/.

For this chapter we will use a TMT (tandem mass tag) example containing the full set of proteins measured in a drought stress time course in rice [16], also present in the GitHub repository above.

2.2 Data Size

Though there are no strict upper and lower bounds for the number of tabs in the input Excel file, running an Excel file with many tabs can take quite a long time. It is recommended to keep the number of tabs to less than 20. The number of proteins (rows) in each tab of the Excel input can be hundreds to thousands (*see* **Note 1**). There is no strict limit for the number of columns to be included in the Excel or CSV files.

2.3 Data Format

The main data is input as an Excel format with one or more tabs, each tab containing a subset of proteins of interests. Very often, the first tab contains the full set of proteins identified in the experiment, to which the other tabs can be compared. The protein identifier column must be present in all tabs and should have the same name in all the tabs, for instance, "Accession", or "Uniprot".

The pathway DB file should be a CSV file and should contain the mapping between all protein accessions and their KEGG pathway IDs. This file will be mapped to the quantitation file using the identifier, hence the two should contain the same type of identifier, usually Uniprot ID.

Optionally, a CSV file containing the protein abundance information can be an additional input file. These files should contain the protein name column similar to the two files above, in order to be matched to the other input files. It can contain other information columns such as protein descriptions which can be ignored, and quantitative data such as log fold changes or protein abundance in each sample which will be aggregated.

2.4 Software Installation

PloGO2 can be installed from Bioconductor [17] or GitHub [15]. While package vignettes and example data sets and scripts are publicly available via both sites, the latter contains more example data sets and codes. Like all other Bioconductor packages, PloGO2 can be installed on Windows, Linux, or Mac platforms; R [18] version 4.0 or later will be needed. Before installing PloGO2, a few dependency packages need to be installed; they can be installed from the R console as shown below:

```
# Install dependencies
install.packages(c('httr', 'openxlsx', 'xtable','heatmap3'))
if (!requireNamespace("BiocManager", quietly = TRUE))
  install.packages("BiocManager")
BiocManager::install(c("GOstats","GO.db"))

# Optionally, install WCGNA if that workflow is desired
BiocManager::install(c("impute","preprocessCore"))
install.packages('WGCNA')

# Finally, install PloGO2
BiocManager::install("PloGO2")
```

3 Methods

3.1 Workflow

We use PloGO2 in two main ways, shown schematically in the workflow Fig. 1. We can either perform various analyses (pairwise comparisons, data partition, etc.) initially, resulting in multiple sets of proteins, then run PloGO2 to annotate and compare them. Alternatively, we can run WGCNA analysis on the full set of experimental proteins or on the differentially expressed proteins only to generate a number of clusters of tightly related proteins, then use PloGO2 on all the clusters to annotate and compare them.

We will use the TMT rice leave data included in our GitHub repository as an example to go through the main steps of PloGO2 for KEGG pathway analysis.

3.2 Load Necessary Packages

1. Run the following command to load the needed packages for running PloGO2:

```
library(lattice)
library(PloGO2)
```

2. Set your working directory, for example, to `C:\temp`:

```
setwd("C:/temp")
```

3.3 Prepare the Input Files

The steps below describe how to generate files such as those already present in the GitHub repository. In order to run the code you could skip the steps and simply use the files provided as examples.

Unlike the publicly available GO annotation databases, KEGG pathway database requires paid subscription for using their APIs. As such, PloGO2 does not provide functions of accessing KEGG pathway annotations on the fly, hence the pathway annotations for all proteins of interest have to be pre-downloaded:

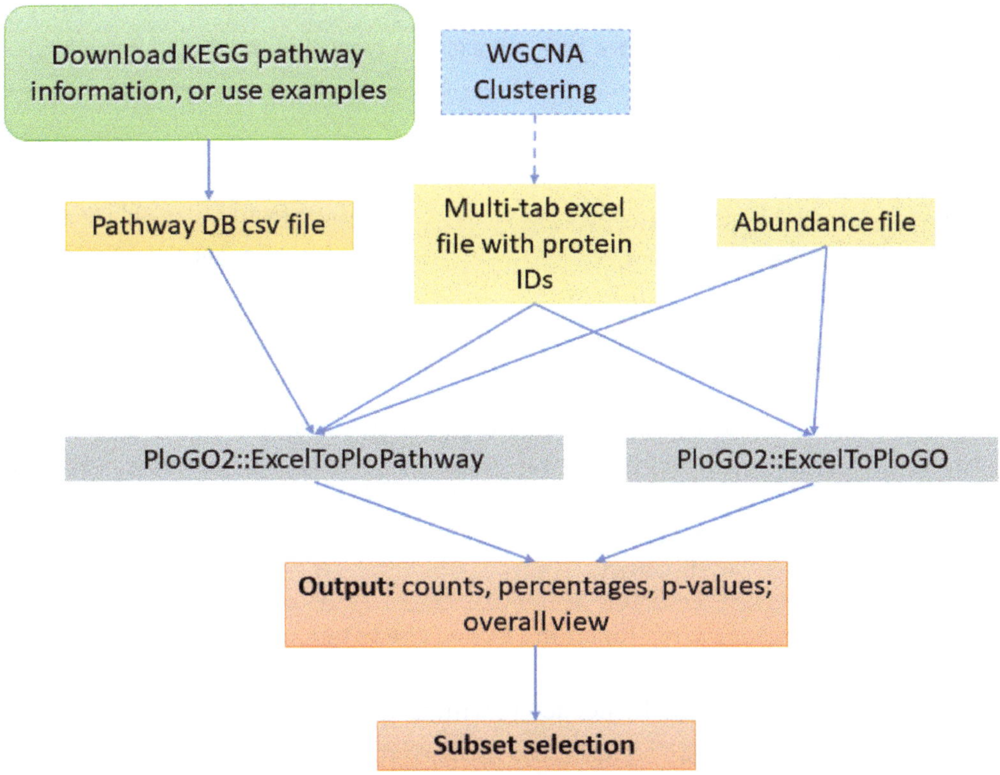

Fig. 1 PloGO2 workflow; input files needed are shown in yellow, and the main top-level package functions are shown in grey

1. Include all protein accessions in the allData tab and retrieve their KEGG pathway annotations (*see* **Note 2**).
2. Save the downloaded pathway annotations in a simple two-column CSV file format. Each row has one protein ID and one or more pathway IDs separated by space (*see* **Note 3**). An example of the pathway DB file is shown in Table 1.
3. Save the file as an CSV file, `pathwayDB.csv`, under your working directory.
4. Create an Excel file to store the subsets of interest of the proteomic data. Each tab contains a subset of the dataset; they could be up and down regulated proteins from various comparisons, or different clusters of correlated proteins, or other experimentally relevant subsets. Usually, the first tab stores the union of proteins in all subsets.
5. Each tab has to contain at least a column of protein IDs. All the tabs should have the same name for the protein ID columns, such as "Accession", or "Uniprot".

Table 1
Example pathway DB file (First few rows only, full example present in the GitHub repository)

Protein	x
A0A0N7KEB1	osa01100 osa01212 osa01040 osa00061
A3C4S4	osa01100 osa01110 osa00520 osa00053
B7E914	osa01100 osa00190
B7EKA4	osa01100 osa01110 osa01230 osa00270
B7ERZ5	osa01100 osa01110 osa00010

6. Besides a mandatory protein ID column, each tab can contain multiple extra columns such as descriptions, annotations, abundance, and so on (*see* **Note 4**).
7. Save the Excel file as `RiceAllData.xlsx` under your working directory.

3.4 Pathway Analysis

1. Use function `ExcelToPloPathway` for KEGG pathway analysis.
2. Set the needed parameters:
   (a) `fname`, the name of the Excel data file
   (b) `colName`, the name of the column containing the protein identifiers
   (c) `compareWithReference`, the name of the tab that serves as basis for enrichment comparison
   (d) `DB.name`, the DB file name for the pathway previously prepared (*see* Subheading 3.3, **Step 3**).
3. For example, the following command invokes the pathway analysis for the example dataset:

```
res <- ExcelToPloPathway(fname="RiceAllData.xlsx",
                          colName="Uniprot",
                          compareWithReference="AllData",
                          DB.name="pathwayDB.csv",
                          data.file.name = "none" )
```

3.5 Results

After running the code a number of files and folders will be created in your working directory:

1. A PloGO2 folder will be created (*see* **Note 5**).
2. By default a PW subfolder will be created to store the transformed annotation files, each one corresponding to a subset

and with the same name as the tab name in the input Excel data file.

3. The main output is an Excel result file, `PloGO2Summary.xlsx`, which will be generated under the PloGO2 folder. This file contains all the initial tabs of the input spreadsheet, to which the pathway annotation has been added as additional columns, to enable easy sorting to select proteins from the pathway of interest. A summary tab is added at the end, containing counts and percentages of each pathway in each data subset, as well as Fisher's exact test *p*-values (*see* **Note 6**) for the comparison of each pathway in each set with the reference subset selected.

4. The results from the summary tab can be used to summarize the pathway percentages in each subset:

   (a) It can be visualized as a stacked barchart using Excel (*see* Fig. 2). This gives the most flexibility, for instance, if the user wants to highlight only a subset of the pathways, or only several data sets.

   (b) Alternatively, the returned results can be used in R, for instance, to select and highlight the enriched pathways only; enriched pathway protein percentage is shown in Fig. 3. To do so, use the following code:

```
# select counts, percentages and p-values from PloGO2 result
counts = res$Counts


perc = res$Percentages
pval = res$FisherPval
# select pathways with > 10 entries in the reference dataset
counts.idx = counts[,1] > 10
# Calculate the fold enrichment compared to the reference set
FE = perc/perc[,1]
sig.idx = !is.na(pval) & (pval < 0.05) & (FE > 1)
sig.index = apply(sig.idx, 1, FUN=sum)
# select enriched pathways only
filter.idx = sig.index & counts.idx
Colors = rainbow(sum(filter.idx))
# print percentages of data
barplot(perc[filter.idx,], col=Colors, las=2)
legend("topright",
       legend=rownames(perc[filter.idx,]),
       fill=Colors)
```

5. Display the enrichment *p* values in R (*see* Fig. 4):

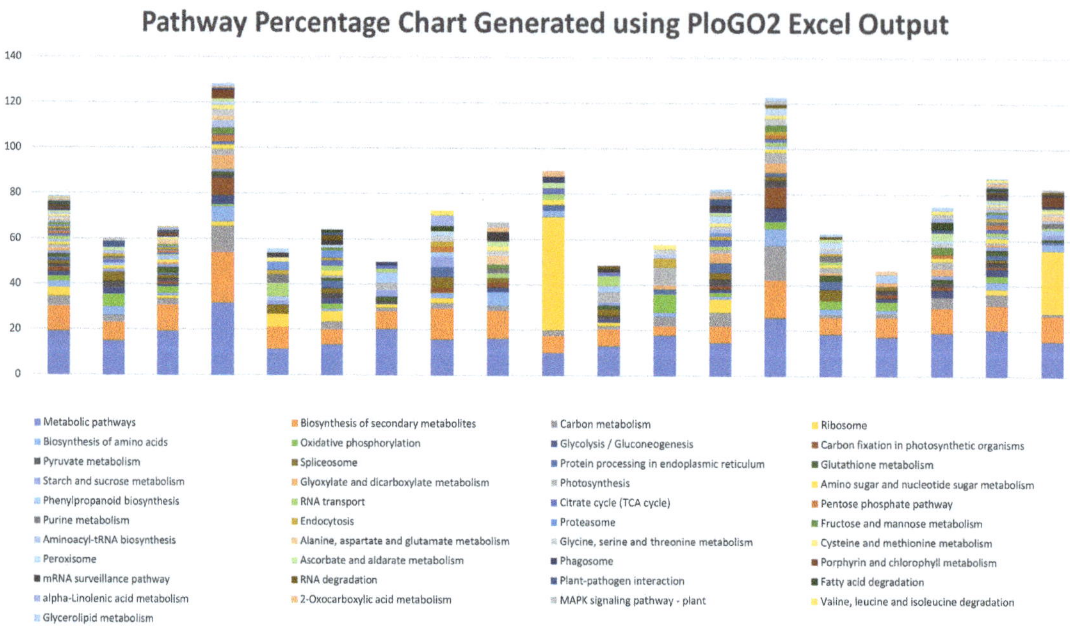

Fig. 2 Pathway stacked barchart generated using Excel showing the percentages of each pathway in each cluster. The two clusters containing most ribosomal proteins (yellow) are visually apparent

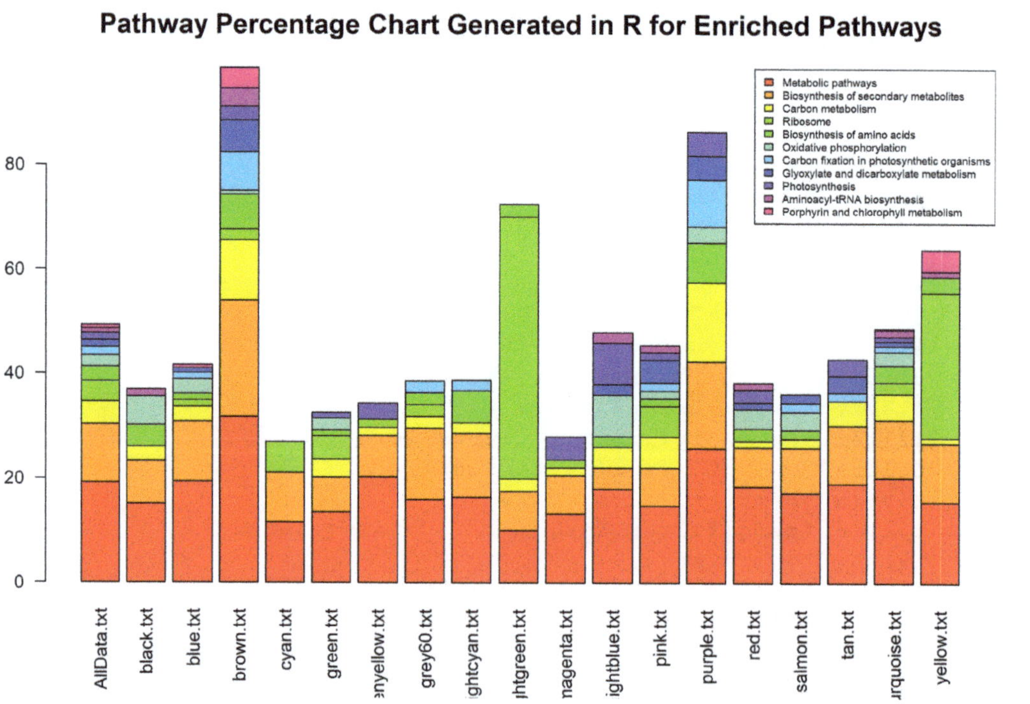

Fig. 3 Percentage of proteins in each pathway, showing the enriched pathways only, based on the comparison with the baseline set of all proteins in the experiment

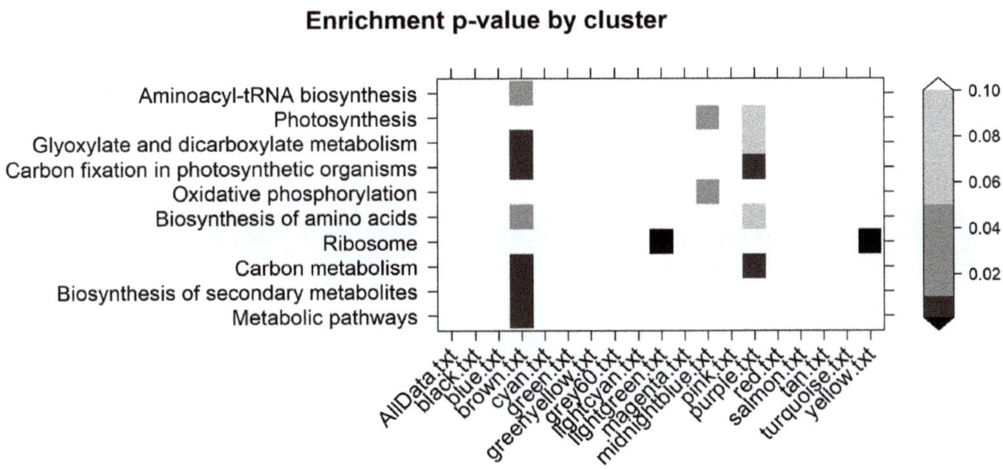

Fig. 4 Fisher's exact test p-values for the enriched pathways are shown, with lower p-values marked with darker color. Five subsets are functionally enriched when compared to the baseline of all proteins in the set. Two subsets clearly contain most of the ribosomal proteins

```
# plot Fisher exact p-values as a levelplot
levelplot(t(pval[filter.idx,]),scales=list(y=list(cex=1),
         x=list(rot=45, cex=1.05)),
         col.regions = gray(c(0:5)/5),
         at = c(-Inf, 0.001, 0.01, 0.05, 0.1, Inf),
         main="Enrichment p-value by cluster")
```

3.6 Adding Abundance

So far, the examples shown have summarized annotation for many data subsets, thus showing how many of the proteins in each subset are present in various pathways, and potentially helping focus on enriched pathways in certain subsets. However, often it is informative to summarize the experimental quantitation from the experiment as well, such as the protein abundance in each sample. Such a summary would give a more complete picture of the dataset, showing, as a hypothetical example, that not only 5% proteins are present in glycolysis, but that they represent 10% of the protein total complement in that sample. In order to do this, we have to add and summarize the quantitative data for the experiment.

The quantitative data is given as a CSV matrix, with rows as all proteins from the experiment and columns as samples (*see* **Note 7**).

The steps below detail the optional abundance part of the workflow. To simply run the code the user can use the "Abundance" data files present in our GitHub repository.

1. The starting point is the total abundance data for the experiment, with rows as proteins and columns as samples. This file is usually generated as part of most proteomics workflows. If the data is additive, such as NSAF abundance data, then it can be used as is. Alternatively, if the experiment has a reference condition such as a no disease group, or a control time point, then

log fold changes from the reference condition can be generated and summarized.

2. Save as CSV file, `Abundance_alldata.csv`, under `C:/temp`.
3. Run `ExcelToPloPathway` with parameter `data.file.name` setting to `Abundance_alldata.csv`:

```
res <- ExcelToPloPathway(
            fname="RiceAllData.xlsx",
            colName="Uniprot",
            compareWithReference="AllData",
            DB.name="pathwayDB.csv",
            data.file.name = "Abundance_alldata.csv"
                        )
```

4. Extract the abundance levelplots for each subset and display them. For example, the following commands will show the abundance levelplots for the fourth subset for pathways which have more than three occurrences (*see* Fig. 5):

```
list.levelplots <- res$list.levelplots
plot(list.levelplots[[4]])
```

3.7 Gene Ontology Analysis

So far, we have highlighted the pathway workflow available with PloGO2; we very briefly give an example of the older GO workflow as well. The GO information is freely available from Uniprot, hence it can be extracted directly. Due to the large number of possible GO categories, for a large experiment with many data subsets it is often useful to simply focus on a limited of sets of GO categories of interest. An example of such subset of categories of interest, `GODefault.txt`, is present on both our GitHub repository (located under folder example data) and PloGO2 R package (located under folder files):

1. Use function `ExcelToPloGO` for Gene Ontology analysis.
2. Set the needed parameters:
   (a) `fname`, the name of the Excel data file; this is the exact same file as that used and described for the pathway analysis part of the workflow
   (b) `colName`, the name of the column containing the protein identifiers, again identical to the pathway workflow
   (c) `compareWithReference`, the name of the tab that serves as basis for enrichment comparison
   (d) `termFile`, The name of the file containing the GO categories. If the default value, NA, is used, the Gene Ontology annotations for "BP" ontology at level two will be extracted.

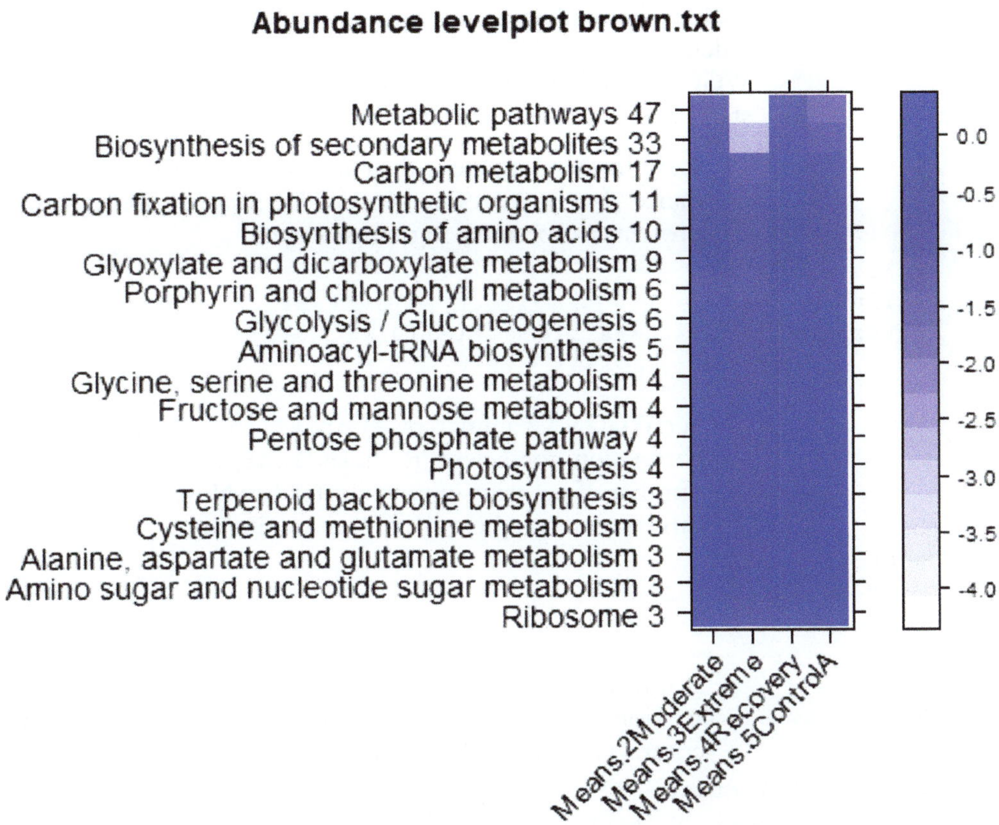

Fig. 5 Levelplot for the abundance data for one of the experimental subset; in this example log fold change data with respect to the control condition has been generated and aggregated for each subset. The pattern of decreased metabolism under extreme drought stress followed by recovery is apparent

3. For example, the following command invokes the GO analysis for the example dataset using an example term file `GODefault.txt` without abundance data (*see* **Note 8**):

```
res <- ExcelToPloGO(fname="RiceAllData.xlsx",
                    colName="Uniprot",
                    termFile= "GODefault.txt",
                    compareWithReference="AllData",
                    data.file.name = "none" )
```

4. Similar results to the previous ones (*see* Subheading 3.5) will be generated for GO analysis.

3.8 WGCNA Analysis for Proteomic Data

As shown in Fig. 1, the workflow can optionally start from a WGCNA analysis, which we use to generate clusters of tightly correlated proteins and identify highly correlated hub proteins within them. WGCNA is a method proposed for gene co-expression network modelling and clustering [1, 2] (*see* **Note 9**). Though it has been widely used in genomics data co-expression analysis, its application in proteomics area is still relatively new. For a detailed WGCNA tutorial, readers can refer to a comprehensive WGCNA online tutorial [19]. Our tailored WGCNA workflow starts from a protein abundance or ratio dataset.

1. Read in the abundance data and only keep the numeric abundance data matrix (*see* **Note 10**). Samples are as rows and proteins as columns; log transformed:

```
library(WGCNA)
allDat = read.csv("WGCNA_input_alldata.csv")
datExpr = log(as.data.frame(t(allDat[,-c(1:4)])))
```

2. Set WGCNA parameters; `RCutoff`, the cut-off value for the correlation coefficient R^2; `minClusterSize`, the minimum number of proteins in each cluster (*see* **Note 11**); `cutHeight` can be used to adjust the merging criterion:

```
RCutoff = .85
minModuleSize = 20
cutHeight = 0.1
```

3. A soft threshold is selected for constructing the weighted correlation matrix by using the approximate scale-free network criteria (*see* **Note 12**). Get the soft threshold power:

```
enableWGCNAThreads()
# Choose a set of soft-thresholding powers
powers = c(c(1:10), seq(from = 12, to=20, by=2))
# Call the network topology analysis function
sft = pickSoftThreshold(datExpr, powerVector=powers,
                        RsquaredCut=RCutoff, verbose=5)
# Get the soft power
softPower = sft$powerEstimate;
```

4. Calculate *signed* correlation network adjacency from expression data (*see* **Note 13**) and calculate the topology overlap similarity (TOM) dissimilarity:

```
adjacency = adjacency(datExpr,power=softPower,type="signed")
# Turn adjacency into TOM distance
dissTOM = 1 - TOMsimilarity(adjacency)
```

5. Build hierarchical clustering proteins using TOM dissimilarity measure, perform dynamic tree cut and merge similar clusters:

```
# Clustering using TOM-based dissimilarity
proTree = hclust(as.dist(dissTOM), method = "average");
# Module identification using dynamic tree cut
dynamicMods = cutreeDynamic(dendro = proTree, distM = dissTOM
                            deepSplit = 2,
                            pamRespectsDendro = FALSE,
                            minClusterSize = minModuleSize);
# Convert numeric labels into colors
dynamicColors = labels2colors(dynamicMods)
# Merge clusters
mergedClust = mergeCloseModules(datExpr, dynamicColors,
                                cutHeight = cutHeight,
                                verbose = 3)
```

6. Get cluster color names and module eigenproteins:

```
mergedColors = mergedClust$colors
mergedMEs = mergedClust$newMEs
# Get the module eigenproteins
MEs = mergedMEs
```

7. Calculate kMEs and hub proteins:

```
# Get kME, module membership
kmes = signedKME(datExpr, MEs)
# Append module colors and kME values for all proteins
dat.res = data.frame(allDat, mergedColors , kmes)
# Get list of data frames for modules,
# order proteins by their kME value decreasingly
list.cluster.dat = lapply(
  levels(as.factor(mergedColors)),
  function(x) {
    dtemp = dat.res[dat.res$mergedColors == x,];
    dtemp[order(dtemp[,paste0("kME",x)==colnames(dtemp)],
          decreasing=TRUE),
        -setdiff(grep("^kME", colnames(dtemp)),
                 which(paste0("kME",x)==colnames(dtemp)))
        ]
      }
            )


names(list.cluster.dat) = levels(as.factor(mergedColors))
```

Object `list.cluster.dat` stores the data frames for the seven clusters. The proteins in each cluster are ordered by their `kME` values decreasingly, and the proteins at the top of the list can be viewed as the hub proteins (*see* **Note 14**). If we print out all clusters into an xlsx file, we will get the example dataset, `RiceAllData.xlsx`, which we have demonstrated previously. The ranking provided by `kME` can give a useful guide for selecting candidate proteins for further analysis or validation. In addition to the `kME` measurement, correlation with biological traits, if available, can also provide suggestions for interesting candidates [20].

4 Notes

1. The datasets should contain at least three proteins in order to have meaningful statistical results.
2. Besides using subscribed APIs, there are a number of ways for downloading KEGG pathway annotations such as using the KEGG website or a third-party tool such as DAVID. An example KEGG web page parser is available at GitHub.
3. No duplication of the protein ID is allowed.
4. All columns will be included in the output.
5. If there already exists a PloGO2 folder in the working directory, this folder will be rewritten when executing `ExcelToPloPathway`. Therefore it is recommended to rename the existing PloGO2 folder before executing `ExcelToPloPathway`.
6. The reported *p*-values were adjusted for each subset by using Benjamini-Hochberg multiple test correction method [21].
7. The rows will be aggregated by summing up, so this data has to be additive for the interpretation to be relevant. Such data could be total protein amounts, as generated, for example, in label free quantitation using NSAF [22], and then the summed up amount would represent total abundance of the proteins in each pathway in the respective sample. Alternatively, the quantitative data could be a log fold change, such as the average log fold change for each condition from control in a time course. In that case, the aggregated amount would represent a cumulative log fold change for the proteins in the respective pathway.
8. This may take a few moments, since the GO information is obtained live from Uniprot.
9. Instead of using the raw pairwise correlation to measure the relationships between genes, it uses a weighted correlation which is modelled as the pairwise correlation raised to a power. For clustering, it uses the topology overlap similarity (TOM) as the distance measurement in the hierarchical clustering algorithm and uses a dynamic tree cutting algorithm to generate an optimal set of clusters.
10. The data should be properly preprocessed and normalized before executing WGCNA analysis.
11. The optimal value for `minClusterSize` depends on the number of proteins in the dataset. Based on our experience, for small to medium proteomics datasets (for example, few hundreds to a thousand), a value between 20 to 30 could

generate a good number of well separated clusters. For a relatively large dataset (say a few thousand proteins), the value might need to be increased to around 50.

12. A scale-free network is one where the topology is dominated by a few highly connected nodes, which link the rest of the less connected nodes to the system [1]; it is assumed that most biologically relevant networks should satisfy this property. By raising the correlation to the selected soft threshold, the correlation network becomes scale-free.
13. WGCNA can build signed or unsigned network adjacency matrix and the default is unsigned which transforms all correlations to positive values. However, we recommend that signed network adjacency matrix should be used for proteomics datasets so that proteins with different regulation trends will be clustered separately.
14. Due to its loose definition, it is hard to give a strict cut-off value for selecting hub proteins, and various papers use the top 10, top 6, or other kME cut-off thresholds.

Acknowledgements

This work was conducted at the Australian Proteome Analysis Facility supported by the Australian Government's National Collaborative Research Infrastructure Scheme (NCRIS). Aspects of this work were supported by the Australian National Health and Medical Research Council (Project Grant GNT1124005 and RD Wright Career Development Fellowship GNT1140386 to AKW), the Ross Maclean Fellowship, and the Brazil Family Program for Neurology.

References

1. Zhang B, Horvath S (2005) A general framework for weighted gene co-expression network analysis. Stat Appl Genet Mol Biol 4(1). https://doi.org/10.2202/1544-6115.1128
2. Langfelder P, Horvath S (2008) WGCNA: An R package for weighted correlation network analysis. BMC Bioinformatics 9(1):559. https://doi.org/10.1186/1471-2105-9-559
3. Consortium GO (2004) The gene ontology (go) database and informatics resource. Nucleic Acids Res 32(suppl_1):D258–D261. https://doi.org/10.1093/nar/gkh036
4. Kanehisa M, Sato Y, Kawashima M, Furumichi M, Tanabe M (2016) Kegg as a reference resource for gene and protein annotation. Nucleic Acids Res 44(D1): D457–D462. https://doi.org/10.1093/nar/gkv1070
5. Schmidt A, Forne I, Imhof A (2014) Bioinformatic analysis of proteomics data. BMC Syst Biol 8(2):1–7. https://doi.org/10.1186/1752-0509-8-S2-S3
6. Carnielli CM, Winck FV, Leme AFP (2015) Functional annotation and biological interpretation of proteomics data. Biochim Biophys Acta (BBA) Protein Proteomics 1854(1): 46–54. https://doi.org/10.1016/j.bbapap.2014.10.019
7. Khatri P, Sirota M, Butte AJ (2012) Ten years of pathway analysis: current approaches and outstanding challenges. PLoS Comput Biol

8(2):e1002375. https://doi.org/10.1371/journal.pcbi.1002375

8. Sherman BT, Lempicki RA, et al (2009) Systematic and integrative analysis of large gene lists using DAVID bioinformatics resources. Nature Protocols 4(1):44. https://doi.org/10.1038/nprot.2008.211
9. Szklarczyk D, Gable AL, Lyon D, Junge A, Wyder S, Huerta-Cepas J, Simonovic M, Doncheva NT, Morris JH, Bork P, Jensen LJ, Mering C (2018) STRING v11: protein–protein association networks with increased coverage, supporting functional discovery in genome-wide experimental datasets. Nucleic Acids Res 47(D1):D607–D613. https://doi.org/10.1093/nar/gky1131
10. Jassal B, Matthews L, Viteri G, Gong C, Lorente P, Fabregat A, Sidiropoulos K, Cook J, Gillespie M, Haw R, et al (2020) The reactome pathway knowledgebase. Nucleic Acids Res 48(D1):D498–D503. https://doi.org/10.1093/nar/gkz1031
11. Ge SX, Jung D, Yao R (2020) Shinygo: a graphical gene-set enrichment tool for animals and plants. Bioinformatics 36(8):2628–2629. https://doi.org/10.1093/bioinformatics/btz931
12. Pascovici D, Keighley T, Mirzaei M, Haynes PA, Cooke B (2012) PloGO: Plotting gene ontology annotation and abundance in multicondition proteomics experiments. Proteomics 12(3):406–410. https://doi.org/10.1002/pmic.201100445
13. Wu JX, Pascovici D, Wu Y, Walker A, Mirzaei M (2020) Workflow for rapidly extracting biological insights from complex, multicondition proteomics experiments with WGCNA and plogo2. J Proteome Res https://doi.org/10.1021/acs.jproteome.0c00198
14. Wu JX, Pascovici D (2020a) Bioconductor plogo2. http://www.bioconductor.org/packages/release/bioc/html/PloGO2.html
15. Wu JX, Pascovici D (2020b) Plogo2 R package. https://github.com/APAFbioinformatics/PloGO2_R_Package
16. Wu Y, Mirzaei M, Pascovici D, Chick JM, Atwell BJ, Haynes PA (2016) Quantitative proteomic analysis of two different rice varieties reveals that drought tolerance is correlated with reduced abundance of photosynthetic machinery and increased abundance of clpd1 protease. J Proteomics 143:73–82. https://doi.org/10.1016/j.jprot.2016.05.014
17. Huber W, Carey VJ, Gentleman R, Anders S, Carlson M, Carvalho BS, Bravo HC, Davis S, Gatto L, Girke T, Gottardo R, Hahne F, Hansen KD, Irizarry RA, Lawrence M, Love MI, MacDonald J, Obenchain V, Olés AK, Pagès H, Reyes A, Shannon P, Smyth GK, Tenenbaum D, Waldron L, Morgan M (2015) Orchestrating high-throughput genomic analysis with Bioconductor. Nature Methods 12(2):115–121. https://doi.org/10.1038/nmeth.3252
18. R Core Team (2020) R: A language and environment for statistical computing. R Foundation for Statistical Computing, Vienna, Austria. https://www.R-project.org/
19. Langfelder P, Horvath S (2014) Tutorials for the WGCNA package
20. Yuan L, Chen L, Qian K, Qian G, Wu CL, Wang X, Xiao Y (2017) Co-expression network analysis identified six hub genes in association with progression and prognosis in human clear cell renal cell carcinoma (ccrcc). Genomics Data 14:132–140. https://doi.org/10.1016/j.gdata.2017.10.006
21. Benjamini Y, Hochberg Y (1995) Controlling the false discovery rate: a practical and powerful approach to multiple testing. J R Stat Soc B (Methodol) 57(1):289–300. https://doi.org/10.2307/2346101
22. Friso G, Majeran W, Huang M, Sun Q, Van Wijk KJ (2010) Reconstruction of metabolic pathways, protein expression, and homeostasis machineries across maize bundle sheath and mesophyll chloroplasts: large-scale quantitative proteomics using the first maize genome assembly. Plant Physiology 152(3):1219–1250. https://doi.org/10.1104/pp.109.152694

INDEX

A

Alternative hypothesis 7, 11, 288

B

Batch effect 179, 197–199, 201, 208, 209, 211, 216, 239, 248, 249, 252, 335
Bayesian model 92
Benjamini–Hochberg .... v, 1, 9, 15, 17, 18,68, 73, 198, 214, 251, 388
Biomarker discovery 197, 198, 200, 222, 238, 239, 316, 317, 344

C

Calibration plot 280, 281
Clustering 95, 172, 199, 208–210, 218, 261, 275–279, 290–294, 298, 299, 358, 375, 385, 386, 388
Connected components 170, 173–175
Crosslink 36, 44–45, 51, 62

D

DAPAR 134, 164, 165, 282
Data integration .... vi, 340, 354
Descriptive statistics 171–173, 177, 179, 186, 193
Differential analysis .... v, vi, 1–23, 131–139, 141–161, 183, 184, 187–189, 245, 251–253
Differential expression analysis, *see* Differential analysis
Dimensionality reduction vi
Dimension reduction, *see* Dimensionality reduction
Discriminant analysis 334, 346

E

Error propagation 91
Exploratory data analysis (EDA) 205–211
eXtracted Ion Chromatograms (XIC) 81, 82, 88, 101,111, 112, 116, 164, 191, 303, 304, 308

F

False discovery proportion (FDP) 4, 5, 11–13, 16–19,21, 26, 73
False discovery rate (FDR) .... v, 1–23, 26, 27, 30–33,42, 48–51, 59, 60, 68, 69, 73, 75, 91, 92, 95, 101–105, 114–116, 142, 149, 150, 152–160, 183, 185, 188, 189, 195, 198, 214, 216–218, 220, 224, 225, 233, 238, 247, 279, 289, 296, 311, 362
FDR, *see* False discovery rate (FDR)
FDR control, *see* False discovery rate (FDR)
Feature selection 13, 238
Filtering .... v, 1–23, 52, 56–59, 68,70, 73–76, 84, 87, 104, 120, 126, 175–179, 187, 188, 192, 193, 220, 222, 224, 225, 233–235, 251, 253, 263, 271, 320, 336
FlashLFQ 53, 54, 56, 63, 303–312
Fold change .... vi, 91, 92, 99, 100, 102–111,116, 137, 142, 150, 155, 157, 158, 160, 183, 200, 210, 240, 245, 255, 258, 309–311, 376, 377, 384, 385, 388
Functional enrichment analysis 383

G

Gene enrichment, *see* Gene ontology
Gene ontology .... vi, 257, 295, 375, 384–385
Generalized linear model (GLM) 198, 199, 201, 209, 211, 213, 236, 237, 239, 245, 251, 253, 263
Gibbs sampler 120, 123
Glycopeptides 45–46, 56, 57
GO analysis, *see* Gene ontology
GO terms, *see* Gene ontology
GSimp 119–128

I

Identification score 73–75
Immunoprecipitation v
Imputation .... v, 18, 92, 109, 119–128, 131–139,181–184, 187, 188, 193, 284, 285, 320
IPinquiry 243–264

K

Knockoff filters 1–23

L

Label-free quantification 36, 43, 53, 56, 57, 63, 67–88, 92, 131, 268, 304, 388

Thomas Burger (ed.), *Statistical Analysis of Proteomic Data: Methods and Tools*,
Methods in Molecular Biology, vol. 2426, https://doi.org/10.1007/978-1-0716-1967-4,

Lasso 10, 12, 13, 18, 22, 23, 315, 316, 339
Left-censored missing value 119–128
Linear regression 328

M

MAR, *see* Missing at random (MAR)
MCAR, *see* Missing completely at random (MCAR)
Meta-analysis .... vi, 361–363, 366–369, 373
MetaMorpheus 35–65, 304–307, 364
MetaMSD 361–373
mi4limma 131–139
Missing at random (MAR) 119, 120, 128
Missing completely at random (MCAR) 119, 120, 128, 181, 193, 284, 285
Missing not at random (MNAR) 119, 120, 181,193, 284, 285
Missing value .... v, 92, 99, 106–108, 111,112, 116, 119–128, 131, 132, 135, 139, 168, 172, 178, 181–184, 187, 188, 191, 193, 194, 283–285, 308, 320, 326, 327, 335, 336, 358
mixOmics 333–358
MNAR, *see* Missing Not at Random (MNAR)
Moderated *t*-test 132, 137, 138
msmsEDA 197–240
msmsTests 197–240, 251, 263, 365
Multicomp 26–29
Multiple imputation...........................132–135, 137, 138
Multivariate analysis 333–358

N

Negative binomial (NB) distribution................ 198, 201, 211, 212, 216, 217, 236, 237
Normalization 18, 56, 63, 80, 141,142, 147–149, 151, 152, 155, 179–181, 187, 193, 199–201, 205, 208, 226–232, 234, 235, 239, 240, 250, 261–263, 282–283, 285, 304, 307–309, 335–336, 365, 388
Null hypothesis 21, 184, 238, 281, 288, 289

P

Parsimony 43, 44, 50, 58
Pathway...................................... 106, 295, 375–384, 388
PCA, *see* Principal component analysis (PCA)
Penalized regression.. 339
Peptide-level analysis... 163–195
Peptide-level validation 73
Peptide spectrum match (PSM) 3, 23, 25–27, 29, 30, 42, 45, 47–51, 55, 57–59, 61, 68, 70, 72–75, 77, 83, 85, 87, 88, 92–99, 101, 104, 143, 190, 304, 305, 307
Peptide-to-protein aggregation, *see* Aggregation
Peptidomic 64, 163–195, 299
PloGO2 375–389
PLS, *see* Partial least square (PLS); Projection to latent structures (PLS)
Poisson model 198, 237
Posterior distribution.....................................92, 99, 100, 102, 104–110, 116
Post-translational modification (PTM)............ 36, 41–42, 45, 49, 51, 54, 60-62, 77, 87, 267–299, 304
Power .... vi, 6, 11–13, 16–18, 40, 101, 106, 111, 115, 279–282, 344, 386, 388
Principal component analysis (PCA)................. 149, 172, 173, 199, 209, 293, 298, 334, 338, 340–346, 358
Prior distribution 102, 110, 111, 142
Probabilistic graphical models.................... 102, 107, 110
Proline............ 64, 67–88, 166, 190, 191, 268, 269, 364
Prostar 163–195
Protein-level analysis 141–161
Protein-level validation 73
Protein roll-up, *see* Aggregation
PSM, *see* Peptide spectrum match (PSM)
PSM-level validation 73
p-value 5, 6, 14, 15, 17, 29,95, 107, 137, 183, 188, 198, 200, 201, 211, 213–222, 224–226, 229, 231, 233, 238, 239, 245, 251, 252, 255, 256, 258, 279–281, 367, 371, 372, 381, 383, 388

Q

QRILC...120, 122, 124, 126
Quality control ....v, 2, 4, 144, 150, 161
Quantification ....v, 36, 43, 50, 53, 56–58, 63, 67–88, 91–116, 120, 131, 135, 136, 141–161, 164, 168, 171, 176, 177, 190, 191, 197, 245–247, 262, 267, 268, 272, 297, 303–312
Quasi-likelihood inference... 198

R

Regression ....vi, 13–18, 120, 180, 198,213, 236, 316, 328, 334, 336, 350, 351
Relative quantification 58, 164, 303
Reproducibility.. 197, 200, 215, 217–222, 238, 246, 268, 274–279, 282
Robust estimation 322
Rubin's rules.............................132, 135, 136, 138, 139

S

seaMass 141–161, 365
Secretome 200, 201, 226–233
Shrinkage 155, 339
Size effect... 201, 222
Spectral count..................... v, 49, 78, 197–240, 243–264
Statistical power, *see* Power

T

Target-decoy competition (TDC) .... v, 2, 3, 21, 22,26, 27, 68, 73
Triqler 91–116, 364, 365
t-test 6, 13–18, 91, 98, 99, 107, 132,137, 138, 155, 160, 161, 187, 279, 304, 364

V

Variance .... v, 9, 18, 21, 22, 40, 122,132, 135–137, 139, 142, 147, 156, 161, 166, 173, 180, 192, 198, 208, 211–213, 216, 237, 277, 279, 280, 289, 298, 334–336, 338, 339, 341, 342, 344, 351, 354, 358
Visualization .... vi, 36, 47–49, 59,171, 173, 179, 222, 243–264, 335, 343, 346, 356, 357, 366, 367
Volcano plot 157, 158, 185, 188,189, 223, 233, 235, 245, 247, 255–259

W

WGCNA 375–389

X

XIC, *see* eXtracted Ion Chromatograms (XIC)

GPSR Compliance
The European Union's (EU) General Product Safety Regulation (GPSR) is a set of rules that requires consumer products to be safe and our obligations to ensure this.

If you have any concerns about our products, you can contact us on

ProductSafety@springernature.com

In case Publisher is established outside the EU, the EU authorized representative is:

Springer Nature Customer Service Center GmbH
Europaplatz 3
69115 Heidelberg, Germany

www.ingramcontent.com/pod-product-compliance
Ingram Content Group UK Ltd.
Pitfield, Milton Keynes, MK11 3LW, UK
UKHW052001051026
15624UKWH00003B/502
* 9 7 8 1 0 7 1 6 1 9 6 6 7 *